House Fires

HOUSE FIRES

Jerry Knapp | Chris Flatley

Fire Engineering
BOOKS & VIDEOS

Disclaimer

The recommendations, advice, descriptions, and methods in this book are presented solely for educational purposes. The author and publisher assume no liability whatsoever for any loss or damage that results from the use of any of the material in this book. Use of the material in this book is solely at the risk of the user.

Copyright © 2019 by
PennWell Corporation
1421 South Sheridan Road
Tulsa, Oklahoma 74112-6600 USA

800.752.9764
+1.918.831.9421
sales@pennwell.com
www.FireEngineeringBooks.com
www.pennwellbooks.com
www.pennwell.com

Managing Editor: Mark Haugh
Production Manager: Tony Quinn
Book Designer: Therman Lee
Cover Designer: Brandon Ash

Library of Congress Cataloging-in-Publication Data

Names: Knapp, Jerry, author. | Flatley, Chris, author.
Title: House fires / Jerry Knapp, Chris
   Flatley.
Description: Tulsa, Oklahom, USA : PennWell, [2018] | Includes
   bibliographical references and index.
Identifiers: LCCN 2017049545 | ISBN 9781593704124
Subjects: LCSH: Dwellings--Fires and fire prevention. | Lifesaving at fires.
   | Command and control at fires.
Classification: LCC TH9445.D9 K63 2018 | DDC 628.9/25--dc23

Printed in the United States of America

2 3 4 5 6   26 25 24 23 22

Contents

How to Use This Book ix

1   The Importance of House Fires to Firefighters 1
     Summary 1
     Introduction 1
     Annual Loss of Life and Dollar Loss to House Fires in the United States 1
     Causes of House Fires 3
     Causes of Fatal House Fires 3
     The Time of Fire Alarms in Homes 3
     House Fires as Mass Casualty Events 3
     Alcohol as a Factor in House Fires 4
     Firefighter Deaths and Injuries 5
     References 6

2   Construction, Types of Houses, and Floor Plans 7
     Summary 7
     Introduction: Reading the House 7
     Home Styles and Floor Plans 9
     House Construction Methods 18
     Construction and Framing Methods 19
     References 41

3   Hazards of House Fires 43
     Summary 43
     Introduction 43
     Common Fires, Uncommon Hazards 43
     Scenario Process 44
     Scenario 45
     Hazards to Firefighters 45
     Rapid Fire Development 49
     Flashover 49
     Backdraft 54
     Fire Spread in Homes 55
     Exterior Fire Spread 55
     Hazards of Utilities 62
     Electrical Hazards 63
     Site Grade Hazards 64
     Energy Efficient Homes 65
     Other Hazards 65
     Scenario Discussion 72
     Additional Scenarios 73
     References 74

4   Size-Up of House Fires 77
     Summary 77
     Introduction 77
     Scenario 77
     General Considerations 78
     Everyone Does Size-Up 79
     Size-Up Process 79
     Existing Size-Up Systems 84
     On-Scene Report 91
     Converted Private Dwellings 92
     Scenario Discussion 95
     Advanced Size-Up Considerations 97
     Additional Size-Up Practice Scenarios 97
     References 99

Contents

5   Command, Control, and Fire Attack Strategies 101
     Summary. 101
     Introduction. 101
     Scenario. 101
     Firefighter Safety 102
     Financial Future 102
     Strategic and Tactical Discussions 102
     Command Principles 103
     Command and Control 103
     Fire Attack Strategies for House Fires 107
     Overall Strategy 108
     RECEO VS. 109
     Four Fire Attack Strategies. 111
     Scenario Discussion 114
     Additional Practice Scenario. 116
     References. 117

6   Search and Rescue Operations. 119
     Summary. 119
     Introduction. 119
     Scenario. 119
     Protecting Life. 120
     Search. 121
     An Effective SAR Plan 127
     Engine Company Search Procedures. 136
     Search Rope. 137
     Search Safety 140
     Moving Victims: Quick Webbing Harness 143
     Search and Rescue at Converted Private Dwellings. 144
     Scenario Discussion 146
     References. 147

7   Preparing for Successful Fire Attack 149
     Summary. 149
     Introduction. 149
     Scenario. 150
     Fire Attack System. 150
     Target Flow. 151
     The Attack Team 156
     Nozzle Selection. 160
     Scenario Discussion 165
     References. 166

8   Ventilation. 167
     Summary. 167
     Introduction. 167
     Scenario. 168
     Ventilation of Modern Home Fires 168
     UL Study Details 170
     Vent for Life 173
     Can You Vent Enough? 174
     Energy Efficient Windows (EEWs) 174
     Traditional Venting 177
     Overhaul Ventilation. 178
     Types of Ventilation 178
     How Much to Vent: When Is Enough, Enough?. 183
     Ladders 183

Chapter Conclusion 184

Scenario Discussion 184

References. 185

9   **Basement and Garage Fires** 187

Summary. 187

Introduction. 187

Important Notes. 187

Basement Fires 188

Garage Fires 201

Basement Fire Scenario. 204

Detached Garage Scenario. 206

Garage Fire Extending to Home Scenario 206

Rapid Fire Development Scenario. 207

Light Smoke Scenario 208

References. 209

10   **First- and Second-Floor Fires**. 211

Summary. 211

Introduction. 211

First-Floor Fires 212

Modern House 216

Traditional House Fire Tactics 220

References. 229

11   **Attic and Exterior Fires** 231

Summary. 231

Introduction. 231

Storage Attic Fires 232

Occupied Attics 242

Exterior Fires. 243

References. 244

12   **Common Nonfire Calls to Homes**. 245

Summary. 245

Introduction. 245

Carbon Monoxide 245

Natural Gas Emergencies 248

Lightning Strikes 258

Leadership: The Most Common Call 260

References. 261

Bibliography **263**

Index **269**

How to Use This Book

This book was written as a comprehensive summary of strategy, tactics, critical background knowledge, skills, and abilities you need to be successful at your next house fire. It is intended for all levels of firefighters from probie through chief officer.

It begins by providing a thorough, national-level understanding of the significance of house fires to America's firefighters, line-of-duty injuries and deaths, civilian injuries and lives lost, and overall cost. It then progresses into logically arranged chapters containing critical background information on construction, search procedures, ventilation, and fire attack plans. Largely these are standalone chapters that can be used for initial learning, future reference on a particular phase of a firefighting operation at a house fire, or general refresher training. Collectively, the book progresses through all the critical skill and knowledge sets necessary for you to be more confident, effective, and safe at your next house fire. This knowledge is an invaluable asset for promotional exams and professional development.

This book will provide you with a number of general rules, strategies, tactics, and other useful fireground hints, tricks of the trade, and so on. It is important to note that many of these are limited to house fires and may not be applicable to other types of occupancies. For the duration of this book, we will deal with single-family dwellings, duplex dwellings, and single-family homes that may have been altered since they were built.

Chapters 1 and 2 provide critical background information. Beginning with chapter 3 we will start the strategic and tactical chapters with a scenario. This exercise is intended to start an open discussion, an interaction between reader and the authors. After reading the scenario, jot down how your department would handle the scenario presented. The scenarios are intended to provide a starting point for the topic in the chapter. After you read each chapter you should have some excellent insight and hopefully some relevant skills and knowledge you can apply to your next house fire. At the end of each chapter are discussions on important points for that particular topic and scenario for your consideration. Compare these to your original answers. The chapters will end with similar smaller-scale exercises based on our fireground experience to convey unique or especially dangerous or common house fires that you will face. You can also use the scenarios as indoor drills for your department or company.

The Importance of House Fires to Firefighters

Summary

This chapter introduces the book and will use national statistical data to paint a detailed picture of the residential fire problem in the United States. Topics discussed include where, when, and how fatal and nonfatal residential fires occur. This will allow the reader to better anticipate and plan for the fires they will respond to. This chapter will detail how impactful house fires are for firefighters in terms of line-of-duty deaths and injuries. The chapter also examines the effect of smoke detectors and other factors such as sprinklers on reducing the current trends.

Introduction

As firefighters, we respond to a variety of alarms, ranging from brush, grassland, and wildland fires to dumpster fires, car fires, vacant building fires, industrial fires, commercial fires, and residential fires. However, it is the residential fire response that is our most important alarm. The following data are presented to provide a historical and overarching perspective on the importance of house fires to the American fire service. In the end, civilian and firefighter lives hang in the balance. Often our actions determine who lives, who dies, or who becomes permanently disabled or injured at the house fire.

We often do not understand the impact of residential fires, especially those in one- and two-family homes. Yet this is where the vast majority of life lost to fire occurs. Individually, the statistics are small and don't make national headlines. Americans die at home, often just one or two at a time. Just one-tenth of 1% of residential fires have multiple fatalities.

Annual Loss of Life and Dollar Loss to House Fires in the United States

The most recent annual statistics prove the modern house fire is our most important alarm. According to the annual report of the National Fire Protection Association's (NFPA) *Fire Loss in the United States during 2015* (Haynes 2016, iii), home fires caused 78% (2,560) of the total (3,280) deaths recorded due to fire in 2015. There was a civilian death from a home fire every 3 hours and 25 minutes. Property loss in house fires accounted for $7 billion, more than half of the total $10.3 billion lost to all types of structure fires (iv). In 2015, house fires were responsible for 71% of all injuries from fire, and a civilian injury occurs from a house fire every 47.5 minutes (iv).

The importance of strategic and tactical competence at house fires for every fire department in the United States cannot be overemphasized. In 2014, 55% of all structure fires were in one- and two-family homes. House fires are the bulk of our work. It is where unacceptable numbers of civilians are hurt and killed and where firefighters risk the most in attempts to reduce those numbers.

In the US Fire Administration (USFA) report *A Profile of Fire in the United States 2003–2007*, the fire problem in the United States is described as follows:

> When compared to other industrialized Nations, the U.S. fire problem is severe. In fact, the United States is ranked as having the fifth highest fire death rate out of 25 industrialized Nations examined by the World Fire Statistics Centre. This general status has been unchanged for the past 27 years. However, the fire and fire loss rate trends for the 5-year period of 2003 to 2007 show, while dollar loss per capita increased by 4 percent, the fire rate per million population declined by 3 percent, the death rate declined by 20 percent, and the injury rate declined by 9 percent. (USFA and National Fire Data Center [NFDC] 2010, 1).

According to the US Census, the majority of Americans (75% or 83 million) live in one- and two-family residences (US Bureau of the Census 2011). It is therefore reasonable to conclude that house fires are and will continue to be our most important alarm. It is the residential fire that has the most impact, both on firefighters and on civilians. In terms of the sheer number of structure fire responses, residential fires outnumber nonresidential fires almost 3:1 in the United States. Fires in one- and two-family dwellings dominate the statistics and account for 66% of all residential fires (National Fire Data Center [NFDC] 2012).

Residential buildings (as defined by USFA) include homes, multifamily buildings, manufactured homes, hotels, motels, dormitories, assisted living facilities, and halfway houses for formerly addicted or incarcerated people. But it is the house fire, the fire occurring in a one- or two-family home, that dominates the fire loss experience in the United States. Figure 1–1 demonstrates the overwhelming number of deaths, injuries, and lost property caused by house fires compared to other residential occupancies.

House fires surface every year as one of our most important fire problems. According to the USFA, "From 2008 to 2010, fire departments responded to an estimated 365,500 fires in residential buildings each year across the Nation. These fires resulted in an annual average loss of 2,560 deaths, 13,000 injuries, and $7.4 billion in property loss" (NFDC 2012, 1). House fire statistics fluctuate a little each year but remain unacceptably high.

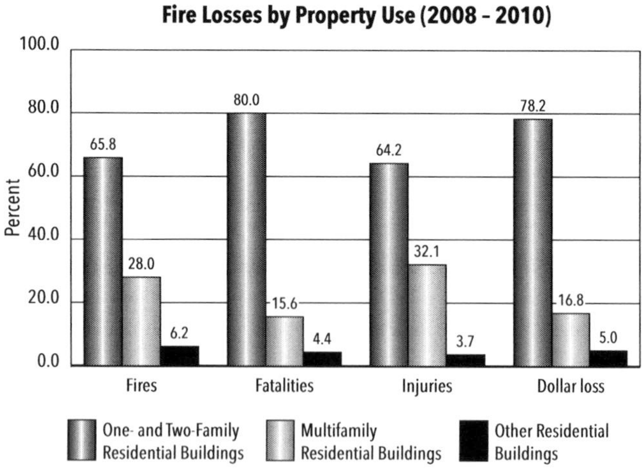

Fig. 1–1. Fire losses by property use (2008–2010). Note: When calculating the dollar losses by property use for 2008–2010, the 2008 and 2009 dollar-loss values were adjusted to their equivalent 2010 dollar-loss values to account for inflation *Source:* NFDC 2012, 3.

Interestingly, while these fire fatalities, injuries, and dollar losses may make big headlines in local news, they occur daily across our country and rarely make the regional or national news. House fires are one of the United States' best-kept dirty secrets. In a broader comparison, according to the US Census, fire deaths in one- and two-family homes far exceed annual deaths from all other natural disasters combined (US Bureau of the Census 2011).

A more personal way to look at the house fires is this: On an average day in the United States, 9 civilians perish in fires and 49 are injured, most of these in their own homes. Fire fatalities in the United States range from a high of approximately 4,080 in 1996 to a low of 2,920 in 1999. NFPA data agree with USFA estimates and interestingly they remain relatively constant (plus or minus a few percent) year after year: 83% of all civilian fire deaths and 77% of civilian injuries from fire occur in residential structures. The largest percentage of these deaths (76%) are in one- and two-family dwellings. Multifamily occupancies account for only 17% of the overall deaths from fire annually. As firefighters, we must be trained, equipped, prepared, and led to be at peak efficiency at house fires; lives, both yours and civilians', depend on your preparedness.

The house fire problem was made painfully clear in the 2010 NFPA report *Home Structure Fires*: "More than 1,000 home structure fires were reported every day. This translates to 43 fires every hour or one reported home fire every 83 seconds. Home fires killed an average of eight people every day. Once every three hours, someone was fatally injured in a home fire. A civilian (non-firefighter) injury is reported every 40 minutes. Home fires cause roughly $200 in damage every second" (Ahrens 2010, 7).

In the 2016 version of the report, the data are very similar: "During the five-year period of 2010–2014, U.S. fire departments responded to an estimated average of 358,300 home structure fires per year. These fires caused an annual average of 2,520 civilian deaths, 12,720 civilian fire injuries, and $6.7 billion in direct property damage. Home fires accounted for three-quarters (74%) of all reported structure fires, 93% of civilian structure fire deaths, 87% of the civilian structure fire injuries, and more than two-thirds (69%) of direct structure fire property loss" (Ahrens 2016, i).

DEFINITIONS

House and home—For the purposes of this book, the terms home and house are used interchangeably and refer to a single-family residence or a duplex, a home where two families live side by side. It is important to note that because of high real estate prices and ebbs and flows in our national economy, many homes that were designed as single-family homes have been converted (legally or illegally) into multiple apartments or single-family homes with a single separate apartment for a tenant.

Residential property—Can mean a wide variety of buildings in which people live, including multifamily housing, row housing, hotels, motels, dormitories, assisted living facilities, etc.

Causes of House Fires

According to a 2008 report by USFA, "Cooking has been the leading cause of residential fires for most of the years since the inception of NFIRS [National Fire Incident Reporting System]" (USFA 2008, 19). Cooking is responsible for approximately 45% of all fires in one- and two-family homes (NFDC 2012, 1) and it exceeds its next closest competitor, heating units (14%; NFDC 2012, 4), by a factor of about two. On heating-caused fires, USFA states, "Heating passed cooking for a few years in the late 1970s when there was a surge in the use of alternative space heaters and wood stoves, but that heating problem has long since subsided" (USFA 2008, 19).

Two groups—electrical malfunctions and carelessness—each account for 7% of one- and two-family house fires, followed by open flame (flame resulting from candles and matches) and intentional ignition, which each account for 5% of residential fires (NFDC 2012, 4). We must always be alert for instances where a house fire has its origins in arson.

Causes of Fatal House Fires

The causes of fatal residential fires may surprise you. Smoking accounts for 13% of all fatal residential fire starts (NFDC 2017, 6). Often (66%), this is associated with alcohol or drug use and involves falling asleep smoking on a bed, chair, or couch. Unintentional careless actions ("misuse of materials or products, abandoned or discarded materials or products, and a heat source being too close to combustibles") account for 16% of fatal fire ignitions (NFDC 2017, 6). Fires of electrical origin come in at 10.4%. In 2005, these four factors accounted for 57% of fatal fires and 59% of the fatalities.

The Time of Fire Alarms in Homes

The time of day you can expect to be called to a residential fire is interesting and important. You will see in the chapter on search and rescue (SAR) that the time of a fire may determine where fire victims are found and hence where you should concentrate your search efforts. Based on the USFA charts that follow here, you have a greater chance of encountering a fatal fire between midnight and 0500 than any other time of day. The greatest number of fires occur between 1500 and 2300. Fires with injuries and fire loss follow closely in pattern to the number of fires distributed throughout the day (fig. 1–2).

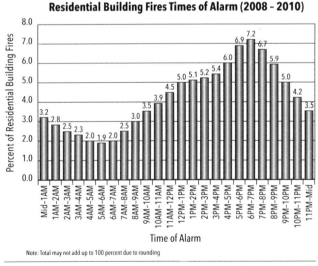

Residential Building Fires Times of Alarm (2008 - 2010)

Note: Total may not add up to 100 percent due to rounding

Fig. 1–2. Time of alarm compared to percent of house fires

House Fires as Mass Casualty Events

According to the USFA (NFDC 2009), each year 250 residential fires result in multiple fatalities. Of these, 81% occurred in single-family dwellings. Interestingly, multiple-fatality fires that result in five or more deaths represent only 3% of multiple-fatality fires in residential buildings. A broader look at national statistics shows that Americans most often die in one- or two-fatality incidents. This is further supported by the fact that 67% of multiple-fatality residential fires result in only two deaths. The next closest is three fatalities, which makes up 20% of multifatality fires.

The slump in the economy and the skyrocketing cost of housing across our country has caused more and more people to occupy individual homes. Headlines in the *Rockland Journal News* (Dec. 12, 1982) reported two fires, in these words: "28 people left out in cold at house fire" and "13 people escape house fire." In the first example, an older Queen Anne–style home had been transformed into multiple apartments. The second headline described a small, straight ranch–style home that was very fully occupied by 13 people.

Take a ride around you first-due area at night. Look for attic and basement lights on. Better yet, look for the blue flickering of televisions indicating that these usually unoccupied spaces have been converted to living spaces. Often, people with limited financial resources will pool those resources and live in very close quarters in what was once a single-family home. Communities with transient populations where seasonal peaks caused by farming, industry, vacation, sporting events, or other events often have extreme problems of overcrowding in houses. College towns characterized by fraternity houses or older homes where students live in every room of a house all possess the potential for multiple-casualty events (fig. 1–3).

Fig. 1–3. Once single-family homes, these may now have numerous apartments and occupancies even in attic and basement areas.

Alcohol as a Factor in House Fires

Many things complicate the American fire problem. Alcohol is one of them. This summary from a report by the Federal Emergency Management Agency (FEMA) and USFA effectively and concisely describes the problem you will face responding to a house fire:

Though the rate has significantly decreased, the United States continued into the late 90s with one of the highest fire death rates in the industrialized world. Given the advancements in fire prevention, including public education, building design, consumer product safety, and sophisticated levels of the fire protection in this country, it is puzzling to many as to why this is so. In an effort to identify the underlying problem(s), researchers have been delving deeper into the extent to which human behavior affects our fire losses.

The connection between alcohol and the ignition, detection, and escape from the fire has been broadly examined by numerous medical and fire protection organization studies. A series of landmark studies undertaken by the Johns Hopkins University and the National Bureau of Standards in the 1970s were among the first to discover a definitive link between alcohol consumption and fire deaths. Many studies have now confirmed their general findings.

Alcohol intoxication may increase the risk of initiating a fire by impairing one's judgment and coordination. An intoxicated individual who is smoking may also succumb to the depressant effects of alcohol, fall asleep and drop a lit cigarette on upholstery or clothing. Intoxication also acutely diminishes one's ability to detect a fire. Under the sedative effects of alcohol, an alcohol-impaired person may fail to notice the smell of smoke, or fail to hear a smoke alarm. Escape from a fire can be hampered by the loss of motor coordination and mental clarity caused by alcohol, even when warning signs are heeded. Furthermore, burns are more physiologically damaging in the presence of alcohol. Several researchers have found that about half of all adult fire fatalities were under the influence of alcohol at the time of the fire. Men have been found to consistently outnumber women among fire casualties and do so with even greater disparity for fire victims under the influence of alcohol. In addition, the younger adult population (ages 15 to 34) seems to incur the greatest number of alcohol-impaired fire casualties. Drinking behaviors that are characteristic of various age groups and sexes may explain these findings.

Studies have also provided conclusive evidence supporting the deleterious effects of chronic and acute alcohol abuse on the occurrence and recovery from burn injuries. Burn injury victims have been found to be disproportionately likely to have been intoxicated at the time of injury or known to be heavy drinkers. From a physiological standpoint, burn victims with histories of alcoholism tend to have longer hospital stays, more complications, and higher mortality rates as a result of their burns.

Questions still remain as to the extent that alcohol affects fire losses. How do we explain the fact that some industrialized countries with some of the highest alcohol consumption rates per capita, e.g. Germany and the Netherlands, have relatively low fire death rates? Researchers have suggested that alcohol-related unintentional injuries have more to do with alcohol drinking patterns than the total amount of alcohol consumed per capita. Who drinks, where they drink, what they drink, and under what social, cultural, and religious circumstances they drink are perhaps more significant factors than the amount of alcohol consumed. A lone drinker at home is probably at greater risk of a fire emergency than a group of people drinking at a bar or restaurant. Moreover, the number of drinks consumed in a single sitting seems to matter a great deal.

Alcoholics have a disproportionately high rate of fire fatalities relative to their percentage of the total population. Non-intoxicated fire victims also may be affected by alcohol: they may have been entrusted to the care of an alcohol-impaired individual. These fire fatalities would not be reported as related to alcohol when blood alcohol levels (BALs) are taken of victims only. As a result, the estimated number of alcohol-related fire casualties as well as the magnitude of the problem may be underestimated. Smoking fires are the leading cause of fire fatalities. The incidence of such fatal fires is higher among those who are under the influence of alcohol and most smoking-related fire fatalities have some connection to alcohol consumption.

In summary, there is a clear connection of alcohol and fire fatalities. Unlike the connection between alcohol consumption and vehicle fatalities, the connection is not often referred to in prevention programs, nor has much been done to address the problem. (NFDC 1999)

Firefighter Deaths and Injuries

According to the USFA report *Firefighter Fatalities in the United States in 2008*, in 2008 there were 21 fatalities where firefighters became ill or injured while on the scene of a structure fire; 71% of these occurred at residential properties (NFDC and National Fallen Firefighters Foundation [NFFF] 2008). In 2011, the USFA reported that 66% of the firefighter fatalities that occurred during fire attack operations occurred at residential fires (NFDC and NFFF 2011). Again, there is a fluctuation in these numbers over the years; however, the statistics remain relatively constant when looking at the overall hazards that house fires present to firefighters (fig. 1–4).

FIXED PROPERTY USE FOR STRUCTURAL FIREFIGHTING DEATHS
There were 21 fatalities in 2011 where firefighters became ill or injured while on the scene of a structure fire. This table shows the distribution of these deaths by fixed property use.

Residential	14
Commercial	6
Other	1

Fig. 1–4. Fixed property use for structural firefighting deaths

In terms of injuries, particularly for firefighters, the impact of house fires is overwhelming when viewed on a national level. In 2004, 76% of firefighter injuries at structure fires occurred at residential properties. Overall, residential properties accounted for 68% of all firefighter injuries. The number of firefighters getting injured at one- and two-family properties is similarly outstanding: 59% of all firefighters injured in 2004 were injured at one- and two-family property fires. Interestingly, apartments and row houses accounted for only 15% of the injuries, a proportion that has remained almost constant for several years according to the USFA.

To really understand the operational hazards of house fires, you should read some of the National Institute of Occupational Safety and Health (NIOSH) firefighter fatality reports. These short but comprehensive reports contain a wealth of information about costly errors or sequences of fireground errors that resulted in firefighter fatalities. The reports are excellent and do not assign blame but rather, after careful investigation, reveal the sequence of events (some that those on the scene may not have even been aware of), which, together with other events, put together the failure chain that resulted in a fatal outcome for one or more firefighters. We must share the experiences of these firefighters to protect our own and to ensure that they did not die in vain.

A particularly poignant report was one done in 1995 by J. Gordon Routley that details the deaths of three career firefighters at a house fire. According to the report, "This incident illustrates the need for effective incident management, communications, and personnel accountability systems, even at seemingly routine incidents. It also reinforces the need for regular maintenance on SCBA [self-contained breathing apparatus], emphasizes the need for PASS [Personal Alert Safety System] devices to be used at every fire, and identifies the need for training to address firefighter survival in unanticipated emergency situations" (Routley 1995).

In essence, this routine house fire, very similar to those faced by departments all across our country on a daily basis, was complicated by several factors that resulted in the deaths of three veteran firefighters. These fatal complications are examples that could be faced by any fire department from the smallest rural district to the largest metropolitan city. This case history clearly shows how complex and demanding the house fire can be. We must never think of a house fire as routine. This fire is an example of what a formidable foe the house fire is and why it requires all the skills and knowledge we can muster. The following chapters lay the groundwork you will need for success at your next house fire.

References

Ahrens, Marty. 2010. *Home Structure Fires*. Quincy, MA: National Fire Protection Association. http://www.ekcjfd.com/Home_Structure_Fires.pdf.

——. 2016. *Home Structure Fires*. Quincy, MA: National Fire Protection Association. http://www.nfpa.org/news-and-research/fire-statistics-and-reports/fire-statistics/fires-by-property-type/residential/home-structure-fires.

Fire Analysis and Research Division. 2006. "U.S. Home Structure Fires." Fact sheet. Quincy, MA: One-Stop Data Shop.

Haynes, Hylton J. G. 2016. *Fire Loss in the United States during 2015*. Quincy, MA: National Fire Protection Association.

National Fire Data Center. 1999. *Establishing a Relationship between Alcohol and Casualties of Fire*. Arlington: TriData Corporation for US Fire Administration.

——. 2009. "Multiple-Fatality Fires in Residential Buildings." *Topical Fire Report* 9 (3). https://www.hsdl.org/?view&did=26259.

——. 2012. "Residential Building Fires (2008–2010)." *Topical Fire Report* 13 (2). https://nfa.usfa.fema.gov/downloads/pdf/statistics/v13i2.pdf.

——. 2017. "Civilian Fire Fatalities in Residential Buildings (2013–2015)." *Topical Fire Report* 18 (4). https://www.usfa.fema.gov/downloads/pdf/statistics/v18i4.pdf.

National Fire Data Center and National Fallen Firefighters Foundation. 2008. *Firefighter Fatalities in the United States in 2008*. Emmitsburg, MD: US Fire Administration.

——. 2011. *Firefighter Fatalities in the United States in 2011*. Emmitsburg, MD: US Fire Administration.

Routley, J. Gordon. 1995. *Three Firefighters Die in Pittsburgh House Fire*. USFA-TR-078. Emmitsburg, MD: US Fire Administration.

US Bureau of the Census. 2011. "Geography and Environment." In *Statistical Abstract of the United States: 2011*, 219–42. Washington, DC: US Bureau of the Census.

US Fire Administration. 2008a. *Fire-Related Firefighter Injuries in 2004*. N.p.: Federal Emergency Management Agency.

——. 2008b. *Residential Structure and Building Fires*. N.p.: Federal Emergency Management Agency.

US Fire Administration and National Fire Data Center. 2010. *A Profile of Fire in the United States 2003–2007*. 15th ed. Emmitsburg, MD: US Fire Administration.

2

Construction, Types of Houses, and Floor Plans

Summary

This chapter provides the reader with the critical background information needed to increase their confidence level for responding to their next residential fire relative to construction and floor plans. If the reader understands basic construction and floor plans and how fire can spread and travel in various types of homes, they will be better prepared to develop strategy, formulate tactics, and understand and execute fireground tasks.

Introduction: Reading the House

A classic World War II story holds that when General George S. Patton saw his enemy, German general Erwin Rommel, far from home in the desert of North Africa, he shook his fist in the air and said, "Rommel, you bastard, I read your book!" Like reading the enemy's book on military tactics, an intimate knowledge of types of homes, their floor plans, their construction, and their built-in fire problems will be an invaluable aid in preparing for your next fire.

If you, the incident commander, company officer, or firefighter, like General Patton, "read the book" on the house before you arrive on the scene, it is likely that you have the prebattle intelligence you need to be safe and successful at house fires. You will understand many critical factors that may mean the difference between life and death for civilians now trapped inside. These same facts will go a long way in both directing and increasing the safety of your members. That is the goal of this chapter: prebattle intelligence. But information is not intelligence until it is analyzed then used on the battlefield. For example, we are all aware of the weakness of trusses under a fire load. The real intelligence is understanding the potential for collapse and rapid fire spread on the particular fireground you're called to, resulting in a plan to keep your members or yourself safe. For example, in figure 2–1, obscured from your command post view by smoke, the roof of this house is supported by lightweight trusses. This is an example of preincident intelligence that is critical to the safety of your members. Some of the critical facts you need to know before you operate at a house fire are as follows:

1. Floor plan—what rooms are where in the house

2. Hazards of that specific type of construction

3. Common routes of fire travel and spread

4. Number of floors

5. Need for ground ladder placement

6. A general fire attack plan

7. A general search-and-rescue (SAR) plan

8. A general ventilation plan

Fig. 2–1. Smoke obscuring intel important for size-up

Single-family dwelling defined

Let's refresh our definition of a house. For the purposes of this book, a house is a single-family dwelling (SFD). An SFD is (or was originally) a house designed for a single family. It may be large or small and vary in layout and number of floors. It is a home.

Many SFDs have been converted into apartments or multiple single-room occupancies. These occupancies have very different fire loads and specific hazards that are not a part of this chapter's scope (but will be covered in chapter 3). These buildings can no longer be called houses and contain deadly general and specific hazards for firefighters. Though many of the rules presented here are applicable generally, we are specifically examining the traditional house designs.

Across our nation, there are many types of homes, with different styles, floor plans, and designs. Regional variations and names abound and it is important to be familiar—very familiar—with specific designs in your first-due area. Across the country, it is impossible to say one design is more dominant than the other. Home designs vary between regions, states, and even down to subdivisions and neighborhoods. Types of homes will vary based on when they were built and the building codes of the time, if any were even in effect. It is very likely that your area contains homes from many different generations of expansion, built by a variety of builders. Thus, it is important to know the types of houses you and your department will respond to in your first-due area or your mutual aid response areas.

Typical room groupings

Like other occupancies, homes are made for a specific purpose to provide for the needs of a group of people, typically a family. This group needs sleeping, cooking, bathing, living, and storage areas, so we can expect these areas in all homes. Most homes have these as separate areas and provide clues to where you may find them during an emergency. Room location and identification is important in developing a strategy and executing tactics and tasks. Curtis Rice, a district chief (retired) from Palm Beach County Fire Rescue (FL), and Elvin Gonzalez, a captain from Miami-Dade Fire Rescue (FL), provide an excellent summary of how and what you can learn from an "exterior 360" of the house (Rice and Gonzalez 2011). A few of their ideas are summarized in the lists below.

Common floor plan concepts:

- There is a natural grouping of rooms that forms the basis for the floor plan
- Kitchen: generally at the rear of the home
- Dining area (room) next to kitchen
- Carport or garage often near kitchen
- Laundry, pantry, or mud room often near kitchen
- Bedroom at opposite side of house (working side, sleeping side)
- If two floors: noisy-type rooms on first floor, quiet areas on second floor
- Even custom-built homes tend to separate functions in a similar way
- Development homes may have common floor plan

Exterior 360 starting at the front ("A") side:

- Front entrance, carport or driveway nearby
- Stairs are generally behind the front door if a multistory
- Large window may indicate a living room
- Tap the glass of the window—hurricane windows make a distinct thudding sound
- Bars or security devices on windows?
- Multistory home: may have a small window at each landing, older homes might have stained glass
- B-side windows: if they are the same, likely both in one or more bedrooms
- C-side windows may indicate bedrooms as well
- Smaller frosted window likely a bathroom, look for plumbing vent through the roof over bath

- Large patio doors likely on C side could be family room, dining room, etc.
- Toward the C-D corner: smaller window, vent stack above, likely over the kitchen sink

Doors:

- If it opens toward you, likely a storage area
- If it opens away from you, generally an area people can occupy
- If you can feel the hinge it opens toward you
- Often an open, nondoor arch between kitchen and dining room

Home Styles and Floor Plans

Straight ranch

This is a common, basic, and simple type of home. Many were built for veterans returning from World War II. It is generally one floor of living space with attic storage. It may or may not have a basement depending on what part of the country you are in. If a basement is present, suspect that it may be occupied due to the limited living space in these homes. Often these are starter homes for first-time home buyers with small and growing families. The house could be rectangular as in figure 2–2 or L- or even T-shaped.

The figures below show examples of this type of home and common modifications or additions you may encounter in your career. In figure 2–2, we can see a basement entrance door just behind the shrubs. This may be living space for one family or it could be a legal or illegal apartment with no access between the basement and first floor. The illegal conversion will present a myriad of challenges and dangers to firefighters. A room has been added onto the straight ranch in figure 2–3, as well as a screened-in porch. The rear door to the house is inside the screened-in porch. The home in figure 2–4 started as a one-and-a-half-story raised ranch. As the family outgrew the 1,100 sq ft, they added a whole additional floor, doubling their living area. The interior stairs to the second floor are just to the right of the front door. Note the area where there are no windows on the first floor, and also the smaller bathroom window and window on the stairwell near the upper right. In figure 2–5, a second floor was added to this previously one-and-a-half-story straight ranch house. Note that there is excellent access to upper floors where the bedrooms likely are. There is no rear door but there is a side door on the B side, hidden by the trees in this photo.

Fig. 2–2. Basement entrance door possibly indicating illegal occupancy

Fig. 2–3. Add-on room and screened-in porch

Fig. 2–4. Additional floor added to one-and-a-half-story raised ranch

Fig. 2–5. Add-on second floor with good access

Raised ranch

One common variety of this type of home is a split level, which has an added upper level or possibly two additional levels. Behind the front door are stairs that go both up and down. Main living areas (all areas not used for sleeping) are usually located on the main floor, with bedrooms upstairs and a large den or open area in the basement or first floor (fig. 2–6).

Fig. 2–6. Common suburban or subdivision raised ranch home. Note that it is essentially a straight ranch floor plan elevated with a basement and garage under it.

Split level. A common variation of the raised ranch is the split-level raised ranch. This design has the raised portion of the home split into two different levels as shown in figure 2–7. In this layout, the bedrooms are most likely over the garage. The basketball hoop suggests that there may be at least one child at home. After you enter the front door, the living room is to the right (larger window) with the kitchen and dining room to the rear. Rooms to the right might be bedrooms, offices, or other similar rooms.

Fig. 2–7. Typical split level. Courtesy of John Mittendorf.

High ranch

This type of home has a garage underneath; inside the garage to the right, a door opens to a mechanical room then to a den or recreational area. Bedrooms are located above the garage with the living room behind the picture window, dining room behind the living room, and kitchen off the dining room. The master bath is opposed to the kitchen between the kitchen and the master bedroom in the rear corner (fig. 2–8).

Fig. 2–8. A high ranch

A high ranch has a major junction of hallways and stairways just inside the front door. It is important to get an interior size-up from this position because this is where you will first get an indication of where the fire is and where it is going. If you are unsure where the fire is in the home, force the door and get a quick look inside. Surely not a place to establish your command post, but this view will provide you critical information such as the location of the fire, where it is going, conditions inside, any victims trapped by fire from their normal means of egress, and so on. Often these homes are called mother-daughter homes. The family lives upstairs or occupies the largest portion of the home and the mother of one of the adults living upstairs lives in a downstairs apartment. Even though there are two separate living units, it is still a single-family home because there is unrestricted access between the living units (fig. 2–9).

Fig. 2–9. A downstairs living unit, just inside the front door, which is open at the left of the photo

Knowing the layout of the home is critical to your safety. At this point in the house in the figures above, a few steps will take you to many different areas. Half the flight of stairs goes up to the living area and half goes down to the recreation area. At the bottom of the stairs you make a left and go through the mechanical room and into the garage. Make a right at the bottom of the stairs and you enter a recreational room or the downstairs apartment. Up the stairs to the right is the living room area, and the hallway to the bedrooms and bathrooms is to the left at the landing in front of the clock. The dining room is to the right, behind the living room, and the door to the kitchen is just to the left of the clock.

Classic two-and-a-half-story wood frame

Architects may know this home as a semi-Victorian or folk Victorian because of its lack of highly ornate detailing both inside and outside. It is large like a Victorian (as opposed to a ranch) but not that large, ornate like a Victorian but not that ornate (fig. 2–10). In the fire service, we know them as two-and-a-half-story wood frames. Two-and-a-half-story wood-frame homes are prevalent across the country, in rural, suburban, and urban settings. Built during the industrial expansion period (typically between 1870 and 1950), they are a blend of new and old construction methods and present a host of issues for firefighters. Typically, they are balloon-frame homes and therefore allow rapid hidden vertical fire spread.

Fig. 2–10. Two-and-a-half-story wood-frame house

Due to the age of these homes in economically depressed areas, they may be run down and are common locations for working fires. In areas where high property values are prevalent, such as the suburbs of major cities, these homes often house large extended families or groups of poorer citizens. Often these large homes will have been converted to legal or illegal apartments or even single-room occupancies. I have been at fires in these types of homes where people had complete (illegal) apartments in both the attic and basement of the same building.

Cape Cod

A Cape Cod is another standard design house that is common across most of our country. They are usually a story and a half, the half being the uppermost portion of the house where the bedrooms are located.

The dormers shown in figures 2–11 and 2–12 are common to Cape Cods. Dormers are windows set vertically in the sloping roof. Dormers have roofs of their own and may be flat or arched but are usually pointed. Similar to a ranch house, they are usually small in square footage and at their maximum height, ground ladders will reach all parts of the building for both vent, enter, and search (VES) and roof ventilation operations. Note the steep roof, which at a minimum will require a roof ladder as footing for the crew cutting the roof.

Fig. 2–11. Dormers on both the front and rear

The dormers provide light and ventilation for the upstairs rooms. Fire departments often mistakenly open the roof over the dormer. Do not use these as ventilation points for roof cuts; only a limited amount of heat and smoke will be released. Cut near the peak of the roof and as close to directly over the fire as possible (fig. 2–12).

Fig. 2–12. Simplicity of design and large rear dormer. Note the air conditioners that will delay entry and ventilation of the windows on the B side of the house.

Cape Cods also have knee walls on the second floor. These walls are 3–4 ft in height and join the steeply sloping roof. These often form large void spaces that are used for storage and are excellent spaces for undetected fire spread.

Contemporary

As the name implies, this is a newer architectural style that varies with the imagination of the designer (figs. 2–13, 2–14, 2–15, and 2–16). Considering that these are still homes, they have many of the same fire problems as other homes and contain the same functional rooms as other homes: living room, kitchen, baths, and bedrooms. The issue is that in contemporary homes, it may be impossible to determine the exact location of these rooms from the outside during an exterior size-up.

Fig. 2–13. Contemporary homes vary widely in appearance, layout, and construction.

Fig. 2–14. A contemporary home

Fig. 2–15. Another design for a contemporary home

Fig. 2–16. Rear view of a contemporary home. Note the lack of window access for VES.

Often these homes have an office as part of their layout. High ceilings can hide high heat buildup from unsuspecting firefighters. Use your thermal imager aggressively to look up, down, and all around for hidden fire.

Large windows probably indicate some type of living or working area in this contemporary home, although it is impossible to tell from outside. However, the ventilation potential of these windows is immense. Should they fail during the fire development, fire will very quickly and violently involve the entire house.

Contemporary homes are generally regarded as those with a modern or futuristic design that doesn't conform to a classic two-and-a-half-story wood-frame private dwelling or duplex. The house in figure 2–17 is contemporary in that it was recently built, but its design is similar to that of an older-style farmhouse. The fireplace, large porch, and pitched roof are characteristics of classic farmhouses, but this house was recently built using lightweight wooden I-joists and roof trusses and will collapse rapidly under a fire load. To some, it may look like an old design but it is more contemporary than classical. One of the tip-offs is that the entire development is made up of similarly styled homes.

Fig. 2–17. Contemporary home designed to look like an old farmhouse

Firefighter Steve Sparks from the Portland (OR) Bureau of Fire shares the following unique tactical challenges presented by contemporary home designs:

First, we are seeing a lot of new homes being built in much older neighborhoods. On the exterior they look exactly like the older homes that were built 60–120 years prior. Older, legacy homes have dimensional lumber with traditional compartmentalization typical of homes of that generation. The new models are built with the same traditional exterior design with dormers, porches, numerous windows, etc. What they conceal from us is that they are built with open floor plans allowing very rapid fire spread and lightweight structural members (manufactured lumber, trusses, etc.). It is very difficult to tell one from another during any fire operation, especially your size up at 0230 hrs with heavy smoke showing.

If the home has a contemporary design, it will likely look much different from the classic two-and-a-half-story wood frame previously described and is easy to size-up as containing lightweight materials. However,

due to desires to keep classic designs coupled with demands to lower costs, lightweight building materials are often used and essentially hidden by cosmetic architectural features (fig. 2–18).

Fig. 2–18. Can you tell which is built with lightweight members that will not bend to gravity under a fire load?

Sparks describes a second problem with a different style of contemporary homes:

They are not classical style homes, they are modern looking, often built with steel frames and structural members combined with lightweight engineered wood and/or trusses. At least we know what kind of tactical issues we may have to deal with at this type house. Frequently these one of a kind (designed) homes that are prevalent throughout the western US; built on steep slopes to take advantage of the spectacular views. The window side is usually opposite the street or driveway side and may have multiple large windows, walk out decks or floor to ceiling windows or a combination of these. Fire Department access is also a frequent problem for search, fire attack and ventilation operations. Another feature of these uniquely designed homes is that they usually have very open floor plans, sometimes with bedrooms that are open in a mezzanine type layout that can be on any of the floors. As a result of the designs it may be impossible or difficult at best to determine priority search areas from the exterior.

The house in figure 2–19 has limited access on three sides. If this is the A side of the home, the C side may have two or more floors below this grade or floor and drop hundreds of feet into a canyon.

Fig. 2–19. Seattle box house, up and down view

Steel structural members are clearly holding up the roof of the huge home in figure 2–20. As this home is vastly larger than anything in the neighborhood, responding firefighters could be surprised when they roll up to what they assumed was a much smaller home. House size has been increasing over the past 20 years in homes all across the United States.

Fig. 2–20. Seattle new steel structural

Saltbox

This is an older style found largely in the northeastern United States. It is a common design in both rural and suburban areas. Although these houses are usually small, they have large-scale safety issues for firefighters.

First, the small rooms will flashover quickly, especially the upstairs bedrooms. These rooms have knee walls in them that make standing up impossible near the roof line of the home at the eaves. The knee wall is a short wall, often only 2–4 ft in height, that meets the sloping roof forming the wall and ceiling of the rooms on the second floor (fig. 2–21). I (Jerry) learned about these knee walls the hard way at a fire

several years ago. We had a good room-and-contents fire and I was searching the second floor. As conditions improved a bit I tried to duck walk rather than crawl. Each time I raised up off my knees, I hit my head on the sloping ceiling just above the knee wall. After three tries, crawling was the least painful option.

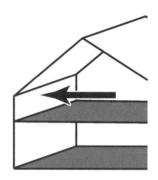

Fig. 2–21. Knee wall in a saltbox

The windows in this style are very important to firefighters. They are approximately 14 in. high and too small to exit from under emergency conditions (fig. 2–22). If fire races up the stairs or the second floor lights up while you are searching it, your only way out is down the stairs. An option to increase firefighter safety is to cut the roof near the eave line to open up a second escape route for members operating on the second floor (see fig. 2–23). This would have to be done preventatively, most likely while members are operating on the second floor. This tactic is time and labor intensive and may not be completed in time to save members if it is done after things go wrong. It is an excellent task for a firefighter assist and search team (FAST) or rapid intervention team (RIT) to accomplish. Another option would be to cut a larger window opening from the window sill downward. This depends on the height of the window in the wall. Again, this is time and labor intensive because members are working near the edge of the roof or off a ladder.

Fig. 2–22. Note the small windows on the second floor.

Fig. 2–23. The rear of this type presents the same firefighter escape issues as the front: windows too small to exit. Judging by the number of exterior doors in this home, it is likely that there is an illegal apartment in the basement in addition to its other occupancies.

A slightly larger version of a saltbox is sometimes called a four square, so named because it has four square rooms above three larger rooms. This particular model has had the porch enclosed, which may be a true porch area or may have been converted into living or sleeping areas. Because of the land-scaping and grades around the house, it may be the equivalent of three or four stories in the rear. Windows off the porch roof may provide access directly into the bedrooms (fig. 2–24). It appears that the addition on the left side of the house was a garage that has been converted to living space (fig. 2–25).

Fig. 2–24. Windows off the porch roof may provide access directly into the bedrooms.

Fig. 2–25. Garage converted to living space

Figure 2–26 shows a western US version of a duplex home. Similar to saltboxes in the east, this home has small windows on the second floor that provide nothing in the way of VES or egress for occupants. Note the second door behind the shrubs on the right side of the house.

Fig. 2–26. Western US duplex home. Courtesy of John Mittendorf.

Bungalows

A bungalow is a 20th-century-style home that is one and a half stories with most of the living space on the first floor. Bungalows were often built during the building boom after World War II. They are characterized by low or steeply pitched roofs. Bungalows may be something as simple as a one-room, studio-type building, the smallest of all their namesakes. Bungalows have a partially occupiable second floor and are often found in seasonal communities but may be used year-round in others. They are typically built with limited insulation and electric service, and when converted to year-round use, improvements may or may not be up to code. The stairs lead up to a partially finished second floor. Knee walls 2–4 ft high run the length of this area.

Deputy Chief Bruce Tenniswood highlights a serious but unique fire problem with these homes. Kitchen cabinets that were installed in the 1950s have likely been replaced with modular, modern cabinetry. During the renovation process, holes in the walls remained where the original cabinets were built in as a part of the structure. This allows fire to spread rapidly into the second-floor knee wall area.

> Kitchen fires in bungalows will normally turn into second floor attic fire unless you understand these characteristics and aggressively pursue the fire through these paths of travel. You need to enter the void space behind the knee walls in the bungalow style house as soon as possible and to find and stop fire spread. This is one time that making an assumption is the correct course of action. Assume that if there is fire behind a cabinet, there is fire behind the knee wall. (Tenniswood 2009)

A second serious path for rapid fire spread is the 2×8 joists that support the second floor. Tenniswood continues,

> These homes normally have 2×8 ceiling joists that span from one outer wall to the other.... Additionally, the builders of these homes normally fastened a 1×6 subfloor in a diagonal fashion to create the floor on the second floor. This floor covers the area between the knee walls only, leaving the areas behind the knee walls open to the roof and floor structures. Any fire that enters the void created by the joists has free run across very dry wood to both sides of the structure. (2009, 105).

Figure 2–27 shows a typical bungalow-style home. Note the proximity of shrubs and trees, which could create a wildland interface issue. The open skylight and tilt-out windows point out living space on the second floor. Figure 2–28 shows the typical floor plan of a bungalow, efficient and economical. Note that there are no large or long hallways; all rooms connect to each other.

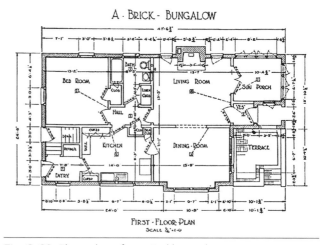

Fig. 2–28. Floor plan of a typical bungalow

Victorian

This term characterizes an architectural style that was popular from 1840 to 1900 and had a wide variety of designs. The industrialization of the United States made new materials and construction methods available to fabricate ornate porches, turrets, and widow's walks. Victorians are often very ornate, with very steep roofs sometimes covered in slate. They usually have a front-facing gable on the third floor or higher, making the house look even bigger.

A Queen Anne is a type of Victorian. It is the most elaborate and is usually highly decorated with fancy railings, trim on eaves, and wraparound porches (fig. 2–29). In the house pictured in figure 2–30, just beyond the front door is a large open foyer and an elegant and wide set of steps to the second floor. We know or suspect that this was previously a large house for a wealthy owner. Chances are, this home now contains several apartments or maybe a bed and breakfast, a law firm, or similar office space. Note the tall and supported chimney on the upper left. Tactically, the front porch offers excellent access to several rooms for VES or other uses.

Fig. 2–27. Typical bungalow. Courtesy of John Mittendorff.

Fig. 2–29. Queen Anne–style Victorian

Fig. 2–30. Large Queen Anne. Courtesy of John Mittendorff.

Colonial revival

Typically built between the 1800s and early 1900s, these are home are similar to classical colonial architecture. They have a rectangular footprint, generally two floors, symmetrical window placement, and a pitched roof. Usually they are clad in brick or wood siding. Bedrooms are on the second floor and the living and kitchen areas are on the first floor. Generally there is a center hallway or entrance (fig. 2–31).

Fig. 2–31. Typical colonial revival home

McMansion

These large homes started showing up in affluent suburbs around 1980. McMansion is a derogatory term used by architects to denote that the building has little architectural value despite its huge size, massive floor areas, and large rooms. The term came from the mass production of fast food, and is entirely apt for this style home: mass produced with little architectural, landscape, or siting care or value. McMansions are often placed on an inappropriately small piece of property and are, from an architect's point of view, tacky. Like

their smaller cousins, in developments they may all have the same floor plan or be limited to several options.

From a firefighter's point of view, these homes present a variety of additional hazards mainly from two sources: construction methods and the sheer size and volume of combustibles. In their favor, McMansions in newer planned communities may have to comply with more stringent building and fire codes. Smoke and fire detection and even residential sprinklers might be present. This is important because responding to an automatic alarm is often the first indication of a working fire. Additionally, the early detection and possibly suppression may reduce your SAR concerns.

As prime land in the immediate city suburbs was built upon, less desirable property was selected to build homes. Sites such as steep slopes, mountaintops, swampland, and even legal or illegal landfills were built upon. It is the steep slopes that provide us with access challenges. Houses built on mountaintops or hills often have steep driveways and long set-backs with multiple layers of retaining walls in the front yard. Essentially, firefighters have to climb several floors worth of height just to reach the front door. Some larger set-backs and the steepest slopes can make even a 100 ft ladder truck unusable, as it will not reach the home (fig. 2–32).

Fig. 2–32. House built on a very steep slope

Often constructed with lightweight building materials (see ch. 4) that span large, open floor areas, these buildings are subject to rapid and localized collapse of floor sections that too many times drop firefighters directly into the fire, causing line-of-duty deaths. The sheer size of these homes causes numerous other problems, such as:

- Flashover: Can occur in high cathedral ceilings and go undetected until it is too late for interior firefighters
- Long stretches to get to the building and move around inside
- Excessive set-backs on steep grades and limited access to aerial ladders

These homes are sometimes built in mountainous terrain far beyond the hydrant district, creating water supply issues. Long supply lines or tanker task forces may be required to establish the required fire flows. Large houses such as these often require multiple lines due to the heavy amount of contents, combustible materials like vinyl siding, and structural members that are burning (fig. 2–33). In the McMansion in figure 2–34, the attached garage was converted to office space, which was where the fire started. An aggressive interior attack supported by roof ventilation contained the fire and saved the remainder of the home. Note that due to energy efficient construction, smoke is not showing from windows or through walls.

Fig. 2–33. Large houses often require multiple lines.

Fig. 2–34. This McMansion fire was able to be contained with an aggressive interior attack supported by roof ventilation.

House Construction Methods

It is absolutely critical that you identify and understand the type of construction that was used to build the house you are conducting SAR or fire suppression operations in. This information only becomes intelligence after it is analyzed, digested, and put into usable form. It will make the difference between life and death for you and your members.

Consider that a military officer would never think of sending their troops into battle without good intelligence on the enemy. The fire officer should never consider sending their firefighters into life-threatening situations without good intel either. This section provides you the intelligence report on the hazards associated with common types of home construction.

Understanding construction

In a conversation the authors had with Frank Brannigan at a Fire Department Instructors Conference (FDIC) in Indianapolis in 2000, he said something like this, referring to a fire and collapse that killed several firefighters: "That building did not just land from Mars yesterday and catch fire today. It had been there for years, decades! What do you mean they did not know the construction weaknesses? Why not?"

This comment highlighted one of our major weaknesses in the American fire service: we don't like to gather preincident intelligence. We focus on a million other things but we don't like to do our critical homework: identifying and understanding construction methods of buildings in our first-due area and analyzing how they react to fire loads.

Preincident intelligence

Every fire department should aggressively go out and document in a useable format information on the houses in their first-due area. Often there are only a few types, and sometimes groups of houses have the same common hazards to firefighters that we examine in this section. The key is to have the information accessible when you need it. The form is not important; the function is.

At the outset, it sounds hard. It is not. It is a bit time consuming. The key is to put the data into some useable and easy-to-understand format. It is the best thing you can do to protect your firefighters' lives. The process is simple if you follow these easy steps:

1. Take a map of your response area and categorize the houses you have by construction type.

2. Document this in a simple, easy-to-use format.

3. If you have computer-aided dispatch (CAD), put this life-and-death information in the system so the dispatcher can refer to it during and after dispatch and warn you over the radio. Better yet, have it available to the officers via a tear-off sheet from the printer so they can read it en route!

It is also important to have this intelligence available to second-alarm companies and covering or mutual aid units. The true value of information and intelligence comes when it is shared.

STOP: Safety Tactics and Operational Plans

The West Haverstraw Fire Department (NY) has distributed critical preincident intelligence on what they call Safety Tactics Operations Plans (STOP). This one-page (front and back) document is in a clear plastic document protector in a three-ring binder, and contains vital information about the homes' built-in construction weaknesses. It is easy to read and conveys critical information (i.e., preincident intelligence) in a very visual way. More importantly, it provides the departments with guidance for rescue and fire attack operations (standard operating guidelines—SOGs) in a visual format. It is an excellent training tool that can be pulled out and turned into or made part of a company drill. It is an excellent refresher for experienced firefighters and a superb training tool for young and developing fire officers as well as a way to pass on critical information about the buildings learned through experience. Remember, experience is our best teacher, but it is also the most expensive in terms of firefighter injuries and deaths (figs. 2–35 and 2–36).

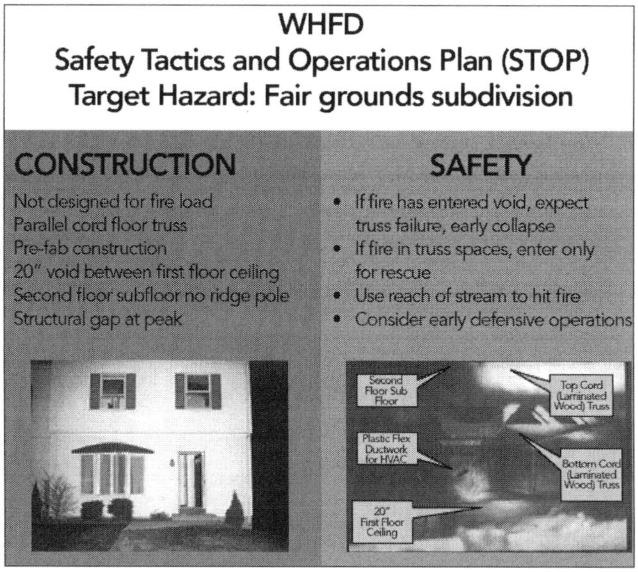

WHFD Safety Tactics and Operations Plan (STOP) Target Hazard: Fair grounds subdivision

STOP PLAN
- If no one is inside, stay outside
- Do not enter for fire suppression if significant fire has reached the truss space
- Stay off the roof

TACTICS
1. Determine if fire has spread/started in void
2. Keep firefighters from above and below fire weakened trusses
3. Use reach of stream to hit fire in involved area
4. Anticipate collapse

Note: Heavy fire involvement of truss

Web of truss that failed due to fire load

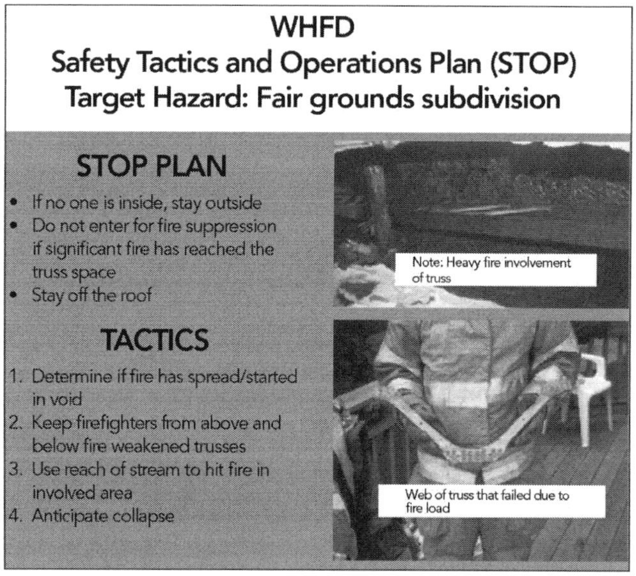

Fig. 2–36. The back page of the STOP document describing strategy and tactics (i.e., the department's policy)

WHFD Safety Tactics and Operations Plan (STOP) Target Hazard: Fair grounds subdivision

CONSTRUCTION
Not designed for fire load
Parallel cord floor truss
Pre-fab construction
20″ void between first floor ceiling
Second floor subfloor no ridge pole
Structural gap at peak

SAFETY
- If fire has entered void, expect truss failure, early collapse
- If fire in truss spaces, enter only for rescue
- Use reach of stream to hit fire
- Consider early defensive operations

Second Floor Sub Floor
Plastic Flex Ductwork for HVAC
Top Cord (Laminated Wood) Truss
Bottom Cord (Laminated Wood) Truss
20″ First Floor Ceiling

Fig. 2–35. Front page of Safety Tactics Operations Plans (STOP) describes building weaknesses

Construction and Framing Methods

There are four main types of construction methods that are widely used across the United States. From the most recent to the oldest they are: lightweight or engineered lumber (and designs), platform or western construction, balloon framing, and post- and-beam framing. Each has built-in strengths and weaknesses that firefighters must be aware of to protect themselves during SAR and fire suppression operations.

Common construction terms

There are several framing members that are found in many if not all types of construction. For example, all walls will have studs; it is the length of the studs that will determine exactly what type of framing was used. Don't be surprised to see a variety of framing methods in the same home. The home may be an older home that was renovated with new framing methods or an addition with different framing. It could be a home that was built before building codes, or built illegally without inspections, or built using a combination of framing methods based on the availability of materials and the skill and knowledge of the builder. Figure 2–37 shows some structural components common to all framing methods. Each particular type of framing may have some of these components and other specific or unique types of structural resistance to the forces of gravity.

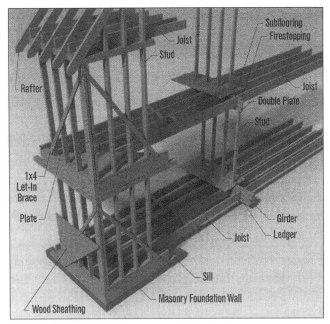

Fig. 2–37. Common terms used in the framing industry

Lightweight or engineered framing

The most important thing you must understand about lightweight or engineered framing materials is this: they are not designed to withstand a fire load. They are designed to stand up to live loads, dead loads, wind loads, and snow loads imposed on them. They are designed to reduce the cost of the building, to reduce construction time, and to span and create large open areas inside the home. Engineered lumber or laminated veneer lumber is stronger than its dimensional counterparts. Often, laminated veneer lumber is used to make lightweight trusses and wooden I-joists. It does, however, trade mass for math, and in terms of fire resistance, mass is very important. Always remember that *engineered lumber and trusses are not designed to maintain their structural integrity when attacked by fire.*

Laminated veneer lumber. Deputy Chief (retired) Gregory Havel from the Burlington Fire Department (WI) and safety director for Scherrer Construction Company is uniquely qualified to describe what engineered lumber really is and how it is made.

Lightweight construction defined. There are two main structural components used in home framing that are referred to as lightweight construction: trusses and wooden I-joists. There are three fire problems associated with both of these. First (for both trusses and I-joists) is that they lack the mass of wood used in traditional or legacy construction. This allows the material to ignite earlier in the fire's life cycle, which obviously adds to the fuel load that is already burning. Wooden I-joists (usually used as floor supports or joists) are made of

plywood and burn through rapidly, which causes the joist to lose its structural integrity, leading to an early collapse.

Second, for trusses, is that the mechanical joints necessary to form the truss fail rapidly during fire exposure. This failure causes weakening of the structural member, which often results in early roof or floor collapse. Both of these events have killed numerous firefighters and continue to do so.

Thirdly, there is interdependency in truss and lightweight framing. Each member must carry its design load, and other members have little excess capacity to share the load, especially if a member fails. In short, it is not overbuilt and failure can cause a cascading effect of other trusses and structural members, resulting in a catastrophic collapse (fig. 2–38).

Fig. 2–38. Legacy construction roof rafters have more mass and fewer connections, limiting the cascading effect when weakened by fire

Parallel cord trusses. Trusses come in mainly two forms for house construction: parallel cord and triangular (fig. 2–39). Parallel cord trusses are used for flooring systems or flat roofs. They are designed to replace legacy joists that may be 2×8 or 2×10 in. in size. Parallel cord floor trusses are engineered with smaller dimensional lumber than legacy construction. If installed properly, they may actually be stronger over the designed span than dimensional lumber under all loads except a fire load.

Fig. 2–39. A triangular truss system holding up a roof in a home

CONSTRUCTION CONCERNS: LAMINATED VENEER LUMBER

Article and photos by Gregory Havel

Laminated veneer lumber (LVL) is one of the types of "manufactured wood" or "engineered lumber" that were developed in the late 20th century. Sawn lumber longer than 24 ft has never been common. Since most sawn lumber today comes from second-growth timber or tree farms, it is difficult to find sawn lumber longer than 16 ft that is straight and strong enough, and of good enough quality, to be used as floor joists and rafters.

LVLs are built up of layers of veneer that are rotary-peeled from logs. These layers of veneer are coated with high-strength glue and stacked in a large billet to the desired length, width, and thickness. Usually, the grain in all of the layers is parallel. The LVL billet is cured in a heated press, which brings it to its final thickness and sets the glue. Then the billet is sawn into standard lumber dimensions for use as beams and joists in construction, and for flanges in wood trusses and wood I-beams. Photo 1 shows a comparison between 2 × 12 LVL beams and the sawn 2 × 6, 2 × 4, and 2 × 12 lumber stored behind them.

Photo 1.

LVLs are advertised to be straighter, more uniform, and stronger than sawn lumber of the same dimensions. Manufacturers' engineering information allows spans 25 percent longer for LVLs than for sawn lumber of the same dimension. Although they are made with water-resistant glue, they are not recommended for continuous exposure to weather without protection. They are stocked by lumberyards in longer lengths than any sawn lumber, and are available by special order in almost any length. Photo 2 shows a large house under construction. The rafters shown in the higher part of the house at the right are 40-foot 2 × 12 LVLs, with a 2 × 18 LVL ridge board. The rafters

shown in the lower part at the left are 26-foot 2 × 10 LVLs with a 2 × 12 LVL ridge board. Both roofs will be sheathed with plywood and covered with slate.

Photo 2.

Photo 3 shows wood I-beams with LVL flanges and high-density oriented strand board (OSB) webs, supported by galvanized steel joist hangers from LVL beams and girders.

Photo 3.

Roof trusses can be up to 24 in. on center and provide a significant amount of material to burn. Note the smaller dimensions. Triangular trusses are used primarily for supporting roofs. They are made from lumber with smaller dimensions and fail rapidly when exposed to fire. It is the connections that cause the rapid loss of structural strength.

Connections used on trusses. Gusset plates hold the parts of the truss together (figs. 2–40 and 2–41). There are no other nails, screws, or mechanical fasteners of any kind. Heat is absorbed by the surface of the plate and transferred into the small fingers that stick into the wood. Char develops, and this weakens the connection until failure occurs. Failure of one member of the truss will result in failure of the entire truss. Note the total failure of the connections and therefore the truss in figure 2–42. This roof is in imminent danger of collapse. In fact, it is a miracle it has not collapsed already, as all its structural elements have failed! This fire burned for only about 12 minutes before extinguishment but had an unlimited amount of air.

Fig. 2–42. Truss failure

Another type of mechanical joint used in trusses is called a finger joint. Like those used in interior moldings, this joint uses friction and glue to hold the connection solidly. Trusses that use finger joints do not have any additional connection. There are no nails, gussets, or other fasteners (figs. 2–43, 2–44, 2–45, and 2–46). Obviously, when the glue melts and the wood begins to char and lose strength, we can expect a total loss of structural integrity and possibly rapid collapse of the load above the truss.

Fig. 2–40. Gusset plate

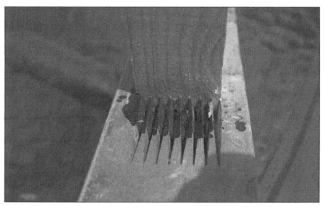

Fig. 2–43. Parallel cord truss with finger joint connections

Fig. 2–41. Parallel cord truss with gusset plate connections

Fig. 2–44. View of the end of a finger joint truss

Fig. 2–45. Cross-section of a finger joint. Courtesy of Rick Strauss.

Fig. 2–46. Parallel cord trusses during construction showing the first floor supporting structure. Courtesy of Rick Strauss.

Fire experience with lightweight construction

Case history. The best way to demonstrate the difference between legacy and lightweight construction is to describe a fire we responded to a few years ago and the case histories that followed it (figs. 2–47, 2–48, and 2–49).

Fig. 2–47. Older, legacy built home renovated by installing a new roof supported by lightweight truss systems

Fig. 2–48. Several years after the renovation we responded and found it fully involved.

Fig. 2–49. Total failure of the roof trusses

Despite an intense fire, the legacy construction maintained some degree of structural integrity. The second floor was still intact, although it was a dangerous place to be because of the collapse of the third floor. The trusses failed quickly and completely under the same fire load. Note that the roof trusses collapsed.

Legacy construction. Legacy construction—also called stick-built because it is built one "stick" at a time—uses dimensional lumber. Stick-built legacy homes use 2×4 in. or 2×6 in. wood for studs and 2×6, 2×8, or 2×10 in. lumber for structural supporting elements such as walls, floor joists, and roof rafters. Under a fire load this dimensional lumber provides more mass to area (fig. 2–50). This means that it will be slower to ignite because as the wood chars, it loses strength slowly. Do not think this means that the collapse will be slow and give you time to escape; it will not. What this means is that even when it is being attacked by fire, it will provide you some degree of structural integrity and safety. You just have to answer the ever-present question: How long has the fire been burning?

Fig. 2–50. These 2 × 10 in. floor joists will maintain their structural stability due to their mass, even as they are attacked and charred by fire.

Hazards of lightweight construction. As a result of the inherent weaknesses of lightweight framing materials, firefighters are killed in residential fires where this type of framing is used when, without warning, the building suddenly collapses. Often firefighters are thrown down into the fiery hell below and are killed or seriously injured. NIOSH reports contain numerous case histories of this exact scenario. You should review these, use them for company and department drills or training and learn from the experiences. The fire service is currently attempting to minimize the dangers to firefighters presented by this type of construction, but the problem is growing as more homes are built with this type of framing.

Case history of firefighter fatalities. According to an article published in *Fire Engineering*, "Between 1997 and 2008, 26 firefighters lost their lives in residential building collapses." Further in the same article, "Recent initiatives such as the National Firefighter Near Miss Reporting system have documented more than 80 reports of incidents involving lightweight and truss construction" (Dalton, Backstrom, and Kerber 2009). To fully understand the hazards of lightweight construction, we must fully comprehend the critical details of incidents involving lightweight construction. This understanding will lead directly to actions that prevent firefighters from being killed at these type homes.

A NIOSH report from 2007 contains the following summary:

> On January 26, 2007, a 24-year-old male volunteer fire fighter died at a residential structure fire after falling through the floor which was supported by engineered wooden I-beams. The victim's crew had advanced a handline approximately 20 feet into the structure with zero visibility. They requested ventilation and a thermal imaging camera (TIC) in an attempt to locate and extinguish the fire. The victim exited the structure to retrieve the TIC, and when he returned the floor was spongy as conditions worsened which forced the crew to exit. The victim requested the nozzle and proceeded back into the structure within an arm's distance of one of his crew members who provided back up while he stood in the doorway. Without warning, the floor collapsed sending the victim into the basement. Crews attempted to rescue the victim from the fully involved basement, but a subsequent collapse of the main floor ceased any rescue attempts. The victim was recovered later that morning. (NIOSH 2007)

It is important to remember two things when reading this report. First, this department did what many other departments across our country would do at this same fire. Second, this analysis must not be viewed as a criticism of their actions, but rather as a teaching tool to save lives in the future. It is critical to learn from these experiences. Experience is the best teacher, but it is also always the most expensive in terms of lives and dollars. So that these firefighters do not die in vain, we must learn from their experiences; read, share, and use them as training tools. Later on in the report, the firefighters on scene that day recount their actions:

> The IC had a crew preparing to make entry through the front door and he went back to the D-side of the structure. The flames were still concentrating and intensifying from the basement garage door. He went to the A-side as a fire fighter ran up to him and told him that the victim had fallen into the basement.

> The fire fighter who was backing up the victim dropped to his stomach and was yelling for the victim to take his hand. The victim replied that he could not reach it. While the fire fighter was on his stomach his low-air alarm sounded forcing him to exit. The IC went to an apparatus and took an attic ladder to the hole where the victim had fallen through. Crew members bumped the victim with the ladder as he crawled, and they yelled for him to try and crawl up the ladder while other members retrieved a roof ladder. The victim was in full gear as he crawled out of view towards the A-side of the structure. The flames were intense as the victim came back into view within 30 seconds and crawled under the ladder. At this point, he did not have on his helmet or Nomex® hood. The members placed a roof ladder into the 10 to 15-foot void in the floor created by the collapse. A Lieutenant immediately entered the fully engulfed basement in an attempt to rescue the

victim. The victim was unresponsive and in convulsions. The Lieutenant grabbed the victim's jacket and tried to move him, but was unsuccessful. He exited the basement as his low-air alarm sounded and requested additional manpower to move the victim. Two other fire fighters entered the basement. They hooked a rope to the victim and were forced to exit due to the extreme heat conditions. Members in the garage attempted to pull the victim up from the basement with the rope, but still could not move him. The remainder of the main floor collapsed as the garage collapsed around them. All the fire fighters were pulled from the structure as it collapsed. The victim was recovered later that morning. (NIOSH 2007)

This case history highlights three critical factors:

1. Preventing yourself from getting caught in the fire is the best option to ensure your survival.

2. A rescue attempt at a situation like this is usually not successful.

3. Even a well-trained FAST or RIT will likely not be able to save firefighters caught in situations like these.

Sadly and simply, once the collapse occurs in this type of framing, it is too late for those involved.

Another case history dealing with lightweight construction was reported in the *National Fire Protection Association Journal* (Earls 2009). You will see this case history again in the size-up chapter. It contains important lessons on construction and size-up learned from fireground experience.

Underwriters Laboratories test results. Underwriters Laboratories (UL), the Chicago FD, and the International Association of Fire Chiefs (IAFC) partnered with a grant from FEMA to conduct testing in order to better understand the hazards of residential lightweight construction. The report (fig. 2–51) was published by UL in 2008 and revised in 2009. The experiment looked at the results of nine fire tests, seven examining floor-ceiling constructions and two looking at roof-ceiling constructions (Izydorek et al. 2008).

The experimental series consisted of 12 furnace fire tests of assemblies representative of typical residential "legacy" and "modern" floor and roof construction. The tests included six structural elements, three ceiling finish configurations, four floor or roof finishes, and one test examining finished ceiling penetrations (Figure 1). All of the test assemblies conformed to the dimensions and span of the available test furnace (14 feet by 17 feet). The structural members tested spanned 14 feet. even though some of the assemblies are capable of spanning greater distances because of either the material strength or joist depth of the floor or roof assembly being tested. Some the spans for these tests are conservative; view resulting failure times with this in mind. Had the structural members been allowed to support loads for a longer span, the resulting failure times would have been potentially accelerated, thus reducing the collapse time for the assembly. Measurements taken during each experiment include observation of the conditions of the ceiling and floor or roof surfaces, temperatures in the concealed space above the ceiling membrane, deflections of the floor and roof surfaces, and failure times of the tested assemblies.

This research conformed to the standard requirements of the ASTM E119, *Fire Tests of Building and Construc-*

Fig. 2–51. Actual UL test report

Figure 2–52 provides a review of key data from the study.

This landmark study had the following conclusions:

1. Floor-ceiling assemblies made with engineered I-joists failed to contain fire 14 minutes sooner than the benchmark, "a combustible floor-ceiling assembly representing typical legacy construction without a ceiling," which lasted 18 minutes (Izydorek et al. 2008, 7).

2. The 12 in. wooden I-joist floor system experienced a total collapse at 6 min and 3 sec, but the floor was unsafe at approximately 4 min (as mentioned in point 1). The plywood panel that provides a great deal of structural stability in this kind of construction burned through quickly, causing the collapse.

3. While intuitively we would think that thermal imaging cameras (TICs) should provide an accurate indication of a weakened floor, this is not true. The study found that a minute before collapse, although average temperatures were almost 1,300°F below the floor, temperatures were only around 100°F on the finished floor level (which had carpet padding and a subfloor below it; Izydorek et al. 2008, 99). Fire suppression water may mask the heat even further. The TIC should not be relied on to detect a weakened floor (figs. 2–53 and 2–54).

Structural Element – Ceiling Finish	Type of Construction	Ceiling Materials	Floor/Roof Subfloor Finish	Collapse Time (min:sec)
2 × 10 Joist Floor – Without Ceiling	Legacy	None	1 × 6 and Hardwood	18:45
2 × 10 Joist Floor – With Ceiling	Legacy	Gypsum Board	OSB and Carpet	44:45
2 × 10 Joist Floor – With Ceiling	Legacy	Lath and Plaster	1 × 6 and Hardwood	79:45
12-inch Wood I-Joist Floor – Without Ceiling	Modern Lightweight	None	OSB and Carpet	6:03
12-inch Wood I-Joist Floor – With Ceiling	Modern Lightweight	Gypsum Board	OSB and Carpet	26:45
14-inch Finger Joint Truss Floor – Without Ceiling	Modern Lightweight	None	OSB and Carpet	13:06
14-inch Finger Joint Truss Floor – With Ceiling	Modern Lightweight	Gypsum Board	OSB and Carpet	26:45
14-inch Metal Gusset Truss Floor with Cord Splices and Framed Stair Opening – Without Ceiling	Modern Lightweight	None	OSB and Carpet	13:20
14-inch-Metal Gusset Truss Floor – With Ceiling	Modern Lightweight	Gypsum Board	OSB and Carpet	29:15
14-inch-Metal Gusset Truss Floor with Cord Splices, Recessed Lights, and Ducts – With Ceiling	Modern Lightweight	Gypsum Board	OSB and Carpet	30:08
Metal Gusset Truss Roof – With Ceiling	Modern Lightweight	Gypsum Board	OSB and Shingles	13:06
2 × 6 Joist and Rafter Roof – With Ceiling	Legacy	Gypsum Board	1 × 6 and Shingles	40:00

Fig. 2–52. Chart of key data

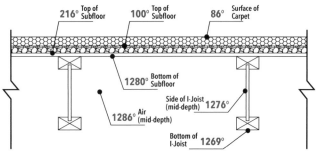

Fig. 2–53. A TIC will not detect unusual heat levels below the floor and thus will not detect a weakened floor. *Source:* Izydorek et al. 2008, 99.

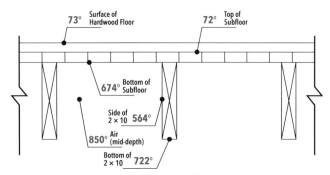

Fig. 2–54. Legacy framing floor and below floor temperatures *Source:* Izydorek et al. 2008, 98.

Another interesting finding was that a plastic ridge vent melted and sagged to the point of sealing itself. This can cause a misleading size-up as very limited smoke will be showing after heat causes the vent to fail. This can not only fool incident commanders but also contain the heat and further intensify the fire, causing earlier weakening and failure of the roof.

Recommendations to minimize the risk of lightweight construction. NIOSH has provided several recommendations to minimize the risk of injury or death involving roof and floor truss systems (NIOSH 2005).

- Know how to identify roof and floor truss construction.
- Immediately report the presence of truss construction and fire involvement to the incident commander.
- Use extreme caution and follow standard operating procedures when operating on or under truss systems.
- Immediately open ceilings and other concealed spaces whenever a fire is suspected of being in a truss system:
  - Use extreme caution, as opening concealed spaces can result in a backdraft.
  - Always have a charged hoseline available.
  - Be positioned between the nearest exit and the

concealed space to be opened.
- Be aware of the location of other fire fighters in the area.

Platform construction

Platform framing became prevalent in our country after World War II. There was less long-length, full-dimension lumber available to build balloon frames, but something had to be done to accommodate the home-building boom caused by returning service members.

By design, this type of framing has built-in resistance to fire extending unchecked in the walls and floor assemblies. On the foundation is the sill plate, usually a 2 × 8 or 2 × 10 that the first-floor joists rest on. On top of the joists is the floor sheathing that extends right out to the wall sheathing. This essentially makes a platform of the first floor. Basement fires often are stopped here in their vertical drive. The first-floor wall studs rise to the ceiling level of the first floor and a horizontal 2 × 4 helps make the assembly rigid.

Second-floor joists sit atop this plate and the second-floor subfloor material is applied on top of the joists. This setup again makes a platform, limiting vertical fire spread within the wall. Boxed-out floor systems covered with plywood create a platform over the foundation that the first floor is built on (fig. 2–55).

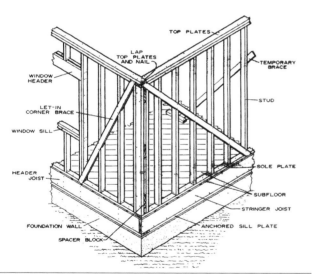

Fig. 2–55. Built-in fire stops created by the floor extending out to the wall sheathing. Note the fire stops between some of the wall studs. *Source*: Anderson 1975, 32.

Balloon-frame construction

Balloon framing is an older type of construction that was common from the mid-1800s up until World War II. You will find balloon frames all across the country, in all settings, from rural to urban. Full-size dimensional lumber was available in long lengths that were a carryover from the wooden ship-

building industry. This was a cost-effective building material for balloon-frame homes.

Balloon framing uses this long lumber as 2 × 4 in. studs (and sometimes larger) that run from the foundation to the eave line. Unlike post-and-beam framing, the structural strength is gained from numerous (usually 16 in. on center) studs in exterior walls. The upper floor joists are supported by a ribbon-ledger board that is notched into and secured to the exterior wall studs. This framing method provides every stud channel as a vertical chimney if fire enters this space. Fire can enter the stud space at the basement level or any floor in between. It is not uncommon for fire to travel up a wall, cross over via the floor joists, and show at an opposing corner or side of the building.

If the house still has the original lath and plaster interior walls, the rough-cut inner surface of the lath has dried and will quickly promote rapid fire spread inside the walls.

Balloon frames are especially susceptible to basement fires. Fire stops are not usually installed, giving a basement fire a direct shot to the upper floors and attic.

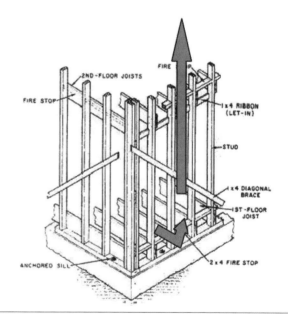

Fig. 2–56. In a balloon-frame home, fire can enter the wall space and travel quickly up and extend into the floor above or directly to the attic. *Source*: Anderson 1975, 33.

Identifying balloon frames from the outside. The age of the home is the first clue. You should suspect that homes built between the mid-1800s and the beginning of World War II have balloon framing, as it was the most common method of framing during that time. The layout of windows is the next obvious clue. Since the studs run the height of the home, it was economical to vertically line up the windows (figs. 2–57, 2–58, 2–59, and 2–60).

Fig. 2–57. Window positioning in a balloon frame

Fig. 2–58. Balloon frame in an urban setting in an old northeastern American town

Fig. 2–59. Large balloon frame. Note there is only one electric meter, which means that this is either a very large single-family home or it contains illegal apartments.

Fig. 2–60. The windows of this balloon frame are not directly over each other but still are lined up so the long studs can provide structural support from foundation to eave line.

Routes of fire spread. Figure 2–61 shows us the basement of a balloon-frame home between the first-floor joists looking toward an outside wall. Note the bottom of the wall stud (1) resting on the sill plate (2) on top of the foundation (3). This home was well constructed. The builders installed bricks (4), called nogging, in the gap to prevent a basement fire from extending up into the walls, which usually results in the total destruction of the home. Notice also the size of the joists, which are full 2×10 in. dimensional lumber. Due to their mass they will take longer to become involved in the fire and will provide structural integrity for some time, even as they burn.

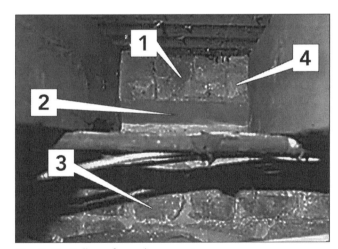

Fig. 2–61. Balloon-frame basement

Despite the excellent fire stopping, fire still has several easy routes into the walls of this home via pipe (supply and return) and electrical chases (fig. 2–62). Note how the lime mortar foundation is deteriorated. Hoselines in the basement can flush this soft mortar out, weakening the foundation and often causing bricks to come out. The pipe insulation contains asbestos, which is a proven carcinogen. The asbestos hazard comes from inhaling airborne asbestos dust. Wetting it greatly reduces the threat, but so does wearing your mask.

Fig. 2–62. Hazards of balloon frames

Balloon-frame case history. In this case history, fire from a child playing with a candle in a closet entered the walls and overran the building before firefighters could open up and extinguish the concealed fire. Often the fire will bypass a floor or two and almost always show in the attic. Typically, the fire department will not be able to open up the interior walls fast enough, stretch lines to those areas, cut the roof, and get lines to the attic simultaneously. The results are evident here, even after an aggressive interior attack. The fire simply has too much of a head start for the attack operation to be successful (figs. 2–63 and 2–64).

Closet where fire started was in the back corner of this room

Fig. 2–63. Aftermath of a fire in a balloon-frame building

Fig. 2–64. This was the primary means of vertical fire spread from the first floor to the attic.

Windows that interrupt the open wall bays act as a fire stop. The default action of the window as a fire stop is evident. This was the exterior wall of the original house. The one floor addition on the B side of the house is the roof you see out the window that is covered with overhaul debris. The cast-iron radiator has partially collapsed through the fire-weakened floor (fig. 2–65).

Fig. 2–65. Cast-iron radiator sunk into the floor nearly through to the floor below

Removing the molding provides access to several channels so you can see if the fire has entered them and, if it has, put water in them (fig. 2–66). Another way to verify balloon framing is to pull the baseboard and probe down with a tool. If it is balloon framing there will be no sill plate and the tool will be able to extend down the exterior wall (figs. 2–67 and 2–68).

Baseboard Molding

Fig. 2–66. Technique to quickly open the stud channels in the wall of a balloon-frame home

Fig. 2–67. One of two routes of vertical fire extension. This was an interior wall.

Fig. 2–68. Attic where the fire raced up the wall and found oxygen and fuel

Once the fire reaches the attic via the stud channels, the underside of the roof provides perfect fuel. It is preheated by the rising heat, which causes the fire to grow rapidly. Balloon frames create fire scenarios where you have not only the origin of the fire to extinguish but simultaneously an attic fire and several void-space fires. All these require attention

immediately, which is usually impossible due to insufficient personnel and untenable conditions inside that must be dealt with first (fig. 2–69).

Fig. 2–69. Old and new create additional fire problems.

Often, older balloon-frame homes have been legally or illegally modified. In this case, thin, non-fire-rated wood paneling caused the fire to spread rapidly on the first floor away from the original fire in the closet, resulting in a well-involved first floor. Decorative tin ceilings concealed the fire as well. Like other opening-up operations, pulling tin ceilings is personnel intensive. This one was filled with sawdust for insulation.

Fig. 2–70. Fire racing through this balloon-frame home caused a total loss.

In figure 2–70, firefighters arrived to find fire on the first floor, fire in the walls, fire in a second-floor bedroom, and a well-involved attic fire. We were not able to open up those areas quickly and extinguish the fire fast enough to save the building and contents. Fire this advanced is almost impossible to stop in a balloon-frame building. To stop the fire, hoselines must go to every floor, ventilation must be swift, and the laborious tasks of opening walls, floors, and ceilings must be done simultaneously if there is any chance to save the building.

Collapse of wood-frame homes

Collapse of wood-frame homes is generally not a serious threat to firefighters. However, when the house is fully involved in fire, has not been maintained properly, or has been vacant and structural elements have been exposed to weather, collapse can be a deadly hazard for firefighters.

Collapse case history. Firefighter Jeff Shupe, formerly of the Cleveland OH Fire Department (former Division Chief for training, North Myrtle Beach Fire Department [SC]), relates the following case history from a fire in 2010:

Engine 11 arrived at a well-involved working fire in a vacant, wood-frame two-family home. We sometimes call them "double deckers" because they are built on top of each other much like the New England triple deckers. *Typically* the attic might be small but this one was a full-height living *area* with finished knee walls and ceilings—plenty of fuel.

Because Cleveland is an old industrial city, most wood-frame structures like this were built using balloon framing techniques, common to that era. On the front of these buildings there are two porches built one on top of the other, 2 stories high; hence the name, double decker. These porches (floors) and roofs are mostly built with old, tongue and groove planking, which provide a large fire load and of course, plenty of weight on the support system. The roof on top of the entire structure was gabled, adding additional weight and fuel.

They are attached using ribbon or "ledger" boards to the front of the houses. They can be pretty good size. In many cases, they are old and rotted from weather and also fatiguing from age and weakening supports. The supports or columns at the corners may be wooden or they may be masonry if that is the veneer of the building. In this case, the railings and supporting columns were thin metal, not a lot of support to begin with. And when I say this structure was well involved, I mean the whole front of the building was ablaze. We spotted the engine to use a hand line and to leave the front for the truck. Our crew stretched a 2 in. handline with a 1 in. solid tip. We thought it would give us the knockdown power needed for this job. I had the nozzle and moved toward a corner of the structure, kneeled down to take a solid stance, and started water. The fire started to darken down but within a matter of seconds the entire porch assembly collapsed and the one corner landed approximately an arm's length in front of me. Our crew was covered in sparks and

embers and displacement knocked me over backward like a toy. This left us behind the eight ball and let the fire spread rapidly throughout the building.

Fig. 2–71. Shows the front of the double-decker upon arrival. Note the lightweight metal supports holding up the front second-level porch. It is likely that the homeowner replaced the original, heavy, more supportive columns sometime during the life of the home. Courtesy of Steve Nedrich.

This case history provides two very important lessons. First, collapse is often instantaneous and catastrophic, as shown in the two photos. Second, when there is heavy fire expect additional problems and hazards such as collapse, downed wires, exposure issues, and so on.

In figure 2–72, we see that the entire second-floor porch came off this home. If an interior attack had been attempted or in progress, the collapse could have killed an entire engine company.

Fig. 2–72. Second-floor porch collapse. Courtesy of Steve Nedrich.

Collapse safety strategies. Collapse of well-involved dwellings should always be a safety concern. Just like in multiple dwellings and commercial or industrial buildings, we must stay out of the collapse zone. The following NIOSH report summary highlights this hazard.

On November 1, 2002, a 36-year-old male volunteer Lieutenant (the victim) died after being crushed by an exterior wall that collapsed during a three-alarm residential structure fire. The victim was operating a handline near the southwest corner of the fire building where there was an overhanging porch. As the fire progressed, the porch collapsed onto the victim, trapping him under the debris. Efforts were being made by nearby fire fighters to free him when the entire exterior wall of the structure collapsed outward and he was crushed. The victim was removed from the debris within ten minutes, but attempts to revive him were unsuccessful and he was pronounced dead at the scene. (NIOSH 2003)

Collapse of burning houses often are quick and catastrophic, giving us no time to react. Firefighters must understand the weaknesses of house construction when they are being attacked by fire. Safety and survival at your next house fire may depend on how well you understand the hazards and develop and execute your strategy.

Post-and-beam framing

This is the oldest style of framing that you will see utilizing wood as structural material. Original homes with this type of framing are rare, and it is sometimes known as braced framing. However, because of its beauty, this type of construction is making a resurgence for specialty homes. It is expensive because it requires large and long heavy timbers; therefore, it is not a common framing method for modern homes.

Resurgence. Computerized cutting and shaping tools have helped reduce the cost of this type of construction. Some new homes are being constructed to imitate this type of construction. Original post-and-beam-framing homes were open throughout the interior to allow the easy spread of fresh air and heating (fig. 2–73). This open design also spreads fire quickly.

Fig. 2–73. New house of post-and-beam construction.

Post and beam is named such because it uses posts, or large timbers, at the corners of the building. Large beams carry the floor loads back to the posts and into the foundation. Walls are built with 2 × 4 in. studs to add structural stability. The house shown in figure 2–74 was designed and the structural members precut and assembled on the homesite similar to a kit. Many of the connections are mortise and tenon joints held together with the wooden dowels, as shown. Figure 2–75 features an assembled joint, which shows the sturdiness of this old but beautiful design. Compare the surface-to-mass ratio of this structural system to that of a parallel cord truss. The larger dimensional members will naturally resist fire longer before they lose their structural integrity.

Fig. 2–74. Sample connection for structural members of a post-and-beam house

Fig. 2–75. Assembled joint

This type of framing is often called braced framing because of the diagonal braces required to keep the frame from racking or tilting side to side from lateral loads (fig. 2–76).

Fig. 2–76. The structural supporting post here is larger than the smaller studs on either side.

In post-and-beam construction, the post is diagonally braced and connected with mortise and tenon connections held together by wooden dowels. Because so much wood is removed to make the joints, these connections can be easily weakened by fire, causing a partial or total collapse

Since building codes did not exist when these original homes were built, always suspect that there is no fire stopping and expect fire to extend quickly once it gains access to the void areas. Also expect surprises in construction methods and insulation methods. I have seen these types of homes insulated with everything from horse hair, bricks, asbestos fibers, and even ground newspaper. In his book *Building Construction for the Fire Service*, Frank Brannigan says, "I have seen houses in which two opposite sides were post and beam and the other two were balloon frame" (1997, 60).

Weaknesses. Deputy Chief (retired) Anthony Avillo of the North Hudson Regional Fire and Rescue (NJ) summarizes the construction weaknesses of braced frame construction in his book *Fireground Strategies*:

> The weakness inherent in this type of wood from construction is the failure of the mortise and tenon joints that hold the walls together. Similar to a truss, or any other type of construction, the point of connection will be the weakest area in terms of resistance to fire. What makes this connection point more vulnerable to failure is that the tenon joint will have a smaller dimension of wood at the end, offering less structural mass at the most critical point, the connection. In addition, consider the fact that this small dimension tenon has dried out over the years, making it smaller and more vulnerable to ignition. As a result of this shrinkage, it may not be as tight a fit inside the mortise as it once was. This creates a heat sink condition inside mortise area. Compounding this issue is the fact that the mortise,

a hollowed out wood member also has less mass. These smaller areas are more vulnerable to destruction by fire. (Avillo 2015, 167)

Chief Avillo continues:

> A concern inherent in all wood frame construction—but even more critical in braced frame construction—is the fact that there is no compensation on the lower floors to support the added load the upper floors place on the weight-bearing structural cage (posts and girts). For instance in a three story braced frame building, the top floor will hold only itself and the weight of the roof. The second floor will hold its own weight plus the weight of the top floor and roof. The most weight and the weakest point will be the ground floor, which will hold the weight of itself plus the weight of the other floors and the roof. The dimensions of the timbers on the first floor will be the same as that of the timbers on all other floors. This places the greatest structural load on the ground floor. That is why when there is a heavy fire on the ground floor of a braced frame building, collapse should be anticipated and companies withdrawn from all areas. Because of this added weight, braced frame buildings of three stories are much more susceptible to collapse than braced frame buildings of two stories. The difference is the added load of the additional floors.
>
> When braced frame buildings fail, they often collapse without warning in an inward-outward fashion. The bottom floor falls outward, while the top floor or floors collapse almost straight downward in an inward fashion. If the building stands alone or is located on a corner, it is even more likely to collapse because the side walls do not receive the same support as an attached building. This is still another indicator of collapse potential in an old, heavily involved building. (Avillo 2015, 167)

Case history and tactics. Let's look at a case history of a fire in a post-and-beam home. An old post-and-beam farmhouse had a basement fire extend into the walls. Upon arrival, firefighters found fire in the basement burning up through the floor. Figure 2–77 shows how, later in the fire, fire took possession of a room on the first floor.

Fig. 2–77. Fire taking over first-floor room. Courtesy of Warren Salle.

This is a common situation and sight upon arrival: smoke showing from many areas, walls, eaves, attic vent, and so on but no visible fire (fig. 2–78). The fire is in the walls or other voids. In this case, the fire was in the basement but was knocked down by an aggressive interior attack. Unfortunately, it had already gotten into the walls and void spaces.

Fig. 2–78. Smoke showing, but no visible fire. Courtesy of Tom Bierds.

The fire will "run the house" quickly via the voids. To save this building you will have to open up quickly wherever you think the fire may be, or may be going, which means practically everywhere (fig. 2–79). It is doubtful the first-alarm assignment will have enough personnel to do this. If fire is in the walls, you must open up quickly to cut off the fire spread. This can be done from the outside, which provides better conditions for members to work. Power tools greatly speed up this process, which is labor intensive by any measure (fig. 2–80).

Fig. 2–79. Fire will "run the house" quickly via voids. Courtesy of Warren Salle.

Fig. 2–80. Power tools are used to quickly open up walls from the outside. Courtesy of Tom Bierds.

Heavy fire is racing up the stud channels and going to be in the attic if it is not already (fig. 2–81). Get lines ahead of the fire spread. Opening up has given the fire lots of oxygen and as you may expect it will light up quickly. Try to get ahead of the fire or at the very least apply water as soon as you can—even if the area is only partially open—to cut off the fire (fig. 2–82).

Fig. 2–81. Fire spreading up stud channels

Fig. 2–82. Apply water as soon as possible, even if the area is only partially open. Courtesy of Tom Bierds.

Direct the water both up and down in the stud channel. Be aggressive with the nozzle; you may have to push it right up under the siding to cut off the fire travel (fig. 2–83). This is one of the rare times where it is okay to put water in from the outside of the fire building. A solid-bore nozzle would be a much better choice to apply decisive amounts of water.

Fig. 2–83. Be aggressive with the nozzle.

Speed in opening up and applying decisive amounts of water quickly are the key to cutting off this fire. Use at least two members to open up while a third member staffs the hoseline. Work in a one-two punch method: open up, apply water quickly (fig. 2–84). This is a labor-intensive operation. Be prepared to have fresh staff to replace those who have worked strenuously. Don't waste time opening below where the fire already is; get ahead of it, which usually means going above where the fire started (fig. 2–85).

Fig. 2–84. Open up, apply water quickly. Courtesy of Tom Bierds.

Fig. 2–85. Get ahead of the fire. Courtesy of Tom Bierds.

Other home construction methods

The kind of house you are in will determine many of the construction features and hazards that are present. We have described many home styles in this chapter. Each house described has its own unique avenues of fire spread and requires its own fire attack plan. Many times, you can determine the style just by looking at it.

There is one construction type that might test some of your assumptions: modular homes. But before we can discuss modular homes we need to define some of the other construction types to which the term may be errantly applied.

Mobile. Mobile homes are built on a chassis with wheels and are designed to move as trailers. This fact limits the design options and floor layouts. Many of these homes are more permanent than mobile, meaning that often the wheels have been removed and the frame and suspension covered by skirting, but the presence of the trailer hitch always indicates a mobile home.

Manufactured. Manufactured homes are built to federal Housing and Urban Development (HUD) standards that require them to have a nonremovable steel chassis. They may be made of one (single-wide), two (double-wide), or in some parts of the country up to four sections. The chassis is supported on concrete piers or placed over basements and braced. They are transported to the site on a trailer and off-loaded or the wheels and hitches are removed and the units are set in place with a crane, creating some of the confusion. Originally regulated as vehicles, they were not required to comply with housing regulations. Federal regulations have increased the safety of this type of construction with the Federal National Manufactured Housing Construction and Safety Standards Act of 1974.

Prefabricated. Prefabricated homes in the traditional sense were "kit homes," where the lumber arrived precut and labeled for its location. In a modern sense, prefabs are often called panelized homes, because the walls are constructed in a factory and shipped flat to the jobsite. Set on traditional foundations, the construction of wall units into panels instead of whole sections allows the walls to be "stood up" in place, reducing build time and cost.

Some panelized construction options deliver panels with the siding and interior wall finishes already completed, providing savings over stick-built homes. (Stick-built refers to the construction method where the lumber is delivered, hand sawn on-site and each stud, rafter, and beam is nailed in place individually, stick by stick.)

Implications for the fire service. The data reported by the manufactured homes industry indicate that manufactured and modular home sales are on the rise nationally. If this is the case, how can we wait to reevaluate our current policies and procedures before adjusting our attack plan for these structures? If modulars and manufactured homes are the new up-and-coming construction method, then it is our responsibility, in a book on modern house fires, to document that trend and provide information on its implications. Deputy Chief (retired) Vincent Dunn (FDNY) wrote for years about the hazards of truss construction while as a group the fire service was burying our own without listening. We feel that is unacceptable and should not be repeated for manufactured homes. Look for these homes in your area and investigate their potential and share your wisdom by publishing your experiences and around the kitchen table in the firehouse.

Modular homes. Modular homes are manufactured in a facility, delivered to the jobsite, and set with a crane. The homes' sections are complete and finished inside and out. Interior walls are finish painted and exterior siding and roofing are applied at the plant. Floor plans and number of sections are determined by the customer's design. The engineering of these homes allows for the separate construction of each section and a crew of workers to build the sections under controlled conditions. When the homes are delivered to the site, the plumbing and electrical connections between sections and any finishing work between the "marriage walls" (walls between the sections) is completed: the exterior siding

that spans the sections will be applied and interior doors and trim will complete the "marriage."

Though the end product may look like a Cape Cod or raised ranch or contemporary, the construction methods used to create these homes may provide new avenues of fire spread. Work with your local building officials and code enforcement officers to visit these sites during the construction phase. In your career, you've probably learned a lot about construction by looking at a building when the walls are open; you'll need to look a lot harder to learn about them when the walls arrive closed!

The information contained in this section is from a 2010 FDIC presentation by Chief Kevin A. Gallagher, Acushnet Fire Department (MA). We are grateful for his willingness to share this lifesaving information, which is used here with his permission.

Construction method. As the name implies, these types of homes are prefabricated. Tradesmen work in an assembly line–fashion building modules (often in 25% increments of the house), which are transported to the site (fig. 2–86). The on-site builder constructs the foundation, assembles the modules, and connects the utilities (fig. 2–87). From the outside, it is impossible to determine if the construction method is modular (figs. 2–88, 2–89, and 2–90).

Fig. 2–86. The modules arrive on a truck. Courtesy of Kevin Gallagher.

Fig. 2–87. A crane picks the module and mates it to the foundation or the top of another module. Courtesy of Kevin Gallagher.

Fig. 2–88. This home is an example of modular construction. Courtesy of Kevin Gallagher.

Fig. 2–89. and Fig. 2–90. Other examples of different modular home styles. Courtesy of Kevin Gallagher.

Advantages over stick-built homes. Modular homes have several advantages over other types of construction:

1. Cost is reduced by 6%–15%.

2. Framing is strong enough to be shipped over the road and lifted into place by a crane.

3. Rapid on-site assembly time, often 1–2 days, with a finished project completed in a few weeks.

4. Affordable.

5. The designs are customizable.

6. The time taken by local inspection is reduced (preinspections are done at factory).

Built-in hazards. There are several built-in weaknesses to modular construction that firefighters must be aware of in order to understand how fire spreads and develop appropriate strategies and tactics (fig. 2–91). It is also vital to understand how quickly fire will spread to prevent firefighters from becoming involved in collapse or falling through fire-weakened floors.

Fig. 2–91. Modular fire

Void between floors. As a direct result of the type of construction, there is an approximately 21 in. high void between the first-floor ceiling (sheetrock) and the second-floor sub-flooring material. When a room-and-contents fire gets into this void, it has the full run of the house. Chief Gallagher refers to this as horizontal balloon framing (figs. 2–92, 2–93, and 2–94).

Fig. 2–92. Void space from the first-floor ceiling up to the second-floor subfloor. Courtesy of Kevin Gallagher.

Fig. 2–93. Another view of the void space. Courtesy of Kevin Gallagher.

Fig. 2–94. Void space from an actual fire in a modular home. Note that the first-floor ceiling joists and the second-floor trusses have burned away due to the intensity of the fire that was in this space. Courtesy of Kevin Gallagher.

Fire spread due to ceiling adhesive failure. Further complicating a fire in the void space is how it got there in the first place. Figure 2–95 shows the adhesive holding the sheetrock to the ceiling of the first floor. There are no nails or screws (mechanical fasteners) holding the sheetrock up. This eliminates the chance for nail pops during transport and construction. The building code allows for deviations like this if they have been successfully evaluated. Unfortunately there is no standard for the glue to be heat resistant. Tests have shown that the adhesive starts to lose its strength and fail quickly, causing the sheetrock to fall and exposing the huge void space to the room-and-contents fire (fig. 2–96).

Crane lift points and fire spread. There are built-in access points for the crane to lift the modules into place (figs. 2–97 and 2–98). These provide another avenue for fire spread. Fire now has access to the void space between floors from an exterior fire or access to the outside of the building from a fire in the void space (fig. 2–99).

Fig. 2–95. A contents fire has caused the adhesive to begin to fail. Courtesy of Kevin Gallagher.

Fig. 2–96. At this fire in Acushnet (MA), investigators determined that failure of the ceiling caused by direct flame impingement on the adhesive led to catastrophic fire spread. Courtesy of Kevin Gallagher.

Fig. 2–97. Boxes being lifted. Courtesy of Kevin Gallagher.

Fig. 2–98. This crane lift point is another easy avenue for fire spread in modular homes. Courtesy of Kevin Gallagher.

Fig. 2–99. Investigators determined that the lift point helped spread the fire in this case. When properly installed, these holes can be sealed using the materials provided by the manufacturer and the hazard reduced. Courtesy of Kevin Gallagher.

Fire spread in marriage walls. Modules have to be built strong enough to withstand transport and placement. Consequently, they essentially are almost completely rectangular boxes. The walls from each box meet, depending on their exact location and design, and form what is called a marriage wall. There are small void spaces at these marriage walls, again providing a great place for fire to spread or burn undetected for periods of time (figs. 2–100 and 2–101).

Fig. 2–100. A typical marriage wall between modules with obvious space for fire spread. Courtesy of Kevin Gallagher.

Fig. 2–101. Fire has spread throughout the structure via the void space between the modular units. Courtesy of Kevin Gallagher.

Wiring. Another characteristic of modular homes is that the wiring may be notched into the outside of the studs (fig. 2–102). Firefighters may not expect wires in these locations and be subjected to dangerous electrical shocks. Wires placed like this also allow fire to run the perimeter of the home for the length of the wire.

Fig. 2–102. In a typical house, you would not expect to find wiring run through the exterior studs, as was the case in this modular home. Courtesy of Kevin Gallagher.

Lightweight materials. Although designs vary, floors may be supported by lightweight trusses or engineered joists (fig. 2–103). As we know, these fail rapidly when put under a fire load. Expect early collapse of these homes.

Fig. 2–103. A finger joint truss was used in this modular home to support the second floor. Courtesy of Kevin Gallagher.

There are no nails or even gusset plates in a finger joint. The webs and cords are held together by adhesive and the friction of the individual wood fingers at each connection. Note also the small dimension of the lumber (fig. 2–104).

Roof construction. To minimize cost, roof sections are pre-fabricated and simply tilted up at the jobsite. A cross member connects each side of the roof, providing some structural stability. Often there is no ridge pole (figs. 2–105 and 2–106).

Fig. 2–104. Complete failure of lightweight structural members caused this home to become a death trap for both civilians and firefighters early in the fire. Courtesy of Kevin Gallagher.

Fig. 2–105. A roof rafter is attached by a mechanical fastener that hinges flat during transport and can swing up at the homesite. This is an obvious structural weak point if attacked by fire.

Fig. 2–106. A finished attic and roof framing assembly

Building inspections. In typical stick-built, on-site construction methods, a building inspector will conduct inspections at these phases of construction: foundation, sheathing, rough

and finish electrical and plumbing, framing, and insulation, and a final inspection at completion. For modular construction, only the following inspections are completed at the homesite: foundation, set, wiring and plumbing final only, and a finish inspection.

Manufactured home builders hire third-party inspectors to provide on-site inspections during the construction phase at the factories. Note that the builder has hired the inspector, making the quality and quantity of the inspection suspect at best. Local building inspectors cannot see electrical, mechanical, or plumbing components because they are already covered with interior finish materials, which also limits the quality of the inspection.

Strategy and tactics. There are several aggressive steps you can do to prevent firefighter injury or loss of life.

1. Identify and preplan for fires in modular homes in your area (see the section in this chapter on STOP).

2. When responding to a modular home fire, conduct a very thorough size-up, including a 360 walk-around, and determine where and how far fire has extended in these buildings.

3. Be aggressive in the use of your TIC to find hidden fire in voids.

4. Keep an eye on the clock and expect early collapse and fire-weakened floors.

5. Expect ceiling failure in a room-and-contents fire.

6. Be aware of the potential for electrical fires in marriage or partition walls.

7. Be aggressive during searches for hidden fire during the fire suppression and during overhaul.

As you can see, it is essential to have in-depth knowledge of building construction, floor plans, and overall home designs long before the fire starts. This preincident intelligence is essential to your survival.

References

Anderson, Leroy. 1975. *Wood-Frame House Construction.* 2nd rev. ed. Washington, DC: Government Printing Office.

Avillo, Anthony. 2015. *Fireground Strategies.* 3rd ed. Tulsa, OK: PennWell.

Brannigan, Francis. 1997. *Building Construction for the Fire Service.* 4th ed. Quincy, MA: National Fire Protection Association.

Dalton, James, Robert Backstrom, and Steve Kerber. 2009. "Structural Collapse: The Hidden Dangers of Residential

Fires." *Fire Engineering* 162 (10). http://www. fireengineering.com/articles/print/volume-162/ issue-10/features/structural-collapse.html.

Earls, Alan. 2009. "It's Not Lightweight Construction. It's What Happens When Lightweight Construction Meets Fire." *National Fire Protection Association Journal*, July/ August. http://www.nfpa.org/news-and-research/ publications/nfpa-journal/2009/july-august-2009/ features/lightweight-construction.

Izydorek, Mark, Patrick Zeeveld, Matthew Samuels, and James Smyser. 2008. *Report on Structural Stability of Engineered Lumber in Fire Conditions.* Project No. 07CA42520. Northbrook, IL: Underwriters Laboratories.

National Institute for Occupational Safety and Health. 2003. *Volunteer Lieutenant Dies Following Structure Collapse at Residential House Fire—Pennsylvania.* NIOSH Report F2002-49. Atlanta: Department of Health and Human Services. https://www.cdc.gov/niosh/fire/ reports/face200249.html.

———. 2005. "Preventing Injuries and Deaths of Fire Fighters due to Truss System Failures. " NIOSH Alert.

———. 2007. *Volunteer Fire Fighter Dies after Falling through Floor Supported by Engineered Wooden-I Beams at Residential Structure Fire—Tennessee.* NIOSH Report F2007-07. Atlanta: Department of Health and Human Services. https://www.cdc.gov/niosh/fire/reports/ face200707.html.

Rice, Curtis, and Elvin Gonzalez. 2011. "Situational Awareness and 'Reading' a House. "*Fire Engineering* 164 (2): 87.

Tenniswood, Bruce. 2009. "Bungalow Fires: Construction Dictates Tactics. "*Fire Engineering* 162 (10). http://www. fireengineering.com/articles/2009/10/bungalow-fires -construction.html.

3

Hazards of House Fires

Summary

Using recent research will help firefighters understand how and why house fires are so dangerous. The facts on modern fire dynamics—the real life cycle of a modern house fire—are not being taught and trained on nearly enough, and the result is that seasoned, veteran firefighters are being killed in flow paths, by flashovers, and burned to death often only feet from safety. The reader will become familiar with interior, exterior, construction, and occupancy-based threats to firefighter safety and gain a thorough understanding of fire growth and spread. More importantly, this chapter will provide an understanding of fire development in residences so firefighters can anticipate and act (either getting out or killing the fire) before the fire kills them.

Introduction

This chapter will explain in detail hazards common to most homes and those "unique" hazards that experience has proven are more frequent than previously thought. Ultimately, the goal of this chapter is to prearm you with the knowledge of these hazards to make you and your crew safer and better prepared for your next house fire. Hazards of construction are discussed in chapter 2, so in this chapter we will discuss the hazards of the occupancy types and methods of fire spread in homes.

Common Fires, Uncommon Hazards

House fires contain unique hazards that too often end tragically in line-of-duty deaths or injuries or, at a minimum, very unpleasant surprises, upon our arrival at the fireground or when we begin operations. At best, these hazards create near misses for firefighters. We often overlook these hazards because we think of homes as safe places. In our minds, your home and my home are safe. If they were not, clearly we wouldn't stay there. This leads to the assumption that other peoples' homes are safe places as well. The result is that even seasoned firefighters will consider these fires *just* house fires.

Scenario Process

There will be a scenario at the beginning of each chapter from now on through the book. We will carry this scenario through the following topics: hazards of house fires, size-up, search and rescue, fire attack, and ventilation. At the end of each chapter will be additional scenarios that will include other types of houses and other important specific fireground situations. To get the most from these scenarios, use the following strategy when reading this book:

1. Read the scenario and answer the questions at the beginning of the chapter.

2. Write your answers on a piece of paper, then fold it and put it at the end of the chapter. Don't ponder and analyze; write down your first thoughts, considerations, and orders. Put yourself in the role of the incident commander (IC), the company officer, a firefighter, or your current position in your department on the first-due engine or truck. As you advance in your career, refer back and reevaluate your decisions based on your new responsibilities and the experience you have gained.

3. Thoroughly read the chapter, *then* return to the scenario.

4. Review your preliminary answers and see if you would change your original decisions based on what you just read.

5. At the end of the chapter, there are short discussions of each question for you to compare your responses.

6. Additional scenarios at the end of the chapter will cover unique examples and more advanced situations that you will face on the fireground.

Through the course of the scenario you will be given a situation report that will give you the address, dispatch information, and some additional information. Use your existing apparatus response type and sequence. Radio reports provided during the scenario will give additional information about the scene and what units are seeing inside or in the rear of the house. For simplicity, the A side is the front, B is the left side as you look at the house from the front, C is the rear, and D is the right side. Assume you have a reliable, adequate, and long-lasting water supply from hydrants or tanker shuttles.

It is important to note that the discussions are just that. There are no exact answers. If there were, firefighting would use a cookbook approach and the strategy and tactics would always result in a successful outcome. You, your department, or your first-due area may have specific reasons to do things differently than presented here. The discussion is designed to mention key points that may or may not be useful to you and your department. It is just one way to do things. Take all of them, some of them, or none of them, modify and use what works for you. It is the application of your training on the fireground, your artful use of the correct principles tempered by your experience, that will make you successful at your next house fire. In short, the purpose of these scenarios is to give you some experience to draw from and apply to your next house fire.

You can also use these scenarios as drill material. Use these same questions to lead group discussions with officers and firefighters. You can also use the scenarios to apply your own department's SOPs as a vehicle to help get your officers and members to understand your strategies and tactics for house fires in your district.

Please remember that *although we are both looking at the same scenario, we will see all of it or parts of it very differently.* This difference will be based on our experiences and training, both of which innocently prejudice our perceptions (what we think, we see), affect our subsequent strategies, tactics, and safety concerns for our members, and ultimately shape our responses to the scenarios.

Using the cookbook analogy, even with the correct recipe, you may not be able to get it right. Too much flour or too much time in the oven makes a bad cake. In firefighting, too little water and too much ventilation will burn down the house. Different geographic areas may require different or altered solutions to the same problem. Conditions, staffing, building construction, and special circumstances vary greatly from Minnesota to Miami.

There is not one right answer. The right answer is the one that works in your community, with your fire department and your SOPs for your members and your customers. The purpose of the scenarios is to provide real-world house fire situations to practice on, think about, discuss, and fine tune for the next fire. The lessons learned here do not cost firefighter or civilian lives. Make decisions quickly during the scenarios presented, then later carefully consider the advantages and disadvantages of these decisions. This may be one of the best ways to become confident at these high-stakes house fires. After all, during the scenarios, the only thing at stake is our ego! On the fireground it is matter of life and death, often ours. And remember General Colin Powell's advice: "Avoid having your ego so close to your position that when your position falls, your ego goes with it."

Scenario

Fig. 3–1. This house fire presents many common and uncommon hazards to firefighters. Courtesy of Brian Duddy.

Situation: It is Wednesday morning, 0223. You are tapped out to a fire at 2 Veterans Street. Caller reported seeing the house across the street on fire. There is no additional information. Upon arrival, the police officer reports to you he heard a dog barking inside.

Resource Report: You are on the scene. Your first-due piece of apparatus is approximately 2 minutes from the scene. There is a good hydrant three houses away from the fire building.

Radio Report: Dispatcher reports multiple 911 calls.

Scenario questions

Based on figure 3–1, answer the following questions and use this scenario as a basis for the topics examined in this chapter.

1. What is your general impression of this house?

2. What specific hazards or threats to you or your members do you see or need to consider?

3. What hazards will you or your members face during an interior search or fire attack?

4. What uncommon or undetected hazards may you face?

Hazards to Firefighters

The universal threat

Fire, of course, is the universal and omnipresent hazard in residential fires. As firefighters we often do not understand the real life cycle of a residential fire. Understanding the real life cycle of residential fires is critical to our safety. In addition to the research we did to develop the flashover survival training program at the Rockland County Fire Training Center (NY), nationwide experience has shown us that experienced firefighters are being trapped and killed in residential fires.

It is important to remember that most homes are open and uncompartmentalized. Fire on the first floor will send deadly heat and products of combustion to upper levels where sleeping areas are usually found. Simply, the fire has an unobstructed path with copious amounts of fuel at every turn. Rapid fire development is an exceptionally dangerous and often overlooked hazard in the common house fire. NIOSH reports are full of firefighter fatalities caused by the rapid fire development found at house fires.

Even a room-and-contents fire presents a serious threat to firefighters. The National Institute of Standards and Technology (NIST) conducted a series of live burn tests on typical living room–type occupancies. In the 16 live burns, most ceiling temperatures exceeded 2,000°F. When we consider that we will experience a first degree burn when our skin reaches 124°F, the fire has a considerable head start and easily has the capacity to injure then kill us by heat alone. Toxic gases, flashover, collapse, and other threats exponentially increase the danger of a house fire.

Training fire development

Why don't we understand the real life cycle of fire? Think about how you were initially trained. You may have gone to your first fire in a concrete burn building like the one in figure 3–2. A few pallets and a pile of hay may have been the training fire. This limited amount of combustibles (as recommended by NFPA) helps keep overzealous trainers from injuring firefighters in training.

Fig. 3–2. Fires in concrete burn buildings do not behave like real room-and-contents fires.

Or you may have been initially trained in a modern, propane-fired structure fire simulator. As you applied water to the propane-fired flames emanating from a simulated bed, couch, or kitchen look-alike, sensors detected the water flow and reduced then shut off the gas flow to the prop. Although both of these training scenarios are necessary to train recruit-level firefighters in critical skills such as SCBA and hose and nozzle techniques, neither actually demonstrate the real life cycle of an uncontained house fire with an unlimited air supply. In a sense we have incorrectly trained ourselves into believing the fire will remain constant, growing slowly and steadily. This is not the case in the real world. Residential fires can quickly flash over, trapping, injuring, and killing firefighters. In the training fire, there is no fire

rolling over your head as you move in and there is no possibility that the room will flash over.

The progression of fire in a concrete burn building is linear. The fire grows steadily but never jumps to a life-threatening situation. The graph in figure 3–3 shows the growth of the training fire over time. Growth of the fire is displayed by temperature over time. Note that it is linear: as time progresses, the fire increases as more fuel is consumed. The building and the atmosphere heat up until the fire is extinguished or heat production is limited by the fuel being consumed, and then the burning rate decreases. This fire grows the same amount in a given time span (figs. 3–4 and 3–5).

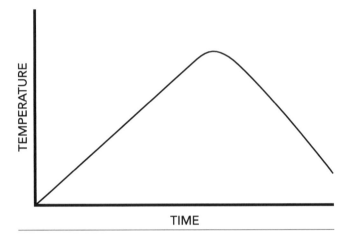

Fig. 3–3. Life cycle of a training fire in a concrete building shown as a time-temperature curve

Fig. 3–4. This training fire, in a concrete burn building, will never go to flashover. There is not enough heat and additional combustibles. It has grown to free-burning stage, with heat and smoke banking down from the ceiling.

Fig. 3–5. Attacking this fire, firefighters will gain valuable experience in SCBA, hoseline advancement, SAR, and ventilation. They will not, however, experience flashover or even realistic preflashover conditions.

Real life cycle of fire

What these training fires don't show us and what we often forget about is the danger of rapid fire growth, commonly known as flashover. The graph in figure 3–6 shows several important areas of the life cycle of an uncontrolled fire. Note the rapid rise in heat over a very short period of time during flashover. Also note the duration of the intense heat: it is not over in a flash.

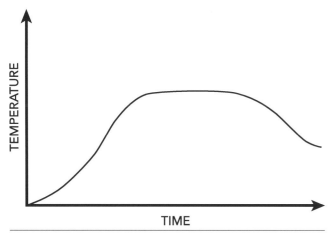

Fig. 3–6. Temperature graph of uncontrolled fire life cycle
Source: Kerber 2010.

First is the incipient or small stage of the fire. At this point in the fire's life cycle it can be extinguished with a pressurized water extinguisher. It is free burning but still small and contained. This stage is characterized by a noticeable increase in heat over a short time.

In what is typically called the free-burning stage, the fire begins to spread to other contents of the room. We call this the "I'm not gonna kill ya" phase because the fire remains fairly constant and does not seem to offer an immediate or

significant threat to firefighters. What the fire is doing, however, is loading the gun (generating flammable gases in the smoke and preheating combustibles) to enter the "I'm gonna kill ya" phase—flashover.

Radiant heat from the fire has now reached sufficient levels to heat up other combustibles in the room near the fire. These as-yet-unignited fuels are giving off flammable gases. This process is called pyrolysis. The fire is loading up the room or building to suddenly take the area to flashover. The term *flashover* is actually a misrepresentation of the event. It is not over in a flash; in fact it lasts several minutes, filling the fire area with flames and aggressively driving fire out windows and doors. Flashover is a nonsurvivable condition for both civilians and firefighters.

It is important to note that time and temperature are not fixed numbers. The time it takes to increase the temperature of the room(s) depends on many variables, such as size, fuel, source of ignition, and most importantly (the controlling factor) the amount of air that is available to the fire.

The flashover in figure 3–7 was obscured by a thick smoke layer and was unnoticed by firefighters. In the photo, it is becoming visible. Firefighters must be aware and look around in a fire room to see the early ignition of flammable gases above the smoke. When you detect sudden increases in heat it may indicate fire has developed over your head and radiant heat is pouring down on you. Now is the time to escape as rapidly as possible or kill the fire with water (fig. 3–8).

Fig. 3–7. This unique photo shows a flashover that began just seconds ago.

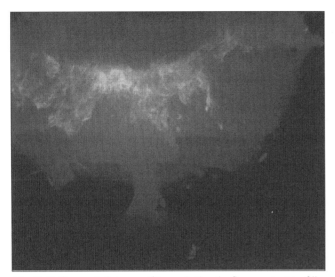

Fig. 3–8. Seconds later, swiftly and silently, flames start to fill the entire room, ceiling to floor.

As flashover silently and rapidly develops, firefighters are being driven to the floor by intolerable heat. Soon they will be enveloped in fire, their gear saturated with heat, and second degree burns will begin to appear in places where gear is compressed by SCBA straps and other areas (fig. 3–9).

Fig. 3–9. Firefighters are driven to the floor. As the pain caused by intense heat becomes unbearable, panic and irrational actions may replace training and rational thought.

In figure 3–10, two chairs are burning at the left side of the photo, a couch is fully involved at the right, and flames fill the room. If you were on this fire floor, you would be enveloped in flames, your turnout gear would not protect you, and, unless you could escape in seconds, you would likely experience severe burns and die immediately or shortly thereafter.

Fig. 3–10. A very rare view of a full-scale flashover from below the fire floor level at a training burn at the Rockland County Fire Training Center (NY)

Flashover continues to push flames from ceiling to the floor level and the room remains filled with fire until the fire runs out of fuel or the flashover is killed by application of decisive amounts of water. Our studies have shown that, despite training and experience, often it is just luck that determines whether firefighters survive being caught in a flashover. Prevention—keeping yourself from being caught in a flashover—is the best route to survival. To successfully prevent yourself from being caught in a flashover at a house fire, you must understand two concepts:

1. The real growth of fire in a compartment (house or room)

2. The flashover development process and firefighters' experience with it

Compartment fire growth

The previous fire growth curve assumed an unlimited amount of oxygen available to the fire. In other words, given an unlimited amount of oxygen, the fire will grow over time as the graph in figure 3–6 shows. However, almost all fires in structures are ventilation limited. This causes the fire to grow as the graph in figure 3–11 depicts. This graph is from the recent study by the UL, *Impact of Ventilation on Fire Behavior in Legacy and Contemporary Residential Construction.* Chapter 8 (ventilation) contains a complete explanation of this study and the changes it will have on our search and rescue and fire suppression operations at future house fires.

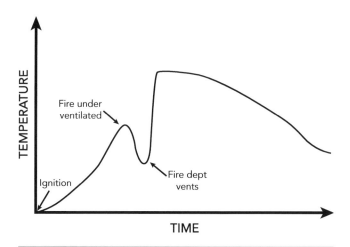

Fig. 3–11. Temperature graph of ventilation-limited fire. *Source:* Kerber 2010.

Rapid Fire Development

In this realistic fire growth scenario, the rapid fire development or flashover is driven by the addition of air after we force the front door of the house to enter for search or fire suppression. For now, in the following pages, gain a thorough understanding of flashover, its cause, and how to protect yourself from getting injured or killed by it at residential fires. The room in figure 3–12 has gone to flashover. Note the aggressive nature of the fire extending out the window. Post-flashover fire spreads rapidly, driven by expanding gases and the additional fuel provided by ignition of the flammable gases in the smoke.

Fig. 3–12. Flashover in a room. Courtesy of Brian Duddy.

To ensure our safety, we must have a thorough understanding of fire development and flashover because residential fires are commonly where firefighters are caught or killed by flashover. Residential fires are extremely dangerous and have been for years. A cover of *Fire Engineering* from 1970 picturing a firefighter jumping out a window to escape flashover, shows that it is nothing new (fig. 3–13).

Fig. 3–13. Flashover is not a new phenomenon.

Flashover

It is no wonder that flashover is not well understood by firefighters; it is not well understood or clearly defined by scientists and engineers. Several definitions are commonly used by the fire service today. According to a handbook published by the Society of Fire Protection Engineers (SFPE), "Flashover is not a precise term, and several variations of the definition can be found in the literature.... Most have criteria based on the temperature at which radiation from the hot gases in the compartment will ignite all of the combustible contents. Gas temperatures of 300–650 °C have been associated with the onset of flashover, although temperatures of 500–600 °C are more widely used" (Hurley 2016, 997).

The NFPA defines flashover as "the stage of a fire at which all surfaces and objects in a room or area are heated to their ignition temperature and all contents and combustible surfaces ignite at once" (Fahy, LeBlanc, and Washburn 1990, 42).

According to Deputy Chief Vincent Dunn of the Fire Department of the City of New York (FDNY),

> The technical definition of flashover is full room involvement caused by thermal radiation feedback (reradiation) from ceilings and upper walls that have been heated by the fire. For example, when an object burns in a room, the resulting smoke and heat rise up to the ceiling. They build and then spread out along the underside of the ceiling to the outer walls, and eventually the confined smoke and heat begins to descend down into the room. As the fire continues to burn in the confined space, the heat from the ceiling and upper walls is radiated back down into the room. If this reradiated heat heats all the combustible in a room to their autoignition temperatures, the room bursts into flame. (Dunn 2015, 77)

Ignition of products of combustion and fire gases from pyrolysis is a driving force in the rapid development of flashover in residential buildings. Consider what is actually happening in the fire growth cycle. It is steady up to flashover, then the room suddenly (in about 5 seconds) and silently fills with flames, and in postflashover, flames are pushing out windows and doors. There was no increase in the fuel. But it looked as if someone threw a can of accelerant on the fire. Where did all that fire suddenly come from?

Types

The late William Clark, in his book *Firefighting Principles and Practices*, explains that this could be a "Type 4 flashover." Clark writes that "it is highly likely that, in current arguments, the term 'flashover' is being applied to what are actually several different processes" (1992, 18).

This is what appears to be happening: As flames reach the ceiling, the CO is ignited as temperatures reach about 1,100°F. The CO_2 in the smoke retards the combustion process, resulting in the lazy ignition of the smoke into a condition widely known as *rollover*. At about 1,200°F, the next step takes place—the CO_2 is broken down into flammable CO. Clark continues:

> At that same temperature, the large amounts of water vapor given off by the fire will also unite with carbon to form hydrogen and more CO. The gases change as follows:

$$C + CO_2 \rightarrow 2CO$$

$$C + H_2O \rightarrow CO + H_2$$

> A large quantity of noncombustible gases, which had been slowing combustion, is suddenly transformed into copious amounts of two highly combustible gases, hydrogen and carbon monoxide, which ignite. (Clark 1992, 18)

This chemical process explains the rapid fire growth. Clark contended that there are three other types of flashover:

> *Type 1*: "Occur when a fire with a continuous supply of oxygen proceeds directly to full development in a straight-line, extremely rapid progression" (Clark 1992, 18). This theory is summarized by the NFPA: "Whenever a fire has been smoldering or burning in an oxygen-starved atmosphere for a long time, there is potential for large quantities of carbon monoxide and other unburned products of combustion to be present. These may not evidence themselves as thick smoke" (Fahy, LeBlanc, and Washburn 1990, 43).

> *Type 2*: This type occurs "when flame ignites unburned fire gases that have accumulated near the ceiling" (Clark 1992, 18).

> *Type 3*: This type happens "when the gaseous products of both pyrolysis and combustion, initially too rich a mixture to burn, suddenly obtain enough oxygen to make a combustible mixture, which ignites vigorously" (Clark 1992, 18).

Controlling factors for flashover

The Rockland County Fire Training Center has conducted flashover survival training since 1996, first using the Swede Survival Flashover Phase 1 unit shown in figure 3–14. These units produce controlled flashovers using particle board lining in the burn module as the main source of fuel. The flashovers are controlled by using a small caliber hose stream inside to cause a smoldering fire and a mechanical ceiling vent, operated inside by an instructor. Opening a rear door varying amounts gives the fire the oxygen necessary to generate several flashovers per burn. Trainees are in the observation module about 3 ft below the burn module's fire floor. This allows the flashover to pass over their heads as they are seated in the observation module. The purpose of this training is simply to have trainees observe the real life cycle of the fire through the flashover phase. An instructor on the nozzle controls the postflashover fire with limited amounts of water, allowing the fire to redevelop into additional flashovers, about six per burn.

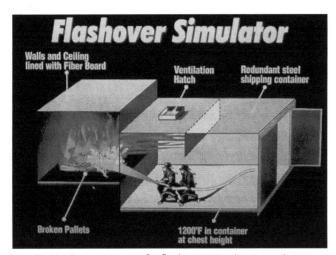

Fig. 3–14. Cross-section of a flashover simulator used to allow firefighters to witness the real life cycle of fire, including flashover and postflashover conditions.

Fig. 3–15. Exterior view of the flashover simulator. The burn module is at the right. Trainees sit on a low bench in the lower observation module and watch as the flashover develops in the burn module and passes over their heads and often exits the rear doors and roof vent.

After conducting this live fire training and witnessing hundreds of burns and flashovers, it is clear to us that Chief Clark is exactly correct with his breakdown of types of flashovers.

While shooting a training video, the photographer insisted we leave the doors open so he could get a good shot. We insisted that there would not be a flashover to video since there was no smoke. We opened the doors wide and began the filming with an unlimited supply of oxygen. Smoke appeared to be very light and flashover occurred quickly, rapidly consuming the fuel load. It is important to note that flashover occurs even with a very light, almost invisible smoke condition. In numerous case histories we studied, many of the firefighters involved said the smoke was light. Remember that these flammable gases are invisible components of the smoke. This was a Type 1 flashover as defined by Clark.

Type 2 flashovers are developed by keeping the doors closed in our simulator, allowing dense smoke to build up. Heat pyrolyzes nearby fuel (particle or oriented strand board), and weak flames of incomplete combustion generate large volumes of flammable gases. Heat is not intense until we open the rear doors (about 30%) to allow the fire to become active and reach up to the ceiling again. Then the flames begin to rollover from the rear of the fire module and ignite the gases at the ceiling of the observation module. The instructor opens the roof vent and flashover is drawn over the trainee's heads and out the vent (fig. 3–15).

Experimenting with different fuel loads (without trainees) in our simulator, we added two couches and an overstuffed chair to make the fuel more realistic. Fires were started by igniting newspaper in a trash can between the couch and the chair. The burning foam rubber and other furniture parts produced dense smoke. Flashover did not occur until we provided significant ventilation by opening the rear doors 50%. Apparently, despite heavy fire involvement of the furniture, dense smoke, and flames reaching into the flammable gases, the gas mixture was too rich to burn. As Clark predicted, a type 3 flashover occurred when conditions inside the simulator were brought into the flammable range by dilution with air from open doors and vents, and it burned vigorously. These were some of the most violent and radiant heat–producing flashovers we have experienced in the simulator.

Flashover of residential occupancies

Watt Street. According to the *NFPA Journal*, on 28 March 1994 the City of New York Fire Department (FDNY) responded to a report of smoke and sparks coming from a chimney at an apartment building. It was a three-story brick of ordinary construction, measuring 20 ft wide by 46 ft deep. Richard W. Bukowski, author of the final NFPA report on the incident, summarizes the incident:

> When firefighters arrived, they saw smoke coming from the chimney but no other signs of fire. The engine companies were assigned to ventilate the roof above the stairs by opening the scuttle and the skylight, and two three-person hose teams were sent to advance hoselines through the main entrance to first- and second-floor apartments.

The first-floor hose team forced open an apartment door and noted a momentary rush of air into the apartment. This was followed by a warm—but not hot—exhaust, which shot from the upper part of the door up the stairway. The first floor team was able to duck down under the flame and run back down the stairs, but the flame filled the entire stairway, engulfing the three men on the second floor. (Bukowski 1995, 86)

Three firefighters were killed at this relatively minor fire at 62 Watts Street. Amateur video showed the flame filling the stairwell and venting out the scuttle for more than 6.5 minutes. Bukowski continues, describing what fueled the fire:

The investigation revealed that the man who lived in the first-floor apartment had left it at 6:25 p.m., after placing a plastic trash bag on top of the gas stove in the kitchen, which he was sure he had turned off. It is reasonable to assume that the pilot light ignited the bag and that the resulting fire involved several bottles of liquor on the counter before it spread to the wood floor and other combustibles....

Clearly, the fire burned for nearly an hour under severely vitiated conditions. The open flue [fireplace] initially provided expansion relief and later vented smoke, as the ceiling layer dropped below the level of the opening. Such combustion produces large quantities of unburned fuel and high carbon monoxide/carbon dioxide ratios.

The report concludes with a description of how computers modeled this fire and proved that the scenario described above is what actually happened.

Was this a flashover or backdraft? It really does not matter. The effect—rapid fire development—caused the death of three firefighters: Captain John Drennan, Christopher Siedenborg, and James Young.

Winnipeg, Canada. Firefighters who have experienced flashover describe the fires leading up to the flashovers as "routine." After the flashovers, however, these fires were anything but. A good example is a flashover than occurred in a row house in Winnipeg, Manitoba, Canada, in October 1994. The original fire involved a couch—a chesterfield specifically. Conditions on arrival included smoke showing out a second-story window of the row house and two occupants on the ledge. Winnipeg firefighter Barry Borkowski describes the interior conditions: "Visibility was probably I'd say 70 percent...nothing seemed to be burning...I saw the chesterfield had been extinguished with the garden hose .... I thought that perhaps another crew had been in it before us; there was so much water; everything was soaked down."

Reacting to reports from bystanders that other civilians were trapped upstairs, Borkowski went to the second floor.

Again, the smoke was very intense, but the heat was not. I got to the window...reached up and slid open the window and smashed out the screen. I had a good opening for ventilation.

I saw Dennis [firefighter Carpentie] coming up the stairs.... He yelled out something about the living room starting on fire. So just as he said that, it became hot upstairs, very, very quickly. . . . In fact, you could feel the pressure differential. Mike Dowhayko was assisting on the second floor, Dennis had gone back down the stairs...and we found ourselves in a very delicate position. Flames were rolling up the staircase.

...Everything in that split second was involved in fire. At this point, I was experiencing what I thought to be thousands upon thousands of bee stings occurring along my arm...my mind would not rationalize it. Mike went to reach a means of egress. I thought I was following him but obviously made a wrong turn. I found myself in the washroom, and I discovered that by the clanking sound my tank was making against the porcelain of the bowl.... At that time I experienced what was the worst fear of my life...it just brought home all those stories you read about, about how firefighters being found in stupid locations like the corner of the bedroom or a bathroom....

Whatever was biting me was getting worse, and I found myself flailing and tugging at my clothing to try to relieve it. At this point, luck or whatever, I heard glass breaking, I heard Mike yell something. I followed the sound, and it brought me to Mike's legs and the same bedroom window I cleared earlier. There was no second thought...the thought did not occur to me, I could be 20, 30, 40 feet; it did not matter if I was on the top of a high-rise...if I stayed there any longer, I was going to be gone...so I flung my legs up, and I managed to clear the window in one shot.... Probably the happiest feeling I ever had in my life was when my feet made contact with the concrete...because I knew 1 was out of that situation. I knew Mike was okay because he was before me...I just hoped Dennis had made it out.

Mike Spaulding's story of being caught and almost killed in a flashover contains many critical life-saving nuggets you can take with you to your next fire (fig. 3–16). Sharing stories like these and other important lessons learned will help prevent you and others from going through similar near-fatal

Never forget that we are working in an uncontrolled environment.

CAPTAIN MIKE SPALDING, INDIANAPOLIS FIRE DEPARTMENT

FIRST-PERSON LESSONS FROM A SURVIVOR

Captain Mike Spaulding, Indianapolis Fire Department

I never thought I'd get caught in a fire. I was too smart for that. I was in good condition ... but I went from a commander to a fire victim.

In February 1992, when I was in my 20th year of service, our department lost two firefighters in a fire in the large, nine-story Indianapolis Athletic Club, which was built in 1922. Both were with me before they died. Their deaths changed me and my fire department.

I responded, as captain of a downtown ladder company, to a normal box alarm. Nothing was showing on the outside when we arrived. Inside, smoke was coming from a grate. We didn't know it at the time we arrived, but the HVAC system was circulating smoke throughout the building. Two other firefighters and I were assigned to search the basement/subbasement area. The fire had a head start of several minutes on us; Athletic Club employees had tried to put the fire out before calling the fire department. There was cool, light smoke in the basement, but no heat. We couldn't find the seat of the fire at first.

One of the firefighters in my crew, John, was inexperienced and had to work alongside an officer. He was an excellent firefighter and worked hard. John and I returned to the lobby, which was filled with smoke.

The chief reported that the seat of the fire was found on the third floor, in a barroom between two large dining rooms. It had started in a small refrigerator and spread up the wall into the concealed space above us. John and I were assigned to hook up with the crew lines and look for ventilation possibilities.

We arrived just below the landing area between Floors 2 and 3. I told John to put on his mask and stay close. I could see the hoseline going round the corner. Smoke banked low to the floors. We crawled and followed the line in. We could hear a lot of noise. We went through the doorway, staying close to the floor with the hoseline. We got behind the firefighter crouched in the doorway.

Somebody from within the room shouted "Woody's in trouble." Woody was a seasoned firefighter. The smoke was heavy and dark. I felt something move. A firefighter was trying to go around me. He had on no helmet or mask. I thought, "It's got to be Woody." I pulled him down to try to get him out of there. He was staggering. I tucked him under my arm.

As John, Woody, and I were crawling out, following the hoseline

in terrible visibility, things changed abruptly—as if a curtain dropped. The heat was unbelievable. The hoseline was burning. At that time, hoods were optional; firefighters used their ears as a barometer for heat. All my unprotected skin was burning. I knew immediately we were in trouble. We were on the third floor of a nine-story building, totally disoriented.

I hit my radio. I was trying to find the emergency button, fighting the urge to get up and run. The radio clip was off. I pulled off my glove to push the button. My hand was seared instantly. I couldn't feel the glove, the hoseline, John, or Woody. I didn't know if they had taken off or if something else had happened because I hit the floor. It was a living nightmare.

I called for help but had the feeling I was all alone. I prayed as I struggled down the hallway. My SCBA straps were burned off my back. The pain was indescribable. If I could have found a window, I would have jumped—and I'm an optimistic person.

I heard a firefighter screaming—a mournful scream for help. There was nothing I could do. I'll never forget that. A lieutenant, who had 30 years on the job, was exiting with his crew through the window. They couldn't take the line any farther. He thought he had heard a human sound and left the line to investigate. He saw my fire boots sticking out. He rolled me over. I hollered. He hosed me down. If it had been anyone else, I never would have been found.

I suffered third-degree burns over 30 percent of my body—face, ears, back, and arms; there was no skin on my right hand. I was in the burn unit for a month and in burn garments for 18 months.

John and Woody didn't survive. I think about that every day.

The fire structure was old and had been remodeled several times. There was no sprinkler system. There were combustible ceiling tiles and poke-throughs from one room to the next. That's what got us. It happened in a moment. The radio system was new. We should have had mandatory straps and hoods.

Buildings like this are all over the country. We can't always pre-plan every one. They hold surprises for all of us. We must not be complacent. There is no such thing as a routine fire. We are working in an uncontrolled environment.

* * *

SAFETY REMINDER

You may have worn your seatbelt 2,000 times in a row. The 2,001st time you may unloosen it because it doesn't feel right [and that's when an accident occurs]. Always put yourself in a position to survive things when they happen. We know they will happen; we don't know when.—*Dennis Compton, chief, Mesa (AZ) Fire Department, "Connecting to the Future"*

* * *

FIREFIGHTER DEATHS AND INJURIES

.... It's almost as though it's acceptable for firefighters to die .... If six were killed in a haz-mat incident, we would have had six meetings. Have you had a structure firefighting meeting after fatalities? There's a difference between what we know and the cultural response we take .... NIOSH devotes 50 pages to what happened and two pages to what to do about it. What's happened that's new? We're still ignoring the lessons of the past 50 years.—*Sacramento Roundtable*

Fig. 3–16. Mike Spaulding

experiences. Here are a few important lessons from Captain Spaulding's story:

1. There was no one single critical event, but rather many small circumstances came together to equal a disaster.

2. It started out as a routine box alarm.

3. The fire was already well-advanced upon arrival.

4. He and other firefighters became disoriented.

5. The fire was in a void space.

6. Sudden heat envelopment.

7. Unbelievable heat, indescribable pain, and extreme action or desire (to jump out the window).

In studying the case histories of firefighters caught in flashovers, we found that many incidents have the following seven events in common:

1. Total fire envelopment: everything goes orange
2. Disorientation
3. Intense pain due to heat penetrating clothing, skin burning
4. Separation from a partner or team
5. Panic caused by thoughts that this event may be fatal
6. Loss of a rational thought process
7. Extreme action

Based on our studies, despite training, experience, or turnout gear, it is often simply luck that may be the key factor in surviving a flashover.

Recognition and survival techniques

- We must train ourselves and our firefighters to observe conditions and avoid being caught in a flashover. Train to recognize the classic flashover warning signs: high heat that forces you to your knees; thick, dark smoke; free-burning fire; fire rollover (a late warning sign); and vent-point ignition (bursts of fire in the smoke from doors or windows). Once you notice these conditions, you must either immediately exit the area or, if you have a hoseline, apply decisive amounts of water to kill the fire.

- Be aware that flashover can occur under a variety of smoke conditions, as described above.

- Recall that our turnout gear will not protect us postflashover. Temperatures will reach 2,000°F+ and our gear will become saturated with heat, transmitting that heat to us and burning our skin. Remember that our skin burns at 124°F.

- Flashover happens rapidly and silently. You must actively look for the warning signs.

- Be aware of flammable products of combustion in void spaces like drop ceilings, cocklofts, and other concealed spaces.

- Many firefighters are killed or injured only a few feet into the fire area. Their bodies are found only a few feet from doors or windows and safety. Dense smoke, intense fire heat, and the chaos surrounding the rescue of an injured or trapped firefighter contribute to the overall danger. You do not need to be deep inside an industrial building to be killed by flashover. Flashovers in residential fires have resulted in many firefighter deaths and injuries.

Dunn, in his writings and videos, argues that a firefighter will not be able to escape flashover if they are more than 5 ft inside the room that flashes over—the point of no return (Dunn 2015). Unfortunately, this is supported by the case histories, particularly a flashover that occurred in 1989 in Oklahoma City. This fire in a one-story wood-frame house claimed the lives of three veteran firefighters. One of these firefighters was in verbal contact with his rescuers after the flashover. The two others were within a few feet of the front door of the house. This is typical of the flashover case histories we studied. Postflashover disorientation, confusion, and extremely low visibility contribute to very low escape and survival rates for firefighters.

- Know your exit and how to get out immediately before or after flashover begins.

- Realize that smoke will obscure your ability to see fire above you, where the flashover may begin.

- Expect flashover at every fire. It is part of the real life cycle of a fire.

Backdraft

Was it flashover or backdraft? This is a common firehouse kitchen discussion. What is most important is that it was rapid fire development and it likely injured or killed firefighters, or at least created a close call. It is sometimes called a smoke explosion.

Backdraft differs from flashover in two ways: the cause and the result. The cause is that the room or building was sealed up and the fire had consumed most of the oxygen. Once oxygen rushes in, the flammable gases in the smoke ignite with almost explosive force. The result is rapid fire development like flashover but with a concussive overpressure similar to what an explosive would produce.

Dunn defines backdraft as "an explosion caused by the rapid ignition of smoke and fire gases occurring in a tightly sealed burning room. The trigger for the backdraft explosion is the fresh air that enters with the firefighters during their initial search and entry to locate the fire…. Firefighters performing initial entry or search and rescue operations can be killed or injured by the blast of a backdraft" (Dunn 2015, 329).

Backdrafts are fortunately rare but can have devastating consequences for firefighters. On 24 July 2010, three firefighters from Tempe (AZ) were injured following a backdraft in a residential fire. A brick duplex home contained heat and smoke. A description of the scene in part said, "When the firefighters opened the door, it just let go." One firefighter was thrown 10 ft, another about 15 ft. The explosion was so powerful that it threw the door across the street, hitting the engine.

Full personal protective equipment (PPE) protected the firefighters, who only sustained minor injuries (Peluso 2010).

Warning signs

It is important to recognize the warning signs of backdraft:

- A closed-up room or house
- Black oily stains on windows
- Lack of visible fire, lots of smoke
- Puffing smoke (in and out) entering and/or exiting the cracks, windows, or door jambs

Vertical ventilation is the key to preventing backdraft explosions, but this is very difficult at house fires. Keeping members in safe areas and conducting horizontal venting may be the best answer.

One alternative is to immediately apply water into the potential backdraft room or house as soon as the door or window is taken. This may prevent backdraft by cooling.

Recognition is the key to backdraft survival.

Fire Spread in Homes

Houses have a variety of unique factors that allow fire to spread rapidly. First and foremost is the open layout of homes. Other occupancies—hospitals, schools, industrial, and some commercial occupancies—have some built-in fire resistance, which often includes compartmentation with fire-resistive doors, smoke doors, fire walls, and so on.

Open floor plan

Houses occupied as homes are generally wide open, which obviously allows a growing fire access to an entire floor (fig. 3–17). In multistory homes, open and unenclosed stairs are a chimney to spread toxic products of combustion to the sleeping area and again allow fire to rapidly spread up and throughout the house. It is critical for both strategic and tactical success on the fireground to anticipate and get ahead of this huge weakness in the building's fire resistance.

Fig. 3–17. Home with an open layout including open and unenclosed stairs

Exterior Fire Spread

The combination of new building methods (modular) and materials (vinyl siding, soffit covers, vinyl framed or sashed windows, and easily ignited foam insulation and sheathing) has created a new and dangerous dimension: extremely rapid fire spread from the *exterior* to the interior. Modern or renovated homes with these features that become involved in fire can have catastrophic progression of fire along several paths at the same time. This rapid fire envelopment from an exterior fire can trap and kill unsuspecting firefighters.

Numerous case histories all show the same sequence. A fire starts in the mulch around the house or on a deck or porch outside the home. The fire ignites the vinyl siding and burns furiously up the side of the house, through the soffit, and into the attic. Along the way, fire enters open windows or fails windows, creating another flow path from exterior to interior for the fire to travel along. If the home is modular, flames can enter the void space between the modules that make up the home and spread rapidly via interior routes. In a very short time, fire has full control of the house. If firefighters are inside on search missions, they will be in extreme danger when the house is enveloped in fire, both inside and out.

It is important that firefighters understand this sequence of events and the important details of its origin and extent of the problem. Tactical fireground successes, safety, and survival will depend on this understanding. Fire envelopment is a relatively new life-threatening sequence for both civilians and firefighters in home fires. Some of the main culprits are discussed below.

Vinyl siding

Hazards of rapid fire spread via vinyl siding were well documented and supported by various testing agencies and researched in detail in a paper by Anthony McDowell of the Henrico County Division of Fire (2009). According to information obtained from the Vinyl Siding Institute, in 2008, 78% of new homes in the northeastern United States and 32% nationwide were clad in vinyl siding. It is a widespread problem that the fire service must deal with both tactically during fireground operations and through code enforcement.

The following is a short list of examples of incidents where vinyl siding directly caused significant amounts of damage.

1. In February 2007, "a rapid exterior fire influenced by high winds, dry conditions, combustible landscaping, and combustible construction" occurred in Raleigh (NC) at a vinyl-clad townhouse complex, destroying or damaging more than 32 townhomes (Bossert 2007, 29).

2. In April 2007, three Fairfax Virginia firefighters were injured during interior operations when vinyl siding burned up through the soffits, resulting in what Fairfax County Rescue Department determined to be a carbon monoxide explosion. Note that CO ignites at about 1,150°F and is very flammable in high concentrations.

3. Also in April 2007, in Prince William County, VA, firefighter technician Kyle Wilson was killed while conducting SAR operations in a single-family home. What began with reports of "heavy fire on the exterior of two sides" quickly became deadly as the exterior fire, fueled by heavy winds and burning vinyl siding, entered the structure at multiple points simultaneously (Naum 2011). Below is a summary of the incident. (For the full report, see NIOSH 2008.)

    > Initial arriving units reported heavy fire on the exterior of two sides of the single family house and crews suspected that the occupants were still inside the house sleeping because of the early morning hour. A search of the upstairs bedroom commenced for the possible victims. A rapid and catastrophic change of fire and smoke conditions occurred in the interior of the house within minutes of Tower 512's crew entering the structure.

    > Technician Wilson became trapped and was unable to locate an immediate exit out of the hostile environment. Mayday radio transmissions were made by crews and by Technician Kyle Wilson of the life-threatening situation. Valiant and repeated rescue attempts to locate and remove Technician Wilson were made by the firefighting crews during extreme fire, heat and smoke conditions. Firefighters were forced from the structure as the house began to collapse on them and intense fire, heat and smoke conditions developed. Technician Wilson succumbed to the fire and the cause of death was reported by the medical examiner to be thermal and inhalation injuries. (Naum 2011)

4. In May 2008, Loudoun County, VA, seven firefighters were injured (some with serious burns) responding to a residential fire when "fire spread along the vinyl siding and combustible sheathing and traveled up through the vented soffits into the continuous attic space" (Bowers et al. 2008, 88).

West Haverstraw fire. The West Haverstraw Fire Department (NY) was dispatched to a mulch fire (as reported by the homeowner) in front of his home in a suburban neighborhood at 0145 hours. At 0147, the police officer on scene reported that "the entire front of the house is involved." A WHFD officer arrived on scene at 0148 and confirmed a fully involved house fire (fig. 3–18). In under 3 minutes, this mulch fire spread so rapidly that when the first apparatus arrived at 0149, the fire had already established flow paths up the exterior of the building that would soon allow it to envelop the entire building. Fire was aggressively pushing out of all the doors and windows on all floors. The roof rapidly collapsed and the attic was burning and showing at the eave line, which was fully involved with fire silhouetted against the night sky. Winds increased the exposure problem on the B side. The first engine arrived on the scene approximately 2 minutes later, established a supply line from a hydrant, and flowed its master stream soon after this picture was taken (fig. 3–19 and 3–20).

Fig. 3–18. Arrival of first WHFD officer. The fire appears to be mainly exterior at this time, but has already extended to the eaves/soffit on the A-B and A-C corners. Courtesy of West Haverstraw Fire Department.

Fig. 3–19. As first-due units protect exposures and prepare for master stream operations, the fire now has complete posession of the house. Courtesy of West Haverstraw Fire Department.

Fig. 3–20. B exposure, the most severe, was held in check with the first hoseline. Before water application, siding and roof materials were vigorously smoking and ready to light up. Courtesy of West Haverstraw Fire Department.

Later investigation revealed that the likely ignition source was a cigarette, carelessly discarded in the landscaping mulch at the front of the house. Another link in the failure chain that caused the ignition of the mulch was the fact that the house faced due south. This orientation (and a lack of shade trees) provided direct sunshine most of the day, drying the mulch to tinderbox conditions. A recent spell of very dry weather completed the mix of dangerous conditions.

Exterior fire envelopment. Occupants first reported a mulch fire, but the mulch burned vigorously and easily ignited the vinyl siding. This is what the police officer saw when he reported that the front of the house was on fire. As the fire burned the siding, it preheated and pyrolyzed the vinyl above it, adding to swift fire advancement up the front of the house.

The pyrolysis also added to the flammable gases in the smoke that entered the attic through the soffits.

As fire moved up the vinyl siding on the front of the house, the next modern house construction material feature to add to this incident came into play: large, vinyl-covered soffits (figs. 3–21 and 3–22). The huge volume of fire racing up the front of the house quickly burned through these thin pieces of vinyl. Soffits are usually covered with perforated sheets of vinyl to use natural convective air currents to ventilate the attic. Air rises, entering the underside of the soffit, enters the attic, and moves along the underside of the roof and out a ridge vent or gable vent, providing natural ventilation for the life of the home. Essentially, this is a preestablished flow path for the fire, drawing it into the unprotected attic space.

Fig. 3–21. A mirror image of the fire building. Note the large soffits on the attic over the windows. Courtesy of Tom Bierds.

Fig. 3–22. Typical perforated vinyl soffit covers

It is uncertain how long the fire remained isolated to the exterior, but the effect of the fire entering the attic was quite clear. Soon after our arrival, the roof collapsed. Fire had gotten in and spread throughout the attic quickly, heavily, and thoroughly. The subsequent roof collapse was actually more of a quick, nondramatic sag.

The complete list of stories on rapid fire spread caused by vinyl siding could fill volumes and the reader is directed to the references at the end of this chapter for more details.

There are, however, two more bits of information that bear discussion. The first is from the report by Madrzykowski, Roadarmel, and DeLauter that compared the combustibility of aluminum siding, T-111 (plywood), and vinyl siding, which found that in the test of vinyl siding, "less than 90 seconds after ignition, the flames began to spread upward and within another 50 seconds the flames were into the attic space" (1997, 347).

The second piece of information comes from McDowell's study comparing exposure fires on T-111 to those on aluminum siding. Referring to the Madrzykowski report, McDowell says, "The results were interesting. A small, focused area of the aluminum siding melted after ten minutes of flame contact, at which point a smoldering fire developed inside the wall, but there was no vertical flame spread. The T-111 allowed burning and flame spread 200 seconds after ignition. The flames then took an average of another 80 seconds to burn to the soffit level" (2009, 15).

Sheathing and insulation

The exterior fire is fueled by not only the siding but also the building insulation (foam or fiberboard) that is directly under the siding. Under the insulation may be plastic house wrap and of course under that is the sheathing of oriented strand board (OSB). The OSB contains large amounts of glue to hold the strands together in a useable board. All of these combustible components play significant roles in the creation of a large body of exterior fire.

In figure 3–23, the number 1 indicates the A side of the house where this residential fire in Rockland County (NY) reportedly started. Note that the siding, insulation, and sheathing were completely consumed by the intense exterior fire. The number 2 on the B side shows where heat caused the vinyl siding to melt and burn away, exposing the combustible foam insulation and subsequently burning the sheathing. After a reliable water supply was established, the exposure line operated on this exterior fire and quickly extinguished it. Compare this to figure 3–24, which shows how the aluminum siding on a different exposure resisted the intense heat.

Here is another clear and dramatic example of exterior fire spread. Truck company members cut a ventilation hole over the fire that has taken control of the office on the side of this home. A line was in place inside and firefighters were beginning the interior attack. The vinyl siding was clearly starting to fail (fig. 3–25).

Fig. 3–23. Exterior fire where the vinyl siding and foam insulation melted and burned, exposing the sheathing. Courtesy of Tom Bierds.

Fig. 3–24. The home on the right is clad with aluminum siding and did not sustain even minor damage despite the severe exposure to radiant and convected heat. Note that the exposure on the B side (far side of the picture) is 75–100 ft away and the vinyl siding is already melting.

In the short time it took to get firefighters off the roof, the siding was close to ignition from radiant heat. One significant clue is that the exposed siding was smoking (fig. 3–26). The siding became involved rapidly when exposed to the heat from the venting fire, creating an intense and fast-moving fire that would have entered the attic through the vinyl gable vent at the peak of the roof (fig. 3–27). In this instance, however, the exterior fire was rapidly extinguished (fig. 3–28). Note that this hoseline was not directed into the building. Interior crews conducted an aggressive interior attack facilitated by effective ventilation and supported by the exterior attack, which prevented the home from being enveloped by fire and endangering interior firefighters.

Fig. 3–25. Firefighters have just finished venting the roof directly over top of the fire. Note the melting of the siding, exposing flammable insulation, house wrap, and OSB board.

Fig. 3–26. Exposed siding smoking. Note the short time from the previous picture.

Fig. 3–27. Involved vinyl siding

Fig. 3–28. Aggressive exterior attack

Vinyl window frames and sashes

Another way the fire could have entered the home in the West Haverstraw case history was via the windows. Live burn tests conducted at the Rockland County Fire Training Center in Pomona (NY) comparing the relative fire resistance of wood-frame single-pane windows versus vinyl-frame, double-glazed, energy efficient windows and sashes provided very interesting and surprising results. Multiple live burns revealed that the energy efficient vinyl window frames and sashes fail quickly under a fire load. This failure of both the frames and the window glazing itself was much more rapid and catastrophic than anticipated and much more rapid than for legacy wood-frame or wood-sash windows. In summary, the legacy windows failed in small amounts over time, and the vinyl (frame/sash) energy efficient windows failed quickly and catastrophically.

These results were surprising, because well-documented research and fireground experience typically show that energy efficient windows withstand heat and a fire load much better than legacy windows. Based on a comparison of our live burns in Rockland, a literature search, and the experiences of fire-fighters in other cities, however, this appears to only be true for aluminum or metal sash and frame energy efficient windows. As you may expect, the vinyl windows fail quickly as the photos below show. The vinyl frames are extruded, which simply means they are made of folded sheets of vinyl, much like the cross-section of an "A" roof post on a car. The vinyl frames and sashes are another modern component that helps fire's extremely rapid fire spread.

Mowrer's (1998, 15) study reports that "vinyl-frame windows faired [*sic*] poorly...at imposed heat fluxes ranging from 0.8 to 1.6 W/cm^2. The vinyl frames and sashes lost strength, sagged and distorted under the imposed heat fluxes, typically within a matter of minutes." This is illustrated in figure 3–29, which shows a wood-frame single-pane window on the left and an energy efficient, double-pane vinyl frame and sash window on the right. Heavy fire inside the test facility is causing the vinyl frame to melt and burn before the wood frame. Later in the test, the vinyl frame and sash completely failed, allowing massive amounts of air into and out of the fire area, which created a flow path and helped drive the fire to the postflashover state shown in figure 3–30.

Fig. 3–29. Comparison of a wood-frame single-pane window and an energy efficient double-pane vinyl frame and sash window

Fig. 3–30. Postflashover state caused by complete failure of vinyl frame and sash

Modular construction

A major component of the exceptionally rapid fire spread in this house was the fact that it was of modular construction. As a reminder, modular homes are brought in to the homesite as nearly finished modules and placed together and secured to each other and an existing foundation. See chapter 2 for details.

Modular construction has its own set of very unique and significant contributing factors to exceptionally rapid fire spread in homes. This was first expertly detected, researched, and presented at FDIC 12 and 13 by Chief Kevin Gallagher of the Acushnet Fire Department (MA) based on a fire his department responded to in 2008 (fig. 3–31). Note the similarities between this fire and the one in West Haverstraw (fig. 3–19): low burn on the lawn, fire from foundation to eaves and roof, exterior surface fire, and total involvement of the interior.

Fig. 3–31. Total involvement of a modular home in Acushnet. Courtesy of Kevin Gallagher.

Chief Gallagher and his department responded to a fire in a modular home and became concerned about how rapidly the fire spread. Researching the causes, he determined that the ceiling sheetrock is often only attached to the ceiling joists with thermal setting glue. There are no traditional sheetrock nails or screws. After careful research and testing, he determined that at approximately 400°F, the glue softens (Gallagher 2009, 162). This obviously causes the ceiling sheetrock to fall, allowing fire to enter the 20 in. void space between the first-floor ceiling and the second-floor subfloor. Something as benign as a small room-and-contents fire can cause this fire barrier failure, leading to rapid fire spread and total loss of the home.

The 20 in. void space is similar to a cockloft and thus has similar fire problems. However, this void is between the floors. It is created as a result of the construction method: stacking the second-floor module on top of the first floor. Clearly, this is another contributing factor to the unusually rapid fire spread seen at this house fire. The danger of fire entering the cockloft is well known, and based on firefighters' recent experiences with the void spaces in modular homes, we should anticipate similar effects at these fires.

Fire department tactical actions

There are several important tactical lessons that have been learned from these exterior fires that can be considered by all fire departments and appropriately applied to other fast-moving house fires.

1. **Houses that have the aforementioned and other "fire-friendly" construction characteristics require new and different tactics.** During size-up and when formulating your SAR, ventilation, and fire attack plans, remember that a building with these dangerous characteristics will likely light up very quickly. Fire can envelop the building via several routes, trapping firefighters. Be conservative in deploying members for interior operations as they may be quickly surrounded by fire from various routes, both internal and external.

   Multiple hose streams (interior and exterior) will be needed to control the fire. If you do not have enough personnel on the scene to accomplish this, consider defensive operations before placing members in the path of the fire. Recall that this fire is advancing in several directions at one time, all contributing to the hazard.

   For fires that begin outside the building (mulch, siding, or decks), starting the fire attack from the outside—cutting off one of the main bodies of fire—can be done by one firefighter in relative safety. This line can be stretched simultaneously with the conventional (in-the-front-door) line if personnel permits. This line, operated by one firefighter, can serve two purposes: protecting the exposure(s) and quickly extinguishing the exterior fire.

2. **These are fast-moving fires that need decisive action.** Use the water you have to deliver a decisive blow to the fire. Saving it in your tank or flowing nondecisive amounts of water in weak, ineffective, and limited-reach streams will not increase the probability of success. This has been the battle cry of Captain Bill Gustin of Miami-Dade Fire and Rescue for years. In a conversation with the authors (June 2014), Bill emphatically sums it up:

   > We have got to deliver water in amounts that make a difference in the fight to protect

exposures or extinguish the fire, no matter what kind of fire it is. We don't want to aggravate it or prolong the fight, we want to kill it, and kill it now, before it can destroy more property or threaten lives of firefighters and civilians. There is almost a paranoia in the fire service about running out of booster tank water before the fire is out. There are firefighters that are more comfortable with fire spreading to an exposure than running out of water; consequently, they do not apply sufficient flow to keep the exposure from igniting. When operating off tank water, you have only two options: First, put the fire out if you have sufficient water. Second, if you do not have sufficient water to put out the fire, use it to protect exposures.

Structure fires are not class-B fires; they don't have to be extinguished completely to keep from reigniting. 300 gallons of water, if properly applied, can reduce a fire's intensity, keeping it from spreading to exposures, until an engine establishes a continuous water supply. Don't forget that it takes a whole heck of a lot less water to keep something from igniting, keeping exposed surfaces moist with intermittent application of water, than to extinguish it once it ignites. There is no option three; saving water in the booster tank is not an option.

3. **These fires require strategic changes to fire suppression plans.** The goal of any fire suppression plan or operation is to either get ahead of the fire and cut it off or to get a hoseline to the seat of the fire and extinguish it. If a large body of fire is on the exterior of the building, it is nearly impossible to extinguish it from the inside with a traditional aggressive interior fire attack.

It is important to consider that there now are two main bodies of fire: one interior, one exterior. The exterior fire has the following advantages. Because it is on a vertical surface, it generates (pyrolyzes) flammable gases that intensify the fire by preheating the uninvolved fuel. This preheating also makes the uninvolved fuel easier to ignite. The exterior fire has layers of fuel that contribute to the fire: vinyl siding and trim, flammable insulation (foam or fiber board), and flammable sheathing (plywood or OSB). All these contribute to a large body of fire outside the structure, which itself is providing a flow path up to the eaves or soffits that offer little or no resistance to the vertical spread. If you don't stop this fire before it reaches the attic, property loss will be exceptionally high, if not total.

This large body of exterior fire must be stopped before firefighters are committed to the interior for any operations. This exterior fire is like an octopus that is wrapped around the building, able to work its tentacles into many different places to slip into the building. Further endangering firefighters is that if the roof structure is lightweight (trusses, I-joists, etc.), it will not withstand a thermal assault very long and will quickly collapse, trapping members inside. A well-involved attic fire can and will extend downward rapidly when sheetrock ceilings are pulled or fail. This scenario often causes a flashover on the second floor, an all-too-common fatal scenario for firefighters, especially if they do not have a hoseline to defend themselves.

4. **Interior and exterior opposing streams need to be included in our strategy.** For years, the simultaneous use of exterior and interior streams has been a major tactical taboo, and rightfully so. Opposing streams pushing products of combustion onto interior firefighters is never good. However, with vinyl siding causing intense fires on the exteriors of these homes, it is imperative that fire be controlled quickly. Obviously, the key is to train members to use that stream only on the exterior fire if other members are working inside, and there must be excellent communication between these lines and command.

5. **Exterior vinyl siding fire can cause the electric service line to fall.** Early in an operation we worked once, the service line from the pole to the house sparked a few times then rapidly came down. The heavy body of fire outside the building caused this major hazard for firefighters.

Hazards of Utilities

Homes, of course, need heating, cooling, electric, and plumbing services. Don't overlook the hazards of home fuels. Natural gas, propane, and oil are common home-heating fuels. The best way to reduce the hazard presented by these substances is to shut them off from the source as soon as you can (fig. 3–32). We conducted a fire attack in a basement fire that was very unsuccessful because the unions holding the gas meter on were damaged by the fire, causing the gas to release and intensify the fire.

Fig. 3–32. An interior natural gas meter that could pose a threat to firefighters

Natural gas

Natural gas is a common home-heating fuel that is very dangerous. Natural gas is flammable when there is as little as 5%–15% in the air. Homes that are damaged by natural gas explosions can present with large amounts of fire or sometimes none at all. A nearby department responded to an incident at a single-story wood-frame home that had been destroyed by a natural gas explosion and it looked like it had been professionally demolished. A pile of shredded wood, furnishings, and belongings were in a neat pile on the footprint of the home.

Natural gas can travel and collect far from the source both inside and outside the home. Natural gas often follows the path of least resistance through electrical conduits, storm and sanitary sewer piping, or any porous route. The odorant, mercaptan, that gives the gas its characteristic smell can be scrubbed out as it passes through soil and pipes. This of course does not make the gas any less dangerous—if anything, it is more dangerous because it is less detectable to occupants' noses.

Propane

Always consider the possibility that a large propane tank may be inside a house under construction (fig. 3–33). Temporary heat is required for painting and sheetrock installers and finishers. Many finished homes are also heated with propane (fig. 3–34).

Fig. 3–33. Propane tank inside a house under construction

Fig. 3–34. Tanks found outside a fully involved rural home

Boiling liquid expanding vapor explosion (BLEVE) hazard. Propane is a good example of a BLEVE hazard you may find at house fires. Propane gas is cooled and compressed into a liquid for economical transportation and use. Fire may impinge on the tank, weakening the metal. The heat of the fire will cause excessive pressure and the relief valve will operate. The relief valve is only designed to reduce the pressure to safe levels in atmospheric conditions. For example, the tank was filled early in the morning when temperatures were in the 20s. The tank is installed in the afternoon sun when temperatures reached 65°F. In this circumstance, pressure caused by heating inside the tank would be relieved by the valve. But the relief valve is not designed to reduce pressure caused by fire impingement. As one relief valve opens and the gas finds an ignition source, a blowtorch-like flame impinges on the metal above the liquid level in the tank. If this continues long enough, the tank metal will weaken and the pressure inside the tank will finally cause the metal to

rip or tear. When the tank is opened to atmospheric pressure, the propane liquid reaches its boiling point and huge volumes of propane gas are immediately generated. The liquefied propane expands and converts to vapor at a rate of 270 times liquid to vapor. Then the gas cloud (100% propane) becomes diluted in the air, causing further expansion until it reaches its flammable range. Combined with an ignition source, the cloud ignites typically in a ground flash (fire) that rises up in a mushroom plume. The tank may shred and explode out in several large and small pieces of deadly shrapnel.

Smaller tanks generally have frangible discs or fusible plugs as pressure relief devices. Once activated, these will not reset and will allow a continuous flow of gas until the cylinder is exhausted. The spring-loaded relief valves found on larger tanks can open and close repeatedly as fire impinges on the tank and increases pressure. The relief valve will open to maintain safe pressure in the cylinder. However, there will be no warning prior to the release of gas, which is an obvious hazard.

Electrical Hazards

(There is a more detailed discussion on electrical hazards in ch. 11.) Overhead electrical wires present the biggest hazard to firefighters at house fires. All homes will be wired with household current (usually 110 volts [V]), and while contact with this current is always unpleasant, it is rarely fatal. The primary wires that transmit electricity along the street poles, however, may contain as much as 13,000 V, and obviously any contact with this amount of voltage will likely be fatal. Firefighters operating in areas with overhead wires must be careful with both aerial devices and ground ladders. The call "clear of wires" should be part of your ground ladder raising steps. In figure 3–35, fire attack operation has to be defensive until the utility company can shut off the electric service to make the area safe for firefighters. Note that the wires on the pole and at the base of the pole are both burning and energized.

Fig. 3–35. Wires down

We responded to an incident at a smart house, one that was wired and had remote controls for everything from heating to intercom systems. Heavy fire in the attic caused the wiring, including the spiral stiffeners from the flexible ductwork, to drop down (fig. 3–36). These wires caused considerable tangle issues even during overhaul when visibility was excellent. During a search or fire attack, this would be a lethal trap. Always keep a set of wire cutters handy (i.e., in an accessible pocket in your turnout gear) to cut your way out of one of these situations.

Fig. 3–36. Wires down in a smart house

Site Grade Hazards

Houses are designed by architects and builders to be pleasing in appearance, especially when viewed from the front. As families expand with children or when elderly parents move in, the home may have additions put on or second floors added to create more living space. The natural or created grade of the homesite often generates the appearance of a one-story home from the front when the rear or sides may actually be three or three and a half stories above grade. The obvious problem is that if firefighters enter from the front of the home and have to make a leap out the rear window, the drop may be lethal. For example, from the front, the home in figure 3–37 looks like a one-story Cape Cod or maybe a ranch home that had dormers added to create living space from what was previously the attic. If you look closely at the rear, you can see that a full-length dormer was added to create living space upstairs. If you look at the same house from the rear, you see that the window on the left has a three-story drop to the ground (fig. 3–38). The basement entrance on the left side may indicate a separate basement apartment. There could be a family living in the basement, another one on the first floor, and a third family or occupant on the second floor.

Fig. 3–37. Front view of what appears to be a one-story Cape Cod

Fig. 3–38. The same house from the rear

According to a USFA report (1995), a fire in a Pittsburgh (PA) house laid out very similarly to the one shown above caused the deaths of three firefighters. A paragraph from the overview section of the report summarizes the hazard:

> This incident also reinforces a concern that has been identified in several firefighter fatality incidents that have occurred where there is exterior access to different levels from different sides of the structure. These structures are often difficult to "size-up" from the exterior and there is often confusion about the levels where interior companies are operating and where the fire is located. In these situations it is particularly important to determine how many levels are above and below each point of entry and to ensure that the fire is not burning below unsuspecting companies. (NFDC 1995, 1)

Two other fires, the Cherry Road row house fire in Washington, DC (1999, two LODDs), and the Diamond Heights fire in San Francisco (2012, two LODDs) are additional examples of the hazards of different grades on opposing sides of the home.

Energy Efficient Homes

As the cost of heating and cooling homes has increased, so has the effort to keep in the heat or cool air. For houses, this has meant three significant changes in construction.

Insulation

The thicker the insulation, the better its insulation value. Homes now have insulation values of R-19 or greater. This increase in insulation in both walls and floors helps keep in the heat generated by a fire. This can reduce the time to flashover and certainly makes conditions more untenable for us.

House wrap

Used to reduce the amount of air both escaping from and entering a home, house wrap has been widely used in both new and renovation construction. House wrap is a thin plastic film that prevents air exfiltration or infiltration into the home. Obviously, this plastic film has the effect of sealing smoke inside the home, preventing us from determining the extent of the fire.

Energy efficient windows

Energy efficient windows are designed to minimize heat or cooling loss from inside the home and to prevent radiant solar heat from entering. Sometimes called insulated glass or thermopane windows, they are made up of two or more panes of glass with either a vacuum or an inert gas such as argon between the layers. The gap provides insulation as there is nothing between the panes to conduct heat through the window. There also may be a film applied to the windows to reflect radiant heat and allow heat to be absorbed in winter.

Fire problems with insulated glass in homes are obvious. They keep in heat longer, again speeding the time to flashover. Energy efficient windows also are well sealed in comparison to their single-pane cousins and can hide fire and smoke from us.

Thermopane windows under a fire load will often have the interior pane of glass fail first. If you are inside searching and hear glass break, you may think a firefighter has taken a window for you from the outside. You'll think the line is moving in and that's why they vented, because the fire is being controlled, so you continue deeper into the home. Suddenly you are caught in a flashover. Due to the pain, your thoughts become only of survival, so logic leaves you and you dive out a window. Thankfully the window was on the first floor and you save your own life, escaping with only minor injuries. What happened here?

Based on live burn tests conducted with energy efficient windows at the Rockland County Fire Training Center in Pomona (NY), it appears that heat from the fire on the interior pane of the window causes the glass to break and fall in. The exterior pane or panes remain intact, allowing heat to build until temperatures reach around 1,100°F, at which point the carbon dioxide in the smoke is chemically changed to carbon monoxide, which lights up at 1,150°F, driving the flashover. Returning to our hypothetical situation, the glass you heard break was what we call a *false vent* and is directly attributable to the energy efficient window (Knapp and Delisio 1997).

Another important finding from the Rockland County tests is that window performance under fire conditions depends on what type of frame the window is installed in. Aluminum frames seem to hold up and maintain window integrity much better than other frames. The integrity of the window is defined by the window's ability to contain the heat and smoke longer, allowing heat and smoke conditions to build faster to deadly conditions. Extruded vinyl frames are actually just plastic frames folded to structural and functional forms. These frames, as one might expect, do not withstand fire conditions very long and perform differently from metal frames. The Rockland tests show that the vinyl frame (extruded plastic) failed even before a single-pane wooden sash window (Knapp and Delisio 1997).

Other Hazards

The population as a whole has a wide variety of interests, jobs, hobbies, obsessions, fetishes, and other strange behaviors. Some of these are legal, some illegal, some suicidal, and some just simply different. For us as firefighters these behaviors manifest as hazards when we respond to someone's home when it is on fire. Here are a few case histories the authors have come across over the years, some in real life and others through newspaper accounts. Everyone thinks their home is safe, and for them it likely is. However, firefighters have been surprised with many dangerous situations during house fires. For instance, we wouldn't normally think of a rollaway bed as a hazard (fig. 3–39). During a search or fire attack operation, if you should get tangled in this, getting your foot caught or your SCBA straps stuck in it, it could become deadly. Bicycles stored in homes also pose a similar hazard. Your SCBA emergency training should involve escape from these hazards.

Fig. 3–39. Rollaway bed hazard

Incomplete or unsafe do-it-yourself construction by the homeowner or occupant presents countless hazards. A firefighter was conducting a search of the home in figure 3–40 while the engine company extinguished a kitchen fire downstairs. Crawling by the sliding glass doors, he thought to himself that the deck would have made an excellent escape route or place to bring the victims. Needless to say, it was very surprising not to see a deck surface. After the fire, we discussed this in the after-action review. When the firefighter saw it from inside during the search, he said he immediately wondered what else could be dangerous in this house.

Fig. 3–40. Do-it-yourself home construction hazard

Businesses operating from homes

Often, people run businesses out of their homes. A fire that a neighboring department responded to became interesting when the drums of seamless linoleum floor solvent caught fire in the attached garage. Other businesses run out of homes present unique hazards. A chief officer in Santa Paula (CA) got an almost lethal dose of cyanide when he inspected a fire suppression operation that was in the overhaul phase. The homeowner ran a precious metal business out of his home that utilized the toxic liquid, which became airborne during the fire.

Crime scenes

Often house fires are crime scenes. It may be simply be an arson job because the homeowner could not pay the mortgage, or it could be arson to cover a burglary or another crime. Arson to cover murder is a possibility and was in fact a reality that the Hillcrest Fire Department (NY) came across. Firefighters responded to a midmorning fire with smoke showing. Search teams discovered what they thought was a fire victim during a routine search. What they'd actually found was a murder victim who had been killed by his former girlfriend. She set a fire in the closet in an attempt to cover the murder.

The Associated Press (AP) reported on 23 July 2010 that a house fire killing a mother and four children in New York City was "being investigated as a possible murder-suicide committed by one of the children, a troubled teenager with a history of setting fires." Three of the children (two girls and a boy, ages 7, 10, and 14 respectively) had their throats slashed. Subsequent investigation reported by the AP on 29 July revealed that the mother killed the children, set the apartment on fire, then sat with her 2-year-old until she and the toddler succumbed to the fire. A remnant of a note with the words "am sorry" was found in the home.

House fires are often set either by inept burglars or efficient murders to cover the scene. What may look like a fire victim may actually be a murder scene with a fire set to cover the crime. In these cases, when you recognize it might be a crime scene, limit overhaul and get the police involved immediately.

Suicides

There have been several successful or attempted suicides in homes with fire as the deadly method for the victim. In one instance, the individual went to the basement and disconnected the gas meter, allowing gas to flow freely. The victim allegedly then stripped down naked and went into the bathtub until the residence exploded, which it promptly did, killing him and presenting first-due firefighters with a fully involved home.

In another incident, a despondent elderly woman attempted to burn down her home by setting numerous small fires throughout the house. Firefighters were not able to save her before she succumbed to smoke inhalation. A neighbor spotted smoke coming from the house. Firefighters were able to extinguish the fires before they extended through the house.

In a suicide attempt in Highland Falls (NY) in 2003, a local man was rescued by police when he was found stuck in the window after trying to escape his burning home. Allegedly, he turned on all the gas burners on the stove and when he thought enough gas had accumulated, he lit off the fireworks he had daisy-chained through the house. Neighbors reported an explosion and fire. The man survived with numerous burns.

Suicides are not only harmful to the person who has decided to end their life; they can also harm firefighters and other emergency responders as well, as the story below clearly points out. Remember that suicides are usually well-thought-out events, often months or years in the planning, and these desperate people often don't care who else they may injure or kill. From the FDNY Watchline:

> Additional Devices Set at Wisconsin Home Explosion—A one-story home in rural Wausau, Wis. was all but leveled in an explosion last Thursday morning in what authorities consider a probable suicide. As fire crews confronted flames and collapse, additional explosive devices were discovered on the property and the bomb squad was summoned. A total of five additional bombs were found, including two in separate cars, one in a shed and one in a camper. All were connected by timed fuses throughout the property. The remains of the homeowner were found in the basement of the home the following day. (*Rockland Journal News*)

Assessment: Explosions in private homes—usually accidental—are not uncommon in the United States, particularly in the winter months where tightly sealed areas can support explosive atmospheres. Intentional explosions, however, as with other forms of arson, may carry added dangers. Responders should be alert to indicators at scenes of intentional explosions and fires, such as the presence of timed fuses or reports of a gunman. Whether the occupant meant intentional harm to others or just a spectacular destruction of their worldly possessions, the danger existed in either case.

Chemical suicides. Initially common in Japan, this method of suicide has been growing in the United States since 2008. Instructions are readily available on the internet and involve mixing common household chemicals to produce deadly quantities of gases, most often hydrogen sulfide but also carbon monoxide and hydrogen cyanide. Typically the suicide takes place in an automobile, but several examples have occurred in small closets in college dormitory rooms and bathrooms in homes. The hydrogen sulfide gas can smell like rotten eggs but easily fatigues the olfactory senses.

Responders to noxious odor calls in homes must be aware that some desperate folks have chosen to end their lives this way. Be vigilant for uncommon odors and ask the person who called in what (if anything) they smelled. Air monitoring is essential, so call the hazmat team quickly. Hydrogen sulfide is deadly even at low concentrations.

Hoarders

Hoarders are people who accumulate mountains of various materials inside their home. Frequently, they feel unable to discard what most of us would consider unusable or trash. Newspapers, magazines, plastic containers, almost anything can be saved and stored in the home in what would normally be living spaces. Hallways, rooms, basements, and attics are often stacked to the ceiling with these collections of everyday items.

For firefighters this results in dangerous conditions created by lack of or blocked exits and lack of access or at best slowed access to the fire area for both engine and truck companies. Fireground experiences abound with dangers created by burning and falling piles of hoarded material. On 19 December 2011, Passaic (NJ) firefighters were confronted with a fire in a hoarder's home. According to the 20 December article by Marlene Naanes and Tariq Zehawi for the *Hackensack Record*, Chief Patrick Trentacost said, "The first responders on the scene of the fire, which erupted just before 6:20 a.m., immediately called a second-alarm because they could not get past the front door of the debris-strewn home on Lexington Avenue. The hallways only had one to two feet to maneuver around, and the stairways had about a foot of room." Captain Luis Sanchez also commented on the fire: "It was wall to wall stuff. It was very, very difficult to make entry. You could step one or two feet inside and there was a mountain of stuff there."

In another case in New York City, the FDNY Watchline daily bulletin reported the following:

> Elderly Couple Rescued from "Collyers' Mansion"—Based on calls from neighbors for a well-being check on a reclusive couple, firefighters forced entry into a home and were met by an overwhelming odor and floor-to-ceiling debris. After donning protective equipment, the firefighters removed the couple to an intensive care unit. A fire department spokesman said the wife may have fallen through the debris first, with the husband becoming trapped in an effort to come to her aid. The wife suffered from multiple rat bites. (*Rockland Journal News*)

Assessment: The term "Collyers' Mansion" is rooted in the 1947 discovery of the pair of eccentric brothers found dead in their NYC brownstone, among 130 tons of debris. First responders should use a deliberate approach when entering and searching hoarder homes, and always maintain a safe egress. Information obtained from neighbors may assist size-up.

Firearms and ammunition

Firearms and ammunition can be a serious hazard to firefighters, as described in this short summary from the FDNY Terrorism and Weapons of Mass Destruction daily bulletin:

> New Jersey Firefighters Impeded by Ammo Fire—Exploding ammunition slowed the efforts of firefighters battling a house fire in Perth Amboy, New Jersey. Firefighters were forced to attack the fire from the outside until the ammunition had cooled down. Three handguns, a shotgun, and a rifle owned by the occupant were later found. "One of the big problems was the son had ammunition on the second floor. That delayed us going in because the ammunition was going off with the large amount heat from the fire," a fire chief said. (*Rockland Journal News*)

Assessment: Firefighters usually have no idea what hazmat or other dangers they will find during a fire. Fire departments may not be notified of legal hazmat if amounts stay below a certain threshold. First responders should be aware that debates on stricter gun and ammunition legislation have caused many people around the United States to hoard bullets in their homes. As per department guidelines, if firearms or explosives are found, notify the IC, evacuate the area, and do not disturb the weapon or device.

Although ammunition in a fire will explode with a popping sound because the round is not contained, it rarely has enough velocity to injure or kill. Rounds in the chamber of a hand gun or long gun, however, can get hot enough to "cook off." This is slang for the moment when the powder reaches ignition temperature, essentially "firing" the weapon without a trigger pull. Obviously, the velocity of these rounds can easily be deadly. Often in situations like this the person will have significant amounts of gunpowder or other explosive hazards.

Pets

We all love our pets; some folks just get out of control. On 14 January 1995, the Associated Press published a photo of a Bengal tiger and 74 other animals that were found living in a home in Syracuse (NY). Included in this menagerie were a wallaby and an African lion. On 6 October 2003, Lydia Polgreen and Jason George wrote an article in the *New York Times* about Ming, a 400 lb tiger police found living in a New York City public housing apartment. In another incident, the headlines for a local paper read, "Animal House of Horrors. Cops say he lived with cobras, wildcats, a gila monster, alligators, foxes, copperheads, boas, wolves, vipers, black widows, pit bulls, a miniature horse, emus, crocodiles…and one chair."

People become very attached to their pets. According to a report in the *Arizona Daily Star* on 8 December 2004, firefighters were endangered during an intense search after a resident claimed her "babies" were still inside the burning house. The "babies" were three cats. The danger of this incident was compounded by a hoarding situation in the home that made the fire burn even hotter, and as Deputy Chief Randy Ogden said, he "wouldn't have committed firefighters to an intense fire situation" had he known that the woman was referring to pets. Sounds crazy, but ask whether the babies are human or animal.

Clandestine drug labs

Often, drug houses have a wide variety of deadly booby traps and improvised security devices. These may include guard dogs with voice boxes removed, firearms that discharge when a door or window is opened, spiked boards that may swing up toward an unsuspecting firefighter, holes in the floor, flammable liquids, and so on.

These illegal home-based drug factories have become an increasing problem in the United States over the past several years. Most produce methamphetamine, which has properties similar to cocaine and is also known as meth, crystal meth, or ice. Meth is one of the most addictive drugs. Long-term use leads to ignoring basic necessities of life, such as food and personal hygiene. The addict will often look emaciated, unkempt, and unclean, with body sores, severely rotten teeth, and so on. Addicts also often have extreme paranoia to the point of misdirected violence.

Similar to a grow house (which we will discuss in the next section), a clandestine lab can be detected by bars on the windows, blacked out windows, chemical odors, unusual chemicals (either in amount or type), unusual stains on walls or floors, and pressurized cylinders.

Common household chemicals and pharmaceuticals are ingredients for meth. Discovery of these products, often in large quantities, is one of the best tipoffs that you have responded to a meth lab. Ephedrine (now strictly regulated) was an over-the-counter medication found in sinus medicine. One pound of ephedrine (usually an abnormal number of small sinus relief pills) will produce ¾ lb of meth. Acids are another common chemical, and can be sulfuric acid contained in traditional acid bottles or something like a drain cleaner. These react violently when they come in contact with water. Muriatic acid or hydrochloric acid are used to generate the hydrochloric gas used in the production process.

Bases are also used in the making of meth. You may find drain cleaners that are alkalis. One meth production method utilizes anhydrous ammonia. Often 20 lb propane tanks are used to steal from legitimate users or stopped railroad tank cars. Often the brass fittings are green in color from the contact with the ammonia.

Ignitable liquids found at meth lab sites include acetone, ether, toluene, alcohol, and xylene. All of these are very flammable.

Lithium or sodium metals are other water-reactive substances required to make meth. Lithium is salvaged from foil in and around batteries. Extracting the lithium leaves a pile of battery parts and residue, which is another excellent clue. Lithium and sodium are water reactive.

Other items indicative of a meth lab are pans and glassware necessary to cook the meth, including blenders, skillets, spoons, coffee makers, pots, filters, mason jars, and syringe-like turkey basters.

It goes without saying that the above mentioned hazardous materials provide a wide variety of hazards—inhalation, fire, explosion, chemical burns, the list is extensive—for responding firefighters. If you discover or find yourself in or near a lab, the best thing to do is get out as quickly and safely as possible. Do not turn anything on or off, knock anything over, touch anything—just get yourself and your crew out and put some distance between it and you.

Grow houses

Another common illegal use for a home is as a grow house for marijuana. While the largest and most lucrative of these operations are located in large warehouses, they can also be just a few rooms in a rented or owned home or apartment (Gustin 2010). The operation often has sealed or blacked out windows and high-voltage intense lighting to stimulate plant growth. According to Captain Bill Gustin from Miami-Dade Florida Fire and Rescue,

> Growing operations are replete with electric hazards because of exposed wiring, terminals, and connections. Artificial light is created by high-voltage mercury vapor or high-pressure sodium lamps, each requiring its own igniter, capacitor, and transformer. Firefighters risk electrocution if they make bodily contact with the metal tool or direct a stream of water on this equipment at close range. The risk of electrocution is intensified when firefighters operate in limited visibility, which may not be improved by a thermal imaging camera (TIC), because the ceilings and walls are commonly covered with reflective insulation board. (Gustin 2010)

Captain Gustin mentions another significant hazard:

> Grow houses commonly have one or more pressurized gas cylinders, which can explode if exposed to fire. The cylinders are used to enrich the concentration of carbon dioxide (CO_2) to hasten growth of the plants.... Propane vapor leaking from a cylinder

connected to a carbon-monoxide generator inside a grow house in Miami-Dade County, Florida, exploded, resulting a partial collapse of the structure and the death of one of the occupants. Fortunately, the explosion occurred before firefighters arrived.

Gustin recommends the following steps when working fires in suspected grow houses (2010):

1. As at any fire, conduct a thorough 360-degree walk-around to detect anything unusual, specifically looking for
   - seemingly excessive security measures,
   - blacked out windows, and
   - unusual hoses or wires entering the building.
2. On the interior, be on the lookout for pumps and hoses to or from a bathroom, kitchen, or even a large garbage can.
3. Always be on the lookout for booby traps.
4. Always assume that the door you entered through is the only point of egress, as other exits may be blocked.

War relics

War relics and soldiers' souvenirs have been found by unsuspecting firefighters. Many of these are high explosives that are live and possibly have functioning fuses. We hastily exited a home a few years ago when we found a grenade during overhaul. Turned out this particular one was a training aid, but it is not uncommon to find live devices, especially from the First and Second World Wars, after veterans have passed away and their homes are being cleaned out or sold.

An excellent example of this is World War II–era first aid kits that contain liquid picric acid, which was used on gauze pads. As the acid dries out and crystallizes, it becomes shock-sensitive. Opening the treated gauze pads could trigger an explosion. An incident of this nature occurred in Boston (MA) in July 2011:

> Decades-old Medical Gauze Discovery Prompts Hazmat Response—Hazmat crews were called to the Boston headquarters of the Girls Scouts of Eastern Massachusetts when it was discovered that first aid kits stored in the building since the 1930s contained gauze pads that were treated with picric acid, which was the definitive burn treatment eight decades ago, due to its antibiotic and anesthetic properties. Similar discoveries have recently occurred in Colorado Springs, Denver and the Washington, D.C. area. No injuries were reported from any of the incidents. (FDNY Watchline)

Assessment: The yellow crystals form when the acid dries (which may be recognizable to hazmat technicians and bomb squad personnel) and are shock-sensitive and explosive. Picric acid is most commonly found in laboratories, but its discovery in out-of-date first aid kits is not unusual. Responders are reminded to always be alert to hidden dangers at incidents.

Security bars

Especially common in high-crime areas, bars on windows and doors present a very serious safety hazard to firefighters. Obviously, barred windows prevent the escape of firefighters during conditions of rapid fire development. Bars often delay entry of rapid intervention teams attempting to rescue firefighters who have declared a Mayday.

If there is any significant fire in the home, remove these bars as quickly as possible. They often can be pried off, with a large tool (Halligan or crowbar) used as a lever. Cutting them with a hydraulic cutter is another option, and if conditions permit, use a winch from a vehicle. However you do it, get rid of this hazard quickly to ensure the safety of firefighters inside.

Fig. 3–41. Bars on windows. Security is obviously an issue for this homeowner. From the front the bars are not visible, but all rear windows are barred. Remove the bars as soon as possible to allow firefighters to escape if needed.

Security bars can be a real threat to occupants as well. An article in the *Detroit News* on 26 January 2004 tells the tragic story of four young children and a woman who perished in Michigan in an early morning fire. The fire started in the kitchen of the rental house where they were trapped behind security bars. The fire blocked the front door, the home's only means of egress.

In December 2011, the Houston Fire Department (TX) was faced with five children and one adult trapped by a fire in a house with bars on the windows. Assisted by first-arriving police officers, firefighters pried off the internal burglar bars and rescued the occupants. This is an important reminder to get the bars off the windows as soon as possible by any means possible.

Security concerns can take other forms. In 2010, firefighters in Tukwila (WA) found a man dead in a house fire. The house was booby trapped with holes cut in the floor and covered with carpet. According to the 4 October report from KIROTV.com, "Fire Marshal Don Tomaso said the victim may have been paranoid about intruders."

FORCIBLE EXIT

Firefighters are taught that breaching walls is an effective means of escaping a dangerous situation if they avoid the strong point such as barred windows or secure doors or doors with multiple locks. Deputy Chief Gregory Havel from Burlington (WI) provides an article with important reminders about breaching (Havel 2008).

> Some of the forcible entry skills we learn during our Firefighter I classes involve breaching walls. These skills could be used to bypass security devices on the doors or windows that we would normally use for entry, firefighter rescue, or self-rescue.
>
> A word of caution, especially when working in residential and light commercial buildings: Not all walls are well-suited for entry or exit through breached openings.
>
> Don't try to breach a curved wall framed with either steel or wood studs. Curved walls have studs spaced more closely than the 16-inch, 18-inch, or 24-inch centers that are common in straight walls to provide support for the multiple layers of drywall board that will be used. The photo below shows a curved wall framed with wood studs on six-inch centers that will be covered with at least two layers of ¼-inch drywall board, which is more flexible and can conform better to the curvature of the wall. Although it is possible to break drywall board and displace a single stud to make an entry or exit opening in a wall, it becomes extremely difficult and time-consuming to break layers of drywall board screwed to studs every six inches, and to displace several studs to make an opening large enough for a firefighter. Look for an easier way.

Photo 1.

In addition, curved walls often conceal void spaces created by the shape of the room on the other side, as in Photo 1.

Breaching a straight wall may not create a usable opening. Photo 2 shows a wood-frame wall in a new home, with studs on 16-inch centers. Only two of the stud spaces shown might be usable for entry or exit openings. The rest contain electric cables and boxes, steel dryer vent pipe, steel ducts, steel gas pipe, and plastic drain-waste-vent pipe. Unless we are very familiar with the building, we won't know this until we have spent a lot of energy and SCBA air in discovery. In addition, we might find that the other side of the wall has a set of cabinets screwed to it, or a row of 250-pound file cabinets sitting against it.

Photo 2.

If we become trapped or disoriented and low on air on the second or third floor of a residence or commercial building, we can create an additional problem for ourselves. Although the wall we choose might be easily breached, we might create our escape opening in an exterior wall at a

height too great to jump, or in a place that can't be reached with a ladder.

We do need to know how breach walls to gain entry to difficult structures. Sometimes we need to breach walls to escape from them—but this must be our last resort, since there is no guarantee of results, and our SCBA air supply will not give us a second chance. Better options are personnel, task, and location accountability; maintaining team integrity; marking our entry route with a hoseline or rope for use in exiting; and establishing secondary (emergency) exits. If we use these options, we are less likely to need to breach a wall for self-rescue or to perform a rescue.

Terrorism

Police need a search warrant to enter someone's home. Firefighters enter buildings to mitigate a wide variety of hazards. Usually this visit is unannounced. As such, you can play a vital role in the war on terror. It is a simple saying but one that makes a great deal of sense: "If you see something, say something!" This slogan is a trademarked program of the NYC Metropolitan Transportation Authority.

The process of gathering intelligence and possibly preventing a terror attack works by gathering many small pieces and bits of information and putting them together, much like a puzzle. What we see inside a home or apartment may be key items in the investigation.

Here is a simple example. To make an improvised explosive device, one needs a power source, an initiator for the explosive, explosive material, and a switch. PIES is the acronym for these four items. Say, for example, your firefighters enter a home for a water leak or to check for fire extension and find a box full of cell phones or timers. These can act as the switch that triggers the explosion. Perhaps firefighters find a white, pasty material in the bathtub or in a container—is it a homemade explosive? If you see something unusual, pass it on through your chain of command to law enforcement. This may be the key break they were looking for, or it could be the first tip that a terror attack is being planned and close to execution.

No matter what kind of illegal lab it is, it is dangerous. There are dangerous and toxic materials in explosives labs just as there are in drug labs. Drug labs may actually be a bit neater because of the value of the product. Material on the floor of an explosives lab can be some of the improvised explosive material itself. It is not expensive to make and may have spilled during the bomb-making process. If you find yourself in a bomb-making lab, retrace your steps during your tactical withdrawal as if you were in a minefield.

A *New York Daily News* article written by Philip Caulfield on 9 December 2010 describes an Escondido (CA) house so contaminated with explosive-making materials that it was intentionally burned down by authorities to reduce the risk of an uncontrolled explosion.

Evidence preservation

Generally, fire suppression and rescue operations focus on the first priority of life safety, but we must also remain cognizant of our duty of evidence preservation. Preserving valuable evidence by not overhauling, limiting access to only firefighters who must be in the building, taking photos, and interviewing first-due firefighters will help in the investigation. Be aware of sights, smells, and the location of items in the area where you are working. Catching the bad guy is a police job.

Scenario Discussion

1. **General impression.** General or gut impressions are important and usually correct. Trust your sixth sense. However, always maintain a degree of suspicion that your overall impression could be seriously misleading. In this case, it looks like a family lives here. It is a good neighborhood. There are several decorations on the lawn and Halloween pumpkins on the front steps. The yard is well kept, and a trash can out front is a good sign that the house has a fairly regular upkeep routine. The police department (PD) reported a dog barking, probably the family pet, maybe behind the fence in the back yard or inside the house. There does not appear to be any half-finished or shabby construction in or around the home. There is no car in the driveway; it could be in the garage, or Dad or Mom may have gone to work on the night shift. General impression here is that this is a single-family home. It is not a converted private dwelling and it appears that occupants are not running a business or workshop out of the home. Hazard wise, we would expect just the regular house fire hazards.

2. **Hazards to firefighters.**

    a. *Construction*—One of the first questions must be: Is this house's framing of lightweight construction? We are all familiar with the hazards of early collapse, floors and roof systems falling and trapping firefighters, floor failures sending entire companies to the basement and into the inferno, and firefighters falling through truss-supported roofs. This must be a primary concern that will determine your overall strategic and tactical possibilities and goals. There is no way to tell if this house contains these deadly traps for firefighters by looking from the outside. You must depend on your preincident intelligence. If you don't have this information about your first-due area, you had better start gathering it right now. Firefighter's lives depend on it. This house looks like platform construction, probably built anywhere between 1960 and the present. It does not contain any of the features of an older balloon-frame home. Therefore, rapid fire spread through the walls and floor joist spaces is not a concern.

    b. *Interior fire spread* appears to be a major concern, possibly *the* major concern at this fire. From this view, it looks like the fire has full possession of one bedroom in the A-D side of the house. Judging by the smoke pushing from the eaves and the intensity and volume of fire in the bedroom, the fire is likely in the attic as a result of the sheetrock failing under the fire load or extension from the autoexposure to the soffits.

    c. *Exterior fire spread* of this fire will be a factor. The fire has extended out the windows with intensity and duration and exposed the thin plastic or plywood soffit, allowing easy access to the attic. Combined with the smoke showing from the attic, it is a good bet the fire has made it to the attic in some significant degree. Although obscured by smoke, fire is showing out the gable vent on the D side. Additionally, smoke is pushing out from the eaves on the A-B corner, indicating that the fire has banked the smoke down in the attic, probably resulting in really strong fire in the attic. Vinyl siding will burn dramatically if ignited but is easily extinguished with a handline. Thick black smoke from the siding may not be indicative of the interior fire conditions. Fast-spreading, spectacular fire on the outside of the building is an attention grabber, but remember the mission of the first hoseline: get to the seat of the fire.

    d. *Rapid fire development* is a hazard during interior operations. It looks like the attic is close to becoming fully involved if it is not already. Rapid fire development in the attic could drop fire down quickly when sheetrock ceilings fail, disorienting, injuring, and trapping firefighters working inside. It is also unclear how much of the inside of the house the fire has possession of. Hopefully (although hope is *not* a good action plan) the bedroom door is shut, containing the fire to the one bedroom. This is likely or at least a possibility, as neither fire nor smoke is showing from other windows or the front door. Energy efficient windows and doors may be preventing heat and smoke from showing at these openings. The initial report from the first-in firefighters will confirm or deny this critical information. Remember the warning signs of flashover: dense smoke, free-burning fire, high heat, rollover, and vent point ignition. All of these are present at this fire and we must be cognizant of the probability of a flashover.

3. **Utilities**.

   a. *Natural gas or oil*—In a suburban or urban setting, the home is almost always natural gas, oil, or electricity heated. Shutting off the gas at the outside meter is a good tactic to reduce the hazard of an explosion or intense fire. Unless the fire is in the basement, the oil storage tank there or in the garage is not a hazard.

   b. *Propane*—In rural settings there may be propane tanks outside the house that need immediate attention if they are exposed to intense heat of the fire. Apply water in copious amounts to prevent a BLEVE.

   c. *Overhead electric service*—Overhead electric service is typical for suburban settings and always a concern for ground or aerial devices.

4. **Hazards to firefighters during interior operations.** Assuming your preincident intelligence tells you that the building is not of lightweight construction and does not contain any other unusual hazards, let's say we proceed with an offensive strategy because the life hazard has not been resolved yet.

   a. *Rapid fire development*—If the fire has extended beyond the bedroom, when you open the front door you will provide it with an unlimited supply of oxygen. It may light up very quickly. As previously discussed, the attic might be close to the same situation. SAR teams will be in extreme danger and should be supported by an engine company (or two) with hoselines with decisive amounts of water.

   b. *Fire-weakened structure*—If there is significant fire in the attic (and this appears to be the case), collapse of the ceilings and possibly some of the attic contents, rafters, and ceiling joists are potential threats. The IC will have to decide if members can safely operate inside, if it is worth the risk for the life

hazard present, and how long firefighters can and should operate inside.

   c. *Uncommon hazards*—This is a home, so be prepared for almost anything. Kids' bicycles in the hallway can trap and kill firefighters. Exotic pets can range from snakes and alligators to lions and tigers. Home businesses can present a variety of hazards. This could be a crime scene, a fire to cover a burglary, an insurance fraud fire, or a suicide or murder attempt.

Additional Scenarios

Fig. 3–42. Practice scenario. Courtesy of Jeffery Arnold.

This fire looks like it is in a second-floor bedroom, and it likely is. The hazard we face at this building is the balloon frame (see chapter 2), which allows fire to spread from basement to attic via exterior wall stud channels if no fire stopping has been installed. Unsuspecting firefighters could go through the front door and fall into a well-involved basement fire while thinking they were going to a one-room fire on the second floor. It is not uncommon for a fire in a balloon-frame home to start in the basement, bypass the first floor, and show and involve a second-floor room. At a balloon-frame building like this one, always check the basement: look into the windows from the outside and then take a small window pane with your hook or your foot. If smoke comes out, there is a basement fire *and* a second-floor fire.

Fig. 3–43. Practice scenario

This scenario provides valuable lessons in both determining the hazards to occupants and firefighters and sizing up, which we will cover in the next chapter. Although the life hazard here looks severe to occupants, it may actually be low for a short time for those in the bedrooms. This observation will play a vital role in your priority, strategy, and tactics for search and rescue. Look at the bedroom windows at the right side of figure 3–43. Little smoke is showing from these windows; occupants may have bedroom doors closed. This may have maintained survivable conditions. Vent, enter, isolate, and search may be a good option here with two separate teams to two windows if you have manpower. The hazard to firefighters could be severe if they try to enter the front door and make a move to the right down the hallway for an interior search.

Fire will soon take full possession of the first floor, as indicated by the heavy black smoke and fire from the rear and attic. Clearly the fire is in the first floor and attic and will be a severe threat to anyone inside. It looks like fire may have entered the attic via the plastic soffit vents that are hanging down near the A-B corner. Heavy fire appears to be venting out the first-floor rear of the structure where you would expect to find a deck with patio sliding-glass doors providing lots of ventilation. Look at the reflection of the flames in the smoke above the roof near the top of figure 3–44.

Firefighters have given this fire what it needs, oxygen, by opening the front door. Smoke is fuel and it lights up here as both vent point ignition (outside) and a ventilation-induced flashover inside in the living room. This is a critical juncture for development of this fire and your search and fire attack. Interior search teams will need to be protected by the hoseline as they try to make the bedrooms to the right. Consideration should also be given to the possibility of bedrooms and surviving occupants in the basement, especially to the left on the A side (front). To the right side may be a garage or garage space that was converted to living or office space.

Fig. 3–44. Practice scenario

Understanding fire dynamics, how the fire will change based on the conditions it has available to it, is critical to everything we do on the fireground. This includes the decisions we make, the risks we take, and the lives we save. We urge you to use the excellent training programs that have been developed by Underwriters Laboratories that are available free on the web at ulfirefightersafety.com.

References

Bossert, Lisa. 2007. *Fire Protection Study: Pine Knoll Townhome Fire*. Presented to the City of Raleigh. http://raleighnc.gov/content/Fire/Documents/Fire%20Prevention/Townhouse_Fire_Full_Report.pdf.

Bowers, Richie, Corey Parker, Jennie Collins, Bill McGann, Justin Green, and Greg Moore. 2008. *Significant Injury Investigative Report: 43238 Meadowood Court, May 25, 2008*. Leesburg, VA: Loudoun County Department of Fire, Rescue, and Emergency Management.

Bukowski, Richard. 1995. "Modeling a Backdraft: The Fire at 62 Watts Street." *National Fire Protection Association Journal* 89 (6): 85–89.

Dunn, Vincent. 2015. *Safety and Survival on the Fireground*. 2nd ed. Tulsa: PennWell.

Fahy, Rita, Paul LeBlanc, and Arthur Washburn. 1990. "Fire Fighter Deaths As a Result of Rapid Fire Progress in Structures: 1980–1989." In *Analysis Report on Fire Fighter Fatalities*, 42–45. Quincy, MA: National Fire Protection Association.

Gallagher, Kevin. 2009. "The Dangers of Modular Construction." *Fire Engineering* 162 (5). http://www.

fireengineering.com/articles/print/volume-162/issue-5/features/the-dangers-of-modular-construction.html.

Gustin, Bill. 2010. "The Hazards of Grow Houses." *Fire Engineering* 163 (6). http://www.fireengineering.com/articles/print/volume-163/issue-6/Features/the-hazards-of-grow-houses.html.

Havel, Gregory. 2008. "Construction Concerns: Breaching Walls." FireEngineering.com (blog), 12 August. http://www.fireengineering.com/articles/2008/08/construction-concerns-breaching-walls.html.

Hurley, Morgan, ed. 2016. *SFPE Handbook of Fire Protection Engineering*. 5th ed. New York: Springer.

Kerber, Steve. 2010. *Impact of Ventilation on Fire Behavior in Legacy and Contemporary Residential Construction*. Northbrook, IL: Underwriters Laboratories.

Knapp, Jerry. 2013. "Rapid Fire Spread at Private Dwelling Fires." *Fire Engineering* 166 (10). http://www.fireengineering.com/articles/print/volume-166/issue-10/features/rapid-fire-spread-at-private-dwelling-fires.html.

Knapp, Jerry, and Christian Delisio. 1997. "Energy Efficient Windows: Firefighter's Friend or Foe?" Firehouse.com, 1 July. http://www.firehouse.com/news/10544872/energy-efficient-windows-firefighters-friend-or-foe.

Madrzykowski, Daniel, Gary Roadarmel, and Laurean DeLauter. 1997. "Durable Agents for Exposure Protection in Wildland/Urban Interface Conflagrations." In *Thirteenth Meeting of the UJNR Panel on Fire Research and Safety, March 13–20, 1996*, edited by Kellie Ann Beall, 345–50. Gaithersburg, MD: Building and Fire Research Laboratory. http://fire.nist.gov/bfrlpubs/fire97/PDF/f97119.pdf.

McDowell, Anthony. 2009. *The Wall of Fire: Training Firefighters to Survive Fires in Vinyl-Clad Houses*. Emmitsburg, MD: National Fire Academy.

Mowrer, Frederick. 1998. *Window Breakage Induced by Exterior Fires*. NIST-GCR-98-751. College Park, MD: Department of Fire Protection Engineering.

National Fire Data Center. 1995. *Three Firefighters Die in Pittsburgh House Fire—Pittsburgh, Pennsylvania*. Emmitsburg, MD: US Fire Administration.

National Institute of Occupational Safety and Health. 2008. *Career Fire Fighter Dies in Wind Driven Residential Structure Fire—Virginia*. NIOSH Report F2007-12. Atlanta: Department of Health and Human Services.

Naum, Christopher. 2011. "Prince William County (VA) Fire Rescue Kyle Wilson LODD 2007; Is This on Your Radar Screen?" Firehouse.com (blog), 17 April. http://www.firehouse.com/blog/10459541/prince-william-county-va-fire-rescue-kyle-wilson-lodd-2007-is-this-on-your-radar-screen.

New York State Office of Fire Prevention and Control. 2012. *Recognizing Clandestine Drug Lab Operations: Student Manual*. Albany: New York State Office of Fire Prevention and Control.

Peluso, Paul. 2010. "Backdraft Throws Arizona Firefighters." Firehouse.com, 28 July. http://www.firehouse.com/news/10465231/backdraft-throws-arizona-firefighters.

4

Size-Up of House Fires

Summary

This chapter provides several time-tested methods for obtaining the necessary facts and describes the assumptions you will need to make to size up a house fire. These facts and assumptions are based on your general knowledge of homes and house fires, fire dynamics, and the information that you passively obtain or aggressively seek out during your initial observations of the scene.

Introduction

A thorough size-up is the foundation of safe and effective strategy, tactics, and tasks that are executed on the fireground. Size-up—asking, "What have we got?"—is a critical first step in protecting firefighter lives, civilian lives, and property. But before we examine this critical task, let's continue with our interactive scenarios to help you build, test, and exercise your skills and knowledge of house fire size-ups.

Scenario

Fig. 4–1. Scenario

Situation: It is Wednesday morning, 0223 hours, and you are tapped out to a fire at 2 Veterans Street. A caller reported seeing the house across the street on fire. There is no additional information. Upon arrival, the police officer reports to you that she heard a dog barking inside.

Radio Report: Dispatcher reports multiple 911 calls.

Resource Report: You are on the scene, your first-due apparatus is approximately 2 minutes from the scene. There is a good hydrant 200 ft before the fire building.

Based on figure 4–1, answer the following questions and use this scenario as a basis for the topics examined in this chapter.

1. What obvious hazards or threats to your members do you see?

    a. Construction type
    b. Fire—extent, intensity
    c. Flashover or backdraft conditions
    d. Utilities—wires, gas, propane, fuel oil, etc.
    e. Unusual hazards
    f. What is the floor plan of the house?
    g. Where are the bedrooms? Stairs?

2. Evaluate the life hazard.

    a. Do you think the house is occupied?
    b. At this time of night, where do you think the occupants are in the house?
    c. Is their exit blocked by fire?
    d. Are there indications that someone may have gotten out? Open doors, windows, etc.?
    e. What are the odds of people inside surviving?
    f. Are there salvageable lives inside?
    g. Where are nonsalvageable lives in the building?
        i. Can you put firefighters inside the building with reasonable risk?
        ii. How many personnel will you need for forcible entry? Are there bars on windows and doors?

3. What is your fire strategy?

    a. Is an offensive strategy appropriate?
    b. Is a defensive strategy appropriate?
    c. Is the fire ventilated?
    d. What type of ventilation is best at this fire (horizontal, vertical, PPV)?
    e. Is one line enough?
    f. Is there a need for a backup line?

As we progress through the book we will systematically address SAR operations, fire suppression, line placement, ventilation, and salvage or overhaul. After you have answered the above questions and you are confident that your size-up is complete, read on.

General Considerations

Size-up is an ongoing process of evaluating a situation to determine what has happened, what is likely to happen, and what resources we need to resolve the situation. This process is carried out continuously by everyone on the fireground during a fire or emergency. Why is this important to firefighters? A firefighter, officer, or commander who can perform a meaningful size-up will be able to recognize potential problems or hazards for trapped salvageable or unsalvageable victims, fire, smoke, collapse, and so on, convey the information to the other members, and (perhaps more importantly) develop strategy and execute tactics for a successful outcome. If you are a thinking firefighter, you will be a safer firefighter or more effective fire officer.

James P. Smith, a 35-year veteran of the fire service and retired deputy chief of the Philadelphia Fire Department (PA) describes size-up in a very practical way: "Size up lets the incident commander gather information for the development of strategic goals. It is a mental process weighing all factors of the incident against the available resources. Size up can be looked at as solving a problem or as a puzzle that requires putting the pieces in their correct place by gathering and interpreting the available information. It is an evaluation process that reviews all critical factors that could have a positive or negative impact on the incident" (Smith 2002).

Smith then utilizes the information from the size-up process to create an action-oriented fireground decision with three simple but critical parts: strategy, tactic, and task or action. For example: The strategy is rescuing the occupants, the tactic is to conduct a SAR operation, and the task is assigned to Truck Company #1.

Your computer-aided dispatch (CAD) should have transmitted your preincident intelligence about the building, its hazards, the neighborhood, hydrant locations and other water sources, and so on. This information may come from a run sheet you tear off a printer or something as simple and cost effective as a three-ring binder with the critical information presented in an easily read format about the specific address or class or group of homes you are responding to. There is no reason to be surprised when you pull up to the address by the construction, water supply, or other firefighter killers that you (ridiculously) did not feel the need to be aware of *before* the alarm. The building did not just sprout up overnight; it has been there for years, and there is no excuse for not knowing critical facts about it and having them in an instantly retrievable format. Firefighters' lives depend on it. It is the officer's job to know this information long before the tones tap you out to the address. It is irresponsible not to.

If you have done your homework, a good part of your size-up is complete before you even leave the firehouse. The preincident intelligence will tell you significant information (size of the home, hydrant location) or special hazards (lightweight construction) related to the house you are about to enter. Just as fire prevention minimizes threats to civilians, preincident intelligence minimizes the dangers to firefighters.

The goal of size-up is simple: to gain as complete an understanding as possible of the building and rescue/fire situation so you can apply and prioritize the appropriate amount of resources to conduct the SAR and fire attack operations in a way that will minimize losses and mitigate the incident. All this while you meet your most important objective: to ensure the safety of firefighters in your command and yourself.

The process of size-up requires significant background knowledge and experience, much of which we covered in the first three chapters. It is the mental agility of the leaders to reach back and pull out these facts based on the current situation that determines the success or failure of rescue and firefighting strategies.

Everyone Does Size-Up

Size-up must be done by everyone on the fireground and it must be a continuous process. Officers on the scene must perform a size-up so they can formulate a strategy supported by appropriate tactics based on assumptions, facts, or considerations from that size-up. Firefighters must do their own size-up so they can properly implement the tactics with the right tools and in the correct order. Firefighters and company officers must also conduct size-ups for two other reasons:

1. Firefighters and company officers may see something that the command officer did not. It may be a severe hazard on the fireground (i.e., downed or energized wires) or a factor that will allow a speedy and effective rescue. Communicate these facts quickly to leaders and other firefighters.

2. The fire scene is dynamic and ever changing, sometimes for the better, sometimes for the worse. For your own safety, maintain your situational awareness 100% of the time on any fireground, but especially at house fires. Your size-up should be continuous with each change of place or function or transition between sectors, divisions, or assigned tasks. As we saw in chapters 2 and 3, houses often contain a variety of deadly surprises ranging from construction hazards to rapid fire development to exotic pets or even deadly chemicals.

Size-Up Process

Let's walk through the mental process of size-up from the time the alarm comes in until you have picked up and are returning to quarters. You will see that the questions generally go from very general to very specific as the incident develops. Think of the size-up process as similar to the steps you would take to build a house. You start off with general questions like: Where are we building the house? What type of house, how many floors, how many square feet? These are all general questions that must be answered prior to construction. As your building plan develops, questions will become more specific: How big will the rooms be? What type of windows? What color siding? Finally, near the end of the project, very specific questions will be asked: What color should we paint the dining room? When the house is built, your after-action review will summarize lessons learned from this experience. Size-up will start with general questions—such as, "What is the location?"—and go to very specific questions: "Where is the master bedroom?"

Playing the odds: Assumptions

Conducting a size-up, developing firefighting strategy and tactics, is one-third art, one-third applied science, and one-third experience. Sizing up a house fire to determine critical factors that the lives of civilians and firefighters depend on is a dynamic and complex skill. Size-up is a skill that you can develop with knowledge, training, and experience. During size-up, commanders and firefighters have to make certain assumptions and then develop their action plans. You never know if your assumptions are correct until the operation is over. This is because your assumptions are based on incomplete information, but that is the best (and in fact all) you have to work with at the time. Indecision is not an option: taking no action is a decision, and one that may tragically limit your options later on in the operation.

Here is a tragic, real-life example of the disastrous effects that incomplete information can have on our survival. In 2006, firefighters responded and entered a house fire in Wisconsin, sounded the floor, found it solid, and began moving forward, only to be dropped into the origin room (in the basement) by a sudden collapse. The incomplete information they were working with was related to the building construction: there was a concrete floor that sounded solid but was supported by "a lightweight wooden parallel-chord truss system and engineered wooden 'I' beams" and catastrophically collapsed, leading to the death of one career engineer and the severe injury of another firefighter (NIOSH 2007).

This is an example of an instance where prefire intelligence on building construction could have changed the strategy of the fire attack. In fact, NIOSH's first recommendation in their report on this fatal incident was that fire departments "conduct pre-incident planning and inspections of buildings within their jurisdictions to facilitate development of safe fire ground strategies and tactics" (NIOSH 2007). Figure 4–2 shows the incident site.

Fig. 4–2. A critical part of your size-up will be to identify as many hazards as possible to ensure the safety of yourself and your crew. *Source:* NIOSH 2007.

This case history makes the point that sometimes your commander's size-up, their decisions, and your ongoing size-up and actions will be flawed by incorrect or incomplete information. The critically important point is to keep updating your initial assessment or previous assumptions with facts as the situation develops. You must also plan for the unexpected events that are always possible (and indeed, probable) on the fireground. Events such as firefighters becoming lost or disoriented or building collapse are threats that can somewhat be mitigated by having a RIT team standing by, but one of your most important size-up priorities is to identify hazards to firefighters and eliminate or minimize those risks.

The proactive approach—conducting a risk analysis, asking, "What am I risking? What am I saving?"—must be taken by every level of firefighter from probies to command officers. When the risks are not worth the gain or reward, it is time to change course (strategy and tactics) immediately.

I (Jerry) trained recruit firefighters for about 20 years at our academy and tried to drum this risk-reward philosophy into their eager minds. The best way to describe it was what I called the Firefighting Conscience. It is a silly but very effective little tactic. Your Firefighting Conscience is a little creature of your own design, floating just off your right shoulder that usually asks things like, "What are you doing?" "What do you expect to save with this action?" "Is it worth the risk?" and "Is this *really* worth the risk?" When we answer these questions, we become safer commanders and firefighters.

What has this got to do with size-up? Size-up is how you see the fire situation and determine what is salvageable (both lives and property), and how you see it will affect how you act. This is true for commanders, line officers, and the newest firefighter.

Receipt of alarm

Upon receipt of the alarm, as soon as you hear the address, you should be asking yourself questions to prepare yourself for anticipated actions on the fireground. Critical questions such as: What do I know about this house? What don't I know about this house? Is it occupied or abandoned?

The definitions we use on the fireground or over the radio during dispatch and operations are very important. According to the National Incident Management System (NIMS), common terminology is critical to a successful operation.

Occupied: Occupied homes are normally inhabited by humans. They are in livable condition and maintained in healthy and sanitary conditions. Example: "Command to dispatch, I am on the scene at 2 Veterans Street and have a working fire in a two-story occupied private dwelling with heavy fire on the A-D corner, second floor."

Unoccupied: A home where residents are not or were not in the building at the time of the fire or emergency. Example: A fire department spokesman says, "The fire destroyed the home but fortunately there were no injuries since the home was unoccupied at the time of the fire."

Vacant or Abandoned: All efforts to maintain the building or property have stopped. The owner has intentionally let the building fall into disrepair. The building may have been sealed by authorities and utility services may also be discontinued. Example: Report from the fire investigator: "The neighbor stated that the family who lived there fell on hard times. The father lost his job and they could not afford to keep the house on the wife's salary alone. One day the neighbor saw a moving truck in the driveway and believes they moved in with the wife's mother. The neighbor has not seen them in months."

General size-up. How big is it? Is it a small straight ranch, two-and-a-half story, Queen Anne, or modern open floor plan with 3,000 or more square feet? This general fact helps define most of the other concerns (search, ventilation, fire control, etc.). The size of the house will also help define how much personnel and water you will need to control the incident.

What house type is it? What is the floor plan or layout? This will tell you where specific rooms are located. How many floors are there? This controls the possible need for and size of ladders. What are the construction hazards? Is it light-weight construction that will fail rapidly under a fire load? Or is it a stick-built house that will usually last longer? Are there overhead wires (prevalent in older sections of town)? Are there large trees or a steep grade that will restrict the use of aerial or ground ladders?

Is the house balloon or platform construction? Balloon-frame fires spread quickly throughout the building. Platform construction generally reduces fire spread with wall studs. Is it prefabricated or modular construction? If it is and fire has entered the void spaces, expect fire spread that is very rapid, especially in the truss spaces between the first and second floors.

Is it a good or bad neighborhood? Could you expect additional hazards associated with illicit drug labs, run down or vacant houses, or boarded-up houses with structural weaknesses? Is security a problem (lots of locks on doors or bars on windows)? This may make forcible entry more challenging, requiring special tools and additional personnel.

Is it likely that someone is home this time of day? We know from fire data that most Americans who die in fires die at night, when they are sleeping, making bedrooms a focal point for search operations. Is it a subdivision with young families with lots of kids? Is it a group home for physically or mentally handicapped people? These may contain several disabled residents with only one caregiver, especially at night. These instances indicate a large search problem. The officer may alter your normal SOGs to focus on the SAR operation.

Establishing a reliable water supply is a critical task and part of your size-up. Can you hit a good functioning hydrant, do you need to draft from a swimming pool or nearby lake, or should you set up tanker shuttles? The size and locale of the fire building and the amount of involvement will determine water supply requirements.

Was the house modified? Original construction may have had lightweight structural members added during renovations that will fail quickly during the fire. Are there additional hazards from illegal occupancies, such as single-room occupancies? What condition is the building in? Is it a single-family home? Is it a legal multifamily? Is it an illegal multifamily?

Windows often indicate the function of the room; for example, the living room is often behind a large picture window. Can you see a smaller bathroom window that may have frosted glass? Off-height windows often indicate stairs or landings. By the process of elimination, you should be able to identify the bedrooms that are a priority to search at nighttime fires.

Based on your specific first-due area, you may have these or other questions that are important in your initial size-up. Discuss them with other members of your department.

Arrival on scene

The next step in the size-up process is to evaluate the building based on what you see upon arrival. In essence, what is the building telling you? Some of these facts are intuitive but important to recall. Additionally, while these next steps will occur almost concurrently, for now let's break it down into two steps: what can you learn that does not involve the fire, and what do you think the fire situation is based on smoke and fire showing.

Common house characteristics. You can pull (with a good degree of certainty) a variety of critical facts about the house from the outside. For example, figure 4–3 provides the following information about the house:

1. Hallway direction. Hallways follow the ridgepole, so the second-floor hallway in this house is an L shape.

2. Room layout. You can guess the room layout based on what type of house this is (see ch. 2). Also, bay windows indicate living rooms, small frosted windows indicate bathrooms, vents indicate kitchen locations, and vent pipes through the roof generally indicate a kitchen or bath below.

3. Possibility of someone living in the attic. This is hinted at by air conditioners, curtains, or larger, newer windows.

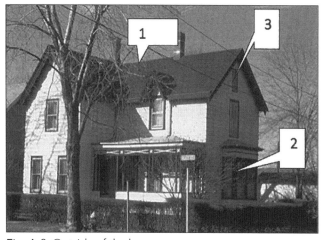

Fig. 4–3. Outside of the home

Common floor plans. The floor plan and room locations for the house in figure 4–4 are easy to figure out:

1. The bay window indicates that the living room is behind it. Based on the general floor plan of this high ranch, the dining room is behind the living room, with the kitchen just off or toward the inside of the house.

2. Smaller windows generally indicate a bedroom, one of three originally in the house.

3. Side windows could be one master bedroom or a master and a smaller bedroom.

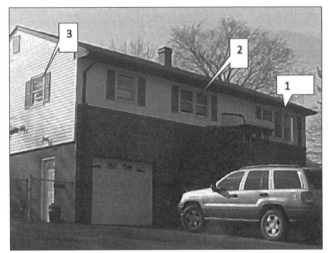

Fig. 4–4. Knowledge of the standard floor plan for each home type can provide valuable information.

The rear of the house has several clues to room layout and access points for firefighters (fig. 4–5).

1. This section indicates a deck that was converted to a sunroom. There usually is a set of sliding glass patio doors leading to the deck or sunroom. Although it seems like it would be a good alternate access point to the front door, it actually brings you through the dining room to the living room at the top of the stairs just up from the front door. If your task was to search the bedrooms, you have not gained a tactical advantage.

2. The kitchen vent fan outlet helps you confirm the location of the kitchen in the house floor plan.

3. The smaller frosted windows confirm the location of a bathroom. Additionally, there may be a vent stack directly above the main bath in the home.

Fig. 4–5. Observations from the rear of the house

Overall impression

From your preincident intelligence and your exterior view of the home you must develop an overall impression to determine your strategy and tactics. Here are some additional items to consider:

- *Does it look occupied or vacant—what is your general impression?* Is there a for-sale sign out front? Is the grass not cut? Has the building been vandalized? Are there children's toys in the yard or driveway? Keep in mind that elderly occupants may lack the resources or ability to maintain their property or home.

- *Multifamily.* This is signified by multiple cars near the house, an enlarged driveway, multiple gas and electric meters, multiple mailboxes, multiple doors either inside a hallway or outside, multiple decks, and so on. This house may have a huge life hazard.

- *Windows.* Are there energy efficient windows that will hold in fire and prevent smoke from showing?

360 size-up

There are six sides to a fire: over, under, left, right, front, and back. For house fires, the back view often provides critical size-up information. We tend to forget the rear view, calling it the C side or the three side, but be sure you get a look at it or get a report from a firefighter there. The fire in figure 4–6 started on the rear second-story deck, so the rear of the house was fully involved and the fire extended into the house through the second-floor rear windows. From the front it is not clear how much fire is in the rear. If you see only the front it looks like an attic fire only. Look at the house on the left: it is illuminated by the light of the fully involved rear. There was a massive fire in the rear. Look to the right of the photo and you can see flames.

Fig. 4–6. Without a 360 walk-around, this house may mislead you and your entire strategy.

Houses usually have a rear or side entrance. Victims may have exited or be unconscious near the exits. Victims may have jumped from rear windows and be lying injured on rear decks or patios or in shrubs. There is an additional spinoff benefit of getting a look at the back, which is that you have to go around one side to get there. So by looking at the C side, you have also looked at the B or D side as well. Going to the rear will also allow you to see if there are any additions, set-backs, or steep landscaping not visible from the front. These surprises can vastly complicate the operation if undetected during size-up.

Rear house entrances, or back doors, are generally not good places for hoselines to access the building. Pushing the line in the rear door requires a long stretch from the front, around the house, then finally into the house to the seat of the fire. There may be fences, shrubs, pools, ponds, vehicles, and kids' playsets in the way. The stretch to the rear door is a very staff-intensive stretch that often is not an efficient use of your crew's time and effort. Of course, never disregard it as an option as the situation dictates: if the house has been divided into apartments or illegal single-room occupancies, this might be the most direct route for the attack hoseline to control the fire. See further discussion later in this chapter.

Fire conditions

The amount of visible fire (or lack thereof) is obviously a huge factor in your size-up. For example, if upon arrival the entire building is fully involved, fire showing out every window and door, and you determine there is no chance for survival of anyone inside, a defensive operation is in order. Protecting nearby exposures will likely be your first goal (fig. 4–7).

Fig. 4–7. Size-up of fire conditions confirms there is no need for interior search operations.

In most home fires, the fire will be limited to one or a few rooms. Your size-up in this case, from the outside, is extremely challenging. Smoke will often obscure your ability to make a solid determination of exactly where the fire is. This is precisely why your SAR crews must be trained to get information about fire location, size, intensity, and spreading out to incident commanders or company officers. Another key skill is to get information about what the other sides and back of the house are doing. The incident commander can do this personally or assign it to a subordinate officer, aide, or firefighter.

It is important to add the fire conditions—size, location, intensity, and what it is exposing—to the other key information about the building you have gotten from other parts of your size-up. When you add all of this information together—floor plan, type of construction, location of the fire, and any information you may have about locations of victims—you can start to develop both strategic and tactical objectives for search and fire attack operations. For example, if you have a significant nighttime fire on the first floor of a two-and-a-half-story wood-frame home with reported victims in the bedrooms, you can and likely will assume the following: the stairs are open and unenclosed so fire and smoke will move quickly up to where the victims are. This will help you design your SAR and fire attack plans.

What does the fire and smoke tell you? The fire and smoke volume, intensity, and speed provide valuable clues to help complete your size-up. These factors, combined with the amount of air that is available to the fire via doors left open by fleeing occupants and windows that have been failed by fire or broken by well-intentioned neighbors or police officers, are key to your size-up. We know from the studies done by Underwriters Laboratories using live, full-scale fire tests in furnished test homes that it is the amount of air available to the fire that will determine how quickly it will flash over. Therefore, you can reasonably assume rapid fire spread if there is a well-developed fire that has a large air supply available to it.

Interior size-up

These days of incident command and fixed command posts can sometimes distract us from critical tasks like interior size-up. The threshold of the front door is not the recommended place for a command post, assuming staffing levels allow you to remain at a fixed command post in your vehicle or another location. But often a quick look inside the house will tell you some important things, especially about fire conditions. A quick look means just that—open the door and crouch low so you can look inside, under the smoke, and gain valuable information. You may see a victim lying on the floor, you will be able to assess the layout of the home, and you may be able to determine the seat of the fire, its intensity, and how fast you anticipate it spreading to other parts of the house. You may be able to determine if the fire has cut off escape routes for occupants. Like all operations at doorways, remember to control the door, even for these quick size-up purposes. An inrush of air may be all the fire needs to light up quickly.

> **Contents Fire**—A fire that involves the material and furnishings in the room(s). A contents fire is contained by the finished walls.
>
> **Structural Fire**—A fire that has extended to the structural members. It has gone past the envelope of the walls and now is impinging on or involving supporting members of the home, such as wall studs, floor framing systems, and members supporting the roof.

Existing Size-Up Systems

In our discussion of size-up systems, it is important to recall that we are dealing with a very specific type of fire: a house fire. Sure, it is a structure fire, a building fire, a private dwelling fire, or whatever you want to call it, but it is a very specific type of structure and occupancy. It is a house fire, not an apartment, not an autobody shop, not a supermarket, not a brownstone, not a railroad flat, and not a row house. It is a house fire. As such it has very specific characteristics that separate it from other types of occupancies. The house fire

is very different from a fire in an industrial occupancy, or a high-rise, or a barn, as we have seen in earlier chapters. First, of all the fires we go to, these have the highest likelihood of containing trapped victims. Second, homes likely have no fire protection system and a completely open path (stairs or open floor plan) for fire to travel throughout the building, endangering everyone in the structure. For these two reasons alone, your size-up and strategy, tactics, and tasks often make the difference between life and death for both occupants and firefighters.

Let's take a look at size-up specifically as part of the house fire response process. But before we do that, remember this: upon arrival, you must deploy your first-due resources and base your strategic and tactical objectives on the information presented, collected, deduced, or assumed during your size-up. Not until the fire is under control will you know if what you observed was realistic, if your preplanning information was correct, and if the assumptions you made to develop your strategic and tactical plans and actions were accurate.

It is often said there is nothing new in the fire service, just new packaging. This rings true in the case of the acronyms we use to help us remember key fireground points.

COAL WAS WEALTH

> **O**ne of the most comprehensive and battle-tested size-up checklists was developed by the FDNY: COAL WAS WEALTH. This acronym takes into account all the factors needed for an excellent size-up. It may seem like too much to remember what each letter stands for, but if you examine several residential fires that injured or killed firefighters, often it is one or more of these key factors that were missed during size-up. COAL WAS WEALTH was the traditional acronym for the 13 items that should be included in a size-up for all types of fires. Many of the items in that list are more intuitive and easy to remember than others, which sometimes gave you the sense that you had considered everything you needed to think about to perform size-up, which is just not true. (Note that the "L" in WEALTH coincides with the "L"—locate the fire—in the newly developed acronym SLICERS.)
>
> | | |
> |---|---|
> | C | Construction |
> | O | Occupancy |
> | A | Area |
> | L | Life |
> | W | Water supply |
> | A | Apparatus/Alarm Assignment |
> | S | Street conditions |
> | W | Weather |
> | E | Exposures |

A	Auxiliary fire protection features
L	Location of fire
T	Time of day
H	Height

RECEO VS

Many of us in the fire service today came up with the acronym RECEO VS: Rescue, Exposure, Confinement, Extinguish, and Overhaul, with Ventilation and Salvage done when possible and appropriate. These actions were to be completed in this order at every fire. With the completion of new firefighting research, however, we now have a new acronym that acknowledges "the importance of controlling ventilation" (Reeder 2014).

SLICERS

SLICERS is a revision of the RECEO VS acronym, developed to "drive us to consider the importance of an awareness of flow path and cooling during fire attack" (Reeder 2014). This acronym was at the heart of the "radical change" mandated by the International Society of Fire Service Instructors (ISFSI) asking firefighters and departments to utilize research in their training and operations. The meaning of the acronym is below (Reeder 2014):

"Size up all scenes

Locate the fire

Identify and control the flow path (if possible)

Cool the heated space from a safe location

Extinguish the fire

Rescue and Salvage are actions of opportunity that may occur at any time"

The acronym you use (SLICERS or RECEO VS) to jog your memory during size-up should remind you of strategic priorities required to save lives and property at the house fire. The subtopics critical to these strategic priorities are built into or assumed to be grouped under a shrink-wrapped and very condensed single letter. Acronyms are a perspective from the 30,000 ft level: big picture, few details. Looking out the airplane window you can see towns or cities linked by major highways. You can't see the fire stations, the water systems, or the local roads, but you know (assume) they are there, within the big overview of the city.

This is the generation gap between SLICERS and RECEO VS: experienced firefighters assume certain critical tasks are built in to the RECEO VS acronym. How could you start a rescue operation (R in SLICERS) until you size up (S in SLICERS) the fire building and locate (L in SLICERS) the fire? Of course this is possible but foolhardy and dangerous. These critical factors are built into the letters of any acronym (in this case, the letter R in RECEO VS). Attempting a rescue is not the first step in RECEO; there is not a human-sized cannon attached to the first-due engine that launches firefighters through a window into a fully involved house as soon as the brakes go on.

Both RECEO and SLICERS are excellent tools if used properly. Neither is right or wrong, they both serve as memory joggers—tools—for the user to fully employ their training, experience, and expertise to conduct a proper size-up and develop a strategic plan. Does it matter if you use a crescent wrench or an open-end wrench to remove a bolt? As long as you have recognized the correct bolt, identified its size, and turned the tool the correct way, what does it matter what tool you used? Clearly, if you try to turn it with a hammer and screwdriver or burr it all up with a set of pliers, the outcome may not be very good. SLICERS is extremely applicable to the modern house fire in that it introduces the concept of the flow path as a main threat to firefighters and is based on excellent UL research.

Let's focus on the strategic objectives and how best to gather size-up information for each of these fireground priorities. Chapters 9–11 will cover specific types of fires in homes (garage, attic, basement, and first- and second-floor fires) and will thoroughly examine the benefits embedded in the SLICERS acronym.

RECEO VS

RECEO VS is short enough to remember and use on the chaotic fireground and covers all key strategic objectives in a reasonable order.

R Rescue
E Exposures
C Confinement
E Extinguishment
O Overhaul
V Ventilation
S Salvage

This easy-to-remember acronym can be used and recalled by incident commanders, veteran firefighters, and the newest probies. It sets the priorities for information gathering in the size-up process and dictates actions required on the fireground in the correct order. It is important to note that this does not mean that these processes must be strictly done in this order. For example, the best way to accomplish the rescue mission in a particular situation might be to extinguish the fire.

RECEO VS is applicable to most house fire situations except for those with unusual hazards such as high winds, illegal single-room occupancies, unusual or unfinished construction, home-run businesses, or of course the ever-present issue of limited apparatus or personnel. It is important to understand that this acronym does not cover all the points that might be relevant to your size-up. It does highlight the key strategic objectives as a memory jogger for the IC. It is relatively simple to remember and is specifically applicable to house fires. It provides a framework or checklist of what your size-up needs to include when gathering information to develop your strategy and tactics.

Rescue. Saving human life is always our top strategic priority. We save lives and property, in that order. Of course, firefighter safety is always assumed to be the key component of planning any strategic or tactical operation. As firefighters, we are paid to take manageable risks when significant gain is probable. We don't trade firefighters' lives for foolhardy attempts at civilian rescues. We evaluate the potential for salvageable human life and act based on that and other conditions, such as structural stability.

One of the best ways we have found to explain this balance is to consider firefighters as building occupants, just like the civilians who might be in the building. If you showed up to a well-involved house fire and were sure no one was inside, why would you let firefighters occupy the building? First, there was no one to save. Second, there was little chance of saving any property. Third, the building was not occupied (no life hazard) until firefighters entered the house.

Gathering size-up information regarding lives in danger comes first because life safety is our absolute first priority. It is important to note that controlling or extinguishing the fire may be the best method to protect lives in danger. Another important factor to consider due to the open and unenclosed stairs of multistory homes is that protecting the means of egress (interior stairs or hallways) with a hoseline is often a huge factor in saving civilian lives and protecting the lives of firefighters searching within. For the purpose of this discussion, we will focus on the information gathered to determine if lives are in danger. The next step is the IC determining how to use this information to develop a strategy and tactics to mitigate what they found out during the life safety (rescue) size-up phase.

Under this major strategic objective, there are a number of key questions that must be answered or assumptions made to establish the strategy, tactics, and tasks.

1. Is there a definite life hazard?

    – Are occupants out on the front lawn confirming that everyone is safe and accounted for?

    – Are occupants telling you that someone is still inside?

    – Are persons in need of immediate rescue visible from windows or porch roofs?

    – Was the alarm transmitted by phone from a passerby? Was it reported by someone inside the house or was it an automatic alarm? These questions will help you in your size-up process. Obviously, a report from someone inside the house has the most credibility while an automatic alarm is among the least credible.

2. Are there salvageable human lives in the house? A fully involved home contains no salvageable human life. Recall that firefighters are occupants also and we don't trade firefighter lives for civilians who are already dead.

3. Is the building structurally sound enough to add the weight of firefighters?

4. Based on the time of day, the condition of the building, and other factors, are the odds good that someone is home or occupying the house? Do these conditions and your department's SOPs prompt you to conduct a primary SAR operation?

5. Where are you most likely to find victims, and can they be safely accessed?

6. Is VEIS a better option than interior access based on fire conditions?

7. Will controlling the fire resolve the life hazard? Can victims wait that long?

8. What is the location of the fire? Is a well-placed hoseline (in the interior stairs, for example) critical to support the self-evacuation of occupants or a firefighter-centric SAR operation?

Gathering this information is not easy. Be aggressive; ask questions of bystanders and police officers on the scene. Be very wary of this information, as it is often false. People get caught up in the moment, thinking lives are in danger, and by the time the situation is processed in their brains and transmitted to you, it has changed from "someone could be home" to "there are definitely people trapped in the house."

HOW DO YOU KNOW?

We pulled up to what was reported as a hydro-fluoric acid spill at a large laboratory facility. This acid is particularly bad since it not only is corrosive and bad for your body, but also seeks out the calcium in your bones, producing very bad effects. There were about 50 employees huddled on the lawn. The fire chief briefed us that there had been a spill in a lab and that it had gotten into the ventilating system, meaning that many people had been exposed and needed to be deconned. There was no one presenting with any severe symptoms of corrosive burns or respiratory distress.

I asked the chief if he had interviewed the person who spilled the acid and he said no. Carefully I approached upwind of the group (there was still no one presenting with the urgency of a corrosive burn) and asked the person who reported the spill to their supervisor to come forward. He did and said he heard that another employee may have spilled hydrochloric acid. The second employee came forward and said that he had *almost* spilled hydrocholoric acid but actually had not. None of the employees ever fessed up to making the call alerting the fire department to the dangerous hydrofluoric spill.

The lesson learned was this: always ask the simple question, "How do you know?" This is especially important when neighbors or others give you "reliable information" about someone inside. During my career I have been involved in nine searches for occupants under heavy and very dangerous fire conditions. Seven of those times, the reliable information was much less than reliable. When bystanders give you information, always ask, "How do you know?" You may get a response that vastly alters your perception of the situation, which may save your life or the life of another firefighter.

It is important to recall that your size-up, as thorough as you think it is, actually is based on incomplete or incorrect information. Always be open to surprises that you may find after your size-up is complete and during your fire suppression operations.

SIZE-UPS OFTEN YIELD SURPRISES

Being a cat lady was the least of this Arkansas woman's issues.

Alice Mildred Barron was arrested after she begged sheriff's deputies to save her cats after her home burned—and they found a stash of drugs instead.

Officers arrived at Barron's house in Greene County on Saturday to find her "in the front yard rocking, uncontrollably screaming to save her babies," KAIT reported.

They managed to calm her down, and she explained her "babies" were, in fact, three cats she feared had died in the blaze.

Greene County Sheriff's Office deputies were able to put out the flames, and then searched the property for the cats.

Instead, they allegedly discovered meth, prescription pills and drug paraphernalia in Barron's bedroom.

The cats survived the fire, BuzzFeed reported. The extent of the damage caused wasn't clear.

During questioning, Barron told deputies she didn't know why she'd started using meth again because she'd been clean for more than a decade.

She was arrested on two felony drug charges. (Moran 2015)

AUTOMATIC ALARM

Improvements in technology have reduced the cost of automatic alarm systems and monitoring. Many homeowners can now afford these systems that detect smoke, fire, carbon monoxide, and natural gas. Although this is good for the life safety of the occupants, the negative side is that there has been a steady increase in false alarms from these systems. *False alarm* is an erroneous term; the system was doing what it was supposed to: detecting, transmitting, and sounding the alarm. The problem is that these alarms are often transmitted as a result of smoke from cooking, steam from a shower, dust from construction or painting, the occupant testing the system without notifying the alarm company, or a host of other benign activities such as cleaning or renovating the home. On the positive side, many fires are now caught in the incipient stage with little property damage and minimal threat to the homeowner.

Size-up of alarm transition data. Passersby may see steam from a dryer vent or smoke from a barbeque grill and report the "fire." The occupant across the street or inside the house generally has correct information about the fire, because they usually verify it before they transmit the alarm. When you arrive, who is giving you information? Is it the neighbor reporting that the family is away on vacation? (How does she know? Is she just well-intentioned or does she actually have good information?) Or is the report coming from a person with soot-stained face dressed in their pajamas saying that someone is unaccounted for? As part of your size-up you must interrogate these people. Ask the hard questions. You will be putting firefighters at risk whenever you conduct a search of the premises based on their information and your own size-up, and it happens all too often that firefighters are hurt and killed looking for persons who are not inside.

POLICE OFFICER INFORMATION

Police officers often arrive on the scene before we do. They have had time to collect and possibly verify reports of people inside the building. A recent experience at a house fire proved this fact. The police officer was on the scene of a working house fire at 0532 hours. As I stepped off the engine, he told me there was a person trapped in the rear bedroom. I asked how he knew and he stated that he had seen a man standing at the closed window and then saw him collapse. With this info in hand, the truck company made a rescue using the VEIS method while the engine company extinguished the fire, allowing the rescuers to exit the building through the newly extinguished (previously fully involved) rooms.

Life hazard size-up is the most important, difficult, and intense part of sizing up house fires. Both civilian lives and firefighter lives are at stake. Often, decisions to commit firefighters to very dangerous search operations are based on incomplete or incorrect information. The overall strategy and tactics chosen by the IC for the entire operation may be based on size-up information regarding the potential life hazard.

Exposures. Evaluation of exposures is the next critical step in the RECEO size-up process. The basic question to be answered here is where is the fire going? For house fires, the term *exposure* has two definitions:

1. **An adjacent room.** These can be exposed to a fire from another area of the house, making them internal exposures. For example, a living room fire on the first floor will expose nearby rooms on the first floor and bedrooms on the second floor via the open interior stairs.

2. **An adjacent home or building.** A fully involved house may expose other nearby houses to radiant or convected heat.

The goal of this step—determining exposures—is to determine where the fire is going and what danger that extending fire presents or will present to occupants, firefighters, structural integrity, and property.

External exposures. Sizing up external exposures is relatively easy. Upon your arrival at the scene there will be a large body of fire with other homes, garages, or outbuildings nearby ready to ignite. Commonly on modern homes, the vinyl siding on the exposed home is melting, dripping, and burning. A time-tested method to determine if a wooden structure is going to light up is if the wood and or paint is smoking or discolored. If it is, be prepared for the fire to extend to the exposure. In figure 4–8, you can see a barn exposed to heavy fire from a fully involved nearby home. Smoking, peeling, and discolored paint are all good signs that it is getting ready to light up. Get water on the exposed surface of this building now.

Fig. 4–8. Exposure getting ready to light up

The best method to protect exposures is to put water directly on the exposed surface. This does not need to be a large caliber stream if you have a limited water supply or personnel. Use a line that provides enough water and reach to hit the entire surface of the exposure. This technique is very effective because the majority of the radiant heat is being absorbed by the film or layer of water running down the exposure. Water curtains are not effective because they are not dense enough to absorb all the heat and so simply delay the exposure's destruction. Using water curtains or protective sprays on the involved structure works because they are usually supplied by a 2½ in. line or greater and some of that large volume of water inadvertently gets on the exposed building. However, these methods are usually water and personnel intensive, two things you will surely lack early on in the firefight.

Internal exposures. At house fires, because of the generally open floor plans of most homes and the open stairs of multistory homes, virtually all of the house is an exposure to even a one-room fire. One of the most important size-up questions is: Is the door to the fire room closed or are other interior doors closed? If the door is closed (and has not burned through), your size-up may indicate you just have a one-room fire. This will dictate your strategy and tactics for the entire operation. If there is intense fire in the room and the ceiling has failed, it will be in the attic as well. In balloon-frame homes, if the fire enters the walls, check the attic right away for extension.

When determining interior exposures, consider what rooms are exposed and the life hazard they contain. For example, there is a good chance a living room fire extending up the open interior stairs at 0200 will present a life hazard to sleeping occupants.

Fire conditions

Fire conditions are probably the most visible and dramatic aspect of size-up. A heavy volume of fire on arrival raises the question, is the building safe to enter? Is there human life that can be saved? Putting firefighters above the fire, as in on the first floor of a basement fire, is especially dangerous in modern construction that does not withstand a fire load. Aggressive interior attack may not be a good option. (See ch. 11 for a complete discussion of basement fires.)

Was the alarm delayed? Does the fire condition, smoke color, or appearance of pressurized smoke indicate something out of the ordinary? If you believe the fire is too large for what you assume is burning based on your size-up, you must ask: Is the fire being fueled by a gas leak? Is it arson? Affirmative answers to either of these questions should cause you to immediately change your strategy and tactics.

Does the smell of the smoke give any clues? The smell of food burning on the stove has a different smell from a room-and-contents fire, which has a different smell from a fire that has extended and is burning the wood framing.

Where the fire is located inside the home and what it is exposing may be the biggest factor in size-up. Hoselines cannot reach the seat of the fire until the exact location of the fire is known. A fire on any floor of the house has its own challenges. Basement fires have limited access via narrow interior stairs, limited ventilation due to small windows, and outside entrances that may be in rear if at all. Utilities, boilers, furnaces, and hot water heaters are often found in basement areas. Penetrations for piping and hot air ducts may not be sealed, allowing fire and smoke to spread throughout the structure. Fires caused by heating equipment are the second leading cause of residential fires.

The leading cause of residential fires is cooking. The kitchen is usually on the first floor, with the cooking appliances set between combustible cabinets that are often involved when something ignites on an unattended stove or in an oven. Voids created to frame out cabinets can extend fire vertically and horizontally. The typical window of time for cooking fires to occur is between 1500 and 2300, which can give the fire a jump on firefighters because it can burn undetected.

Living rooms full of flammable furnishings may be just inside the front door. A fire on the first floor can cut off access to the front door or second floor, if bedrooms are upstairs. Second-floor bedroom fires can easily expose the attic to a fast-moving fire when fire vents out the window and penetrates the already porous soffit covering.

Confinement. Once you have determined the external and internal exposures, the next step is to figure out how to confine the fire. At this point, your goal is to keep the fire from gaining control of any additional real estate or square footage and taking over more of the neighborhood or more of the house. In essence, the concept is to surround the fire and cut it off from extending to new areas. Although this process is conceptually simple, the challenge is in the details. How many personnel do you have available to get multiple hoselines in place? Is forcible entry required to get lines inside? At a minimum you need a primary and a backup line on most working fires. What is the best route to get these lines to points where they can limit fire spread? With a fully involved house, exposures getting ready to light up, and limited personnel, the exposure line will need to be big enough and agile enough to protect the exposure by flowing water onto it and if possible controlling the exterior fire. There are a million different details and possibilities but the goal remains the same: confine the fire to already burned areas.

Size-up involves piecing the available information together to create a picture of what you have, from which you can develop strategic and tactical objectives (what you are or will be doing and what additional resources you will need to do it). During this step, the IC will determine where the initial hoselines will be placed to confine the fire. As part of the size-up process, this step must be related to the life hazard that is present. Confinement of the fire, for example, away from the interior stairs to allow the interior teams to make the stairs and rescue occupants may be the priority confinement action.

Extinguishment. Size-up information will determine which and how many resources you will need to extinguish the fire. Will it be an offensive or defensive operation? Size-up information will also determine how many lines of what size and estimate the approximate necessary flow. Essentially, in this step, you are making the final push on the fire to evict it from the home.

Overhaul or salvage. The size-up process continues into the final information-gathering step, overhaul. With the main body of fire extinguished, the question becomes, "Has the fire entered concealed or void spaces?" Property conservation is our second priority. Sure, we have a duty to open up and make sure the fire does not rekindle and destroy additional property or threaten additional lives, but we also have a responsibility to the homeowner to protect as many of their possessions as possible. Size-up considerations during overhaul should include: How can we overhaul this home without causing unnecessary damage to the structure or its contents? Can we use salvage tarps to protect furniture and carpeting? Can we place debris in tarps, large buckets, or trashcans to limit damage to this or other sections of the home? What valuables can we secure with the help of law enforcement for the homeowner? Salvage should be an ongoing process. Size-up to salvage what we can for the homeowner or occupant should be done continuously to help us execute our second most important mission: protecting property.

Another consideration for size-up during overhaul is finding the cause of the fire. Some states have legal requirements that the chief or IC determine the cause of the fire. If members do a thorough job of overhaul, critical evidence of the fire origin might be lost or destroyed. In this case, you may have to revisit information gained in your size-up to help you determine the cause.

Ventilation. In this section we examine the information you need for size-up to develop ventilation tactics as part of your overall strategy. Successful ventilation operations are often the key to the success of the entire rescue and fire attack operations at house fires. Proper ventilation is a critical supporting task for rescue, exposure protection (especially interior), confinement, extinguishment, and overhaul or salvage. Proper ventilation begins with proper size-up. Let's look at some of the important things you should be looking for during size-up relative to your planned ventilation operation.

Los Angeles City Fire Department Battalion Chief (retired) John Mittendorf states the following regarding the critical factors and background information required for size-up for ventilation operations in his classic book, *Ventilation Methods and Techniques*:

> A working knowledge of building construction not only provides the necessary expertise to conduct a quick and accurate size-up of a structure, it also provides the foundation for effective, timely, and safe ventilation operations in the following areas:

> Size-up—your action plan based on all perceived factors.

> Ladder Placement—Ladders should be placed to the strong side of the building. This results in strength and stability for ladder operations.

> Structural integrity—The effects from fire on a particular type of construction, building or roof should be evaluated to determine the remaining time necessary to conduct a safe ventilation operation.

> Ventilation feasibility—can a ventilation operation be safely conducted?

> Operations—offensive or defensive

> Structural members—Safe routes of travel for personnel must be determined. The direction of structural members (rafters/joists) should be determined to enhance cutting operations. (Mittendorf 1988)

The aforementioned critical construction and ventilation considerations build the foundation of the detailed information you will gain observing the fire building during your size-up.

There are three types of ventilation possible at house fires: horizontal (taking windows), vertical (roof-cutting operations and positive pressure), and positive pressure ventilation (PPV; using large fans to control the movement of products of combustion).

Horizontal ventilation. "Taking the windows" is firefighter slang for this type of ventilation. With respect to size-up, here are some factors that you should consider when gathering information to formulate your ventilation plan:

- Location of the fire
- Location of the windows relative to the flow path of the fire—will they draw the fire to undesirable areas?
- Location of the search teams and victims—will taking specific windows make the fire situation worse for interior crews and victims?
- Size of windows—are they large enough to remove significant amounts of heat and smoke?
- Window glazing (hurricane windows are hard to take quickly)
- If fire vents out the opened window, will flammable siding and foam insulation ignite?

Vertical ventilation. Cutting the roof is a bit more complicated than taking the windows. Roof operations put firefighters above the fire and in a much more dangerous environment. Due to the additional dangers, there are many more factors to consider in addition to those for horizontal ventilation:

- Is fire in the attic? (If so, roof opening will be necessary early on.)

- Do you anticipate the fire extending to the attic?

- Has the fire attacked the structural members of the roof support system?

- Are truss or other lightweight supports used in the roof support system?

- Is the slope of the roof safe for firefighters to operate on?

- Is the roof surface slippery (ice, metal, slate, or tile)?

- Will falling roof coverings (slate or tile) endanger firefighters operating below or within?

- Is there an excessive weight load on the roof from snow, water, heavy tiles, or slate?

- Are there enough ladders, roof ladders, and personnel to safely conduct a roof operation?

Positive pressure ventilation. PPV is generally used in support of an interior fire attack. Some departments use it concurrently with hoseline advancement and others only after the fire is controlled to clear out lingering products of combustion. During your size-up, if you are considering using PPV, there are several questions you should ask first:

- Is there an appropriate location for a blower outside the building?

- Is there an appropriate vent location to drive products of combustion from the house?

- Will the positive pressure and induction of air push the fire into concealed areas?

- Will the induced air cause a ventilation-limited fire to free burn?

- How quickly will water be applied to the fire relative to the blower being used?

Clearly, gathering information during size-up to plan your horizontal, vertical, or PPV operation is complex and will dramatically affect the overall success or failure of the fireground strategy selected.

On-Scene Report

Now that the information gathering is complete for your size-up, you need to package the information in a usable form for incoming units. This is commonly referred to as the on-scene report. The on-scene report is a short description of the address, fire building type, fire situation or location, and any other pertinent information. There are a thousand different variations and styles used to accomplish this important task. The real bottom line is that your report should paint the picture of the situation for incoming units:

Chief 1 to Dispatch: I'm on the scene at 20 Main Street, we have heavy fire on the A-B corner in the kitchen of a 20 × 60 straight ranch. Occupants report one child is unaccounted for and possibly inside.

This report is very valuable to incoming units and provides the following key points:

- It confirms that the address you're responding to is correct: Main *Street*, not road, avenue, lane, or court. It also confirms that there are no homonyms ("soundalikes"), like with *beach* and *beech*.

  The address provides a clue to the type of home you may be responding to, as long as you've done your homework and know what kind of houses are found in the neighborhood. This might also provide clues about water resources and life hazards. Incoming engine companies should be searching preplan data for the nearest reliable water sources.

- It confirms a working fire. This may generate a box alarm assignment or automatic mutual aid in your department. It may also automatically bring resources such as utility companies, EMS, and so on.

- It describes the type, size, and height of the house. Engine companies can now tentatively plan how long the stretch may need to be based on the address and size of the home. Priority for line placement will be to support the search operation, not the fire attack. Truck companies now know they will (probably) be conducting an aggressive search operation for the missing child and can prepare themselves for it. Ventilation operations may have to be delayed if initial personnel is limited on the truck company. Horizontal ventilation of a one-story straight ranch can be accomplished by a single firefighter with a pike pole due to its low height, so the use of aerial apparatus may not be necessary. Portable ladders may be available from engine apparatus parked closer to the scene. Does this report describe a home with a walkable roof pitch, indicating that roof operations may be possible? Laddering potential for this building may be limited due to its low one-story height; however, landscaping and other factors may necessitate the use of ladders.

- All units know the general location of the fire in the building and the potential associated hazards. The kitchen fire may have started with deep frying so it's possible that a pot of oil is on the stove and that cabinets and furniture are involved. Natural gas or propane could be fueling the stove. Firefighters can likely rule out other more hazardous conditions that fires in a garage or basement may present, such as flammable or compressed gases or liquids.

- The style of the building provides general information about the saturation. This house is long and narrow and the bedrooms are likely at the opposite end of the house from the kitchen (main body of fire). This gives the victims a better chance of survival. The location of the fire and the possible location of victims mean that there is the potential to use the two-team approach. It may be possible to perform the vent, enter, isolate, and search (VEIS) operation from the outside under these conditions.

- You have a heads-up that this will be a physically and emotionally intense fire operation for everyone. We have reliable reports of a person trapped. Truck companies will conduct rapid primary search operations supported by a swiftly placed hoseline delivering decisive amounts of water to protect egress routes for search teams. The reports of persons trapped—especially children—ramp up firefighters' emotions, but this is when firefighters must remain calm and consistent. As search team members, be cognizant of your surroundings and be alert for signs of flashover and developing fire conditions. As company officers, communicate conditions to command so they have a complete understanding of the interior situation. As a command officer, provide resources in advance of the need (firefighters for relief or FAST or RIT duties, medical personnel for victims or firefighters).

This 10-second radio transmission tells the radio dispatcher that the chief and members will likely need additional resources and support. It is a working fire, so dispatch may be authorized to send additional fire units, EMS, command, and gas and electric crews to reduce risks to firefighters. The dispatcher now knows that they need to pay close attention to the radio for this call.

For incoming officers, this report paints the initial picture of a serious operation where members will be taxed to their limits and require relief sooner and more often. Officers may decide to tap out additional resources based on local needs. This could be a second or third alarm for personnel, a request for a water shuttle or drafting or tanker operation, and so on.

How the on-scene report is transmitted is another often-overlooked facet of the size-up process. What you say is as important as how you say it! If you have just pulled up and begin screaming into the radio at a pitch and volume that make the transmission incomprehensible, the only message the incoming units will get is "Holy crap!" Don't give a report when you are out of breath from making the 360 survey either; this will also not provide units with the correct information. Think about what you need to say, take a deep breath, and say it in a calm, collected manner. The report paints the picture; you may be viewing bedlam and chaos but the report must convey the message that things are under control.

Converted Private Dwellings

Size-up of converted private dwellings is a major problem for fire departments all across our nation. Converted private dwellings (CPDs) are a pervasive problem today and will continue into the foreseeable future. They are just as much a problem for the biggest urban department as they are for midsize suburban departments and even the smallest rural departments. They can be single-family dwellings (SFDs) where extended families live together, farm communities where migrant workers temporarily reside, college towns with houses converted for students—really, anywhere that has high housing prices. CPDs are everywhere and must be correctly sized up in order for the operation to be successful.

Private dwellings can be converted into apartments and even rooms (single-room occupancies, or SROs) for a variety of occupants. They are common in college towns where larger homes are now divided to house several students. This was the case at a residential fire just outside Boston in March 2013. John Zaremba, Colneth Smiley Jr., and Laurel J. Sweet of the *Boston Herald* report it this way: "The packed Allston home where a 22-year-old Boston University student died in a raging fire Sunday is an illegal rooming house that hasn't been inspected in 21 years, frustrated city officials said yesterday as firefighters sifted through the ashes and tenants salvaged anything the blaze left behind." The news article goes on to say that at the time of the fire, there were 19 tenants living in the building, which was listed as a two-family home. The landlord had other similar properties, which were obviously a very lucrative source of income. Although there are ordinances that require reinspection upon change of tenants, the landlord has to request it.

CPDs defined

In essence, the building looks like an SFD from the outside but further detailed size-up reveals a very different occupancy with very different hazards that require different strategies and tactics from you would apply to an SFD house fire. CPDs will fool you, as they may be illegal conversions, in which case they will be intentionally disguised and hide deadly hazards. These dangerous buildings are constructed in violation of local and state fire and building codes. There may be a mix of apartments and SROs in the same home. Proper size-up of converted private dwellings is the key to success and safety.

Figure 4–9 shows broken glass from a venting operation covering the mattress on the floor of a very small SRO after a fire downstairs. One side of the room is mattress; the opposite side has an electric hot plate on a small table and several piles of clothing. Note the broken deadbolt from the search team's forcible entry on the mattress.

Fig. 4–9. Some hazards of SROs

Origins

CPDs have their origins, usually based on financial considerations, in the following circumstances:

- Families and extended families may be forced to live together as a result of low income levels or recent losses or reductions in income.

- College students often live in SROs in larger converted homes.

- Migrant workers occupy CPDs during certain seasons.

- Lower socioeconomic level populations may overoccupy a house by living in basements, attics, and garages.

- Unscrupulous building owners often convert large homes into SROs, multiple apartments, or combinations of both for profit.

Hazards

The unique hazards of CPDs are why we must identify them quickly during our size-up. Undetected, a CPD will have many deadly surprises for firefighters:

- Overcrowding caused by occupants' belongings (storage boxes, bikes, and bulky children's toys in hallways)

- Hallways too narrow for a firefighter with an SCBA on to turn around in

- People living in unexpected and dangerous locations (attics, basements, hallways)

- Small rooms created for privacy, often deadbolted where only a privacy lock would be found in an SFD (fig. 4–10)

- Small rooms crowded with the same amount of furniture normally found in larger rooms (figs. 4–11 and 4–12)

Fig. 4–10. A deadbolt on what would typically be a bedroom door is a good clue that this is a CPD containing SROs

Fig. 4–11. A crib, a full-size bed, and a single bed leave little floor space for firefighters to operate. At this fire, bunk beds also blocked a back door and window in the corner of the house and a refrigerator blocked a kitchen window.

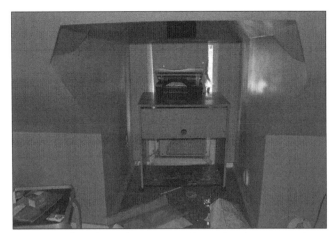

Fig. 4–12. Illegal attic living space. Note the dormer window is blocked by both an air conditioner and a computer table. The only exit is down two flights of very narrow winding stairs that would have acted as a chimney during a fire on a lower floor.

Fig. 4–14. To allow heat to flow passively up stairs, the building owner replaced the double layer plaster-on-lath transom over the door from the first-floor kitchen to the hallway with this single-thickness wood paneling. Numerous holes were drilled in it to heat the upstairs rooms. The closed door below the transom held the fire to the kitchen but the holes allowed smoke and heat to heavily charge the upper floors.

Building hazards include

- Substandard construction methods used to convert the home, often destroying fire barriers (plaster or sheetrock walls) and replacing or covering them with combustible paneling or leaving open voids for fire travel (fig. 4–13)

- Overloaded or substandard electrical service

- Lack of proper egress from attics or basements, no fire escapes

- Ventilation openings in walls, floors, and ceilings to allow vertical spread of ambient heat and cooling, which also allow for extremely rapid fire spread throughout the dwelling (fig. 4–14)

- Windows covered over from the inside (often with sheetrock) to hide the occupancy from neighbors

- Division into multiple apartments, limiting our access in the building

As previously stated, CPDs are a nationwide problem for the fire service, especially during size-up. It is important and valuable to learn from other fire service leaders regarding CPDs. The following comments were originally published in *Fire Engineering* in an article on CPDs. These three extremely valuable and insightful comments from experienced leaders provide a thorough understanding of the problem and how important it is to conduct a proper size-up. Additionally, they provide a look into possible alternative strategies when you discover a CPD during your size-up.

Captain Bill Gustin, Miami-Dade Fire and Rescue, has the following to say on the dangers of CPDs:

Illegal conversions of single-family homes in my district are a big problem. It is important to get the

Fig. 4–13. Just inside the front door to the left should be the living room. To make this an apartment, a wall was made of very combustible paneling (replacing the original fire resistant plaster or sheetrock). This room is now a bedroom that contains both a full-size bed and a crib, which creates a narrow hallway to the rear of the house.

hoseline in place by the most direct means possible. In my district, fire showing at the rear of a house can often not be reached by a hoseline advanced through the front door because the house has been divided, creating a separate occupancy in the rear. In the neighborhood where conversions are prevalent, we take the line to the doorway or stairway that is closest to the fire because there is a good chance that it is the only path to reach the fire.

Similarly, firefighters stretching a hoseline through the front door and up an interior stairway to reach a fire on the second floor are likely to find their ascent blocked by plywood—the stairway was closed off to create a separate occupancy on the second floor that can be reached only from an outside rear stairway.

In terms of forcible entry, consider taking bolt cutters, as SROs are commonly locked with a padlock and chain that passes through an opening in the door, usually where the door knob was removed, and a hole in the adjacent wall. It is critical to get a hose line in position and operate on the fire as quickly as possible; this is the most life-saving function at any fire where there will be a delay in searching all areas.

Your search of upper floors will fail if the fire is not rapidly confined and firefighters conducting the search will be placed in grave danger. Firefighters ascending above the fire floor are depending on the engine company operating the hoseline to protect the stairway. (Knapp and Zayas 2011)

Tom Labelle, executive director of the New York State Association of Fire Chiefs relates this story of a converted home in the Albany (NY) area: "We had a second-floor fire in a legacy two-story wood-frame 'farmhouse' with a scissor stair to the right of the front door. When I took the line the top of the stairs, I was looking at gypsum board! They had made it into two apartments. I could not get to the fire" (Knapp and Zayas 2011).

Jeff Shupe, a very experienced and now retired member of the Cleveland Fire Department, describes the danger of CPDs as "a changing urban environment." He discusses the need for a change in the tactics used at these homes: "It is almost impossible for firefighters to 'adapt and overcome' problems presented in converted homes on the fireground because they are in the middle of a rapidly deteriorating condition bordering on or going to untenable, and they likely do not know what is happening to them and why." Shupe continues, agreeing with the statements of Captain Gustin: "Another thing we have to get away from is always trying to hit the fire

from the unburned side. It does not hold true anymore. Back in the old days, that was fine, but now fires burn with more speed and intensity. We must get water on the fire in the quickest way possible" (Knapp and Zayas 2011).

Size-up

One of the best ways to size up CPDs is to drive around your first-due area at night. Look for lights on in attics or basements. CPDs often have some obvious exterior clues that give them away:

- Multiple mailboxes
- Unusually high number of cars in the driveway or near the home
- Multiple satellite TV receivers on the roof
- Windows that are blacked out, either from inside or out
- An attic air conditioner
- Enlarged basement stairs (exterior)
- Makeshift driveways or parking areas
- An excessively large number of trash bags or containers in front of or near the home

Notice that we did not include multiple utility meters as a clue for your size-up. Utility companies will only install meters on code-compliant buildings. These CPDs are anything but code compliant.

Reaction

Once a CPD has been detected and confirmed by interior teams, it is critical to get the word out to the IC immediately. CPDs require more personnel for forcible entry operations due to the locked doors inside. Estimation of the search resource requirements will likely change along with the staff needed to stretch lines to the fire if normal access routes (through the front door) are blocked. When your size-up reveals a CPD, it will often change your entire strategy. In the SAR and fire attack chapters, we examine specific considerations for these dangerous house fires.

Scenario Discussion

1. What obvious hazards or threats to your members do you see?

    a. Construction type: Because you have done your prefire planning, you know the home was stick-built and does not contain any lightweight construction materials.

    b. Fire extent and intensity: Obviously heavy fire spreading to the attic, possibly causing a collapse potential.

c. Flashover or backdraft conditions: The fire in the room of origin has gone to flashover and the remainder of the house is probably heavily charged with smoke, providing backdraft and flashover potential.

d. Utilities (wires, gas, propane, fuel oil): Again, from the preplan information, we know there is natural gas service in the area and underground electric, neither of which provide a significant or immediate hazard to firefighters at this point.

2. Evaluate the life hazard.

a. Do you think the house is occupied? The house appears well kept and occupied as opposed to vacant or abandoned. There is no for-sale sign, the grass is cut, and there are Halloween decorations in the yard, supporting this assumption. The home is in a neighborhood that typically contains young families with children. Considering the time, it is likely they are home and probably sleeping. It is likely that at least one parent is home as well.

b. At this time of night, where do you think the occupants may be in the house? It is likely they are in the bedrooms.

c. Is their exit blocked by fire? It appears that the bedroom door is closed and still holding inside. This makes escape from the bedrooms adjoining and opposite the fire room possible. However, with the intensity of fire, the exit down the hallway will not remain tenable for very long.

d. Are there indications someone may have gotten out (open doors, windows, etc.)? Nothing from the A side, and the 360 report is the same, with no open windows or doors.

e. What are the odds of the people inside surviving? Good except for the fire room. There is no smoke showing from other windows or the front door. This could mean that the house is sealed well for energy efficiency or that the door to the fire room is still mostly intact.

f. Are there salvageable lives inside? There may be salvageable lives in many areas of the home, including the basement, which might be a separate apartment as in a mother-daughter home. Additionally, occupants may have been incapacitated by smoke or fire and collapsed in egress routes from where they were to a door or window.

g. Where are nonsalvageable lives in the building? In the fire room. Sadly, it is too late for these occupants.

h. Can you put firefighters inside the building with reasonable risk? It looks like it at this point, although fire will extend rapidly to the attic via the exposed soffit and probably directly through the

ceiling of the fire room(s) when the sheetrock collapses. Observation on the D side shows fire pushing out of the gable vent through the smoke.

3. Fire strategy

a. Is an offensive strategy appropriate? Yes, again probably for a short time since the risk appears to be justified by the potential reward of several rescues.

b. Is a defensive strategy appropriate? No, a rapid interior search must be conducted with support from a hoseline in the hallway. It appears that this fire has not moved into other living areas of the home where salvageable victims can be saved during an offensive operation. Once the primary search is completed and the first-floor fire is knocked down, it may be necessary to go to a defensive operation. This decision will be driven by two factors on this fireground. First, fire may have taken complete control of the attic, creating collapse concerns for members operating inside. Second, depending on your available resources, you may not have enough personnel on the scene to conduct a sustained interior fire attack, pulling ceiling and extinguishing fire in the attic from below.

In terms of sizing up this fire using RECEO VS, the following considerations are appropriate:

Rescue. It appears that, barring any additional solid information to the contrary, an aggressive SAR operation is possible and appropriate at this incident. The primary search will locate immediately salvageable victims. A secondary search must be conducted as soon as it is safe and staffing levels permit.

Exposures. There are two important exposures: the interior of the house and the attic. The house interior may contain salvageable victims. The most important exposure is the means of egress for any trapped occupants, and it needs to be maintained for the SAR crews. The attic is severely exposed by this intense, well-ventilated, and free-burning fire. Although not significant for civilian lives, heavy fire in the attic could result in collapse onto firefighters inside and may be the limiting factor in keeping this operation an offensive attack.

Confinement. Confining the fire starts with the first hoseline supporting the rescue, likely going through the front door, up half a flight of stairs, then down the hallway to the right to keep the hallway safe for victims and rescuers. This line may push into the fire room and greatly reduce hazards on the fire floor by controlling the fire in that area. The attic still remains a threat to be dealt with.

Extinguishment. The first line should be able to maintain the egress route and knock down the fire in the room of origin.

Additional lines and personnel will be necessary to control the attic fire.

Overhaul. Perhaps this fire took place in one of the states where the law requires the fire chief to determine the cause of the fire, or maybe this was a fatal fire or one where firefighters or others were injured, making evidence protection vitally important. It could also be a crime scene: often burglars set fire to cover their crime, and frequently owner or occupants will set fires for revenge, love triangles, or financial reasons. If the fire cause is not immediately apparent, consider limiting overhaul until the arson investigators can get a look at the scene and conduct their investigation.

Advanced Size-Up Considerations

In figure 4–15, a typical home in the Chicago area provides a size-up challenge. A 360-degree size-up (or as close to it as you can get) will provide the critical information for your strategic and tactical plans for rescue, ventilation, and fire attack. From the front, this appears to be an average split level. But the lower windows at the A-D corner and the absence of a garage where you would typically find it give you a clue that this may not be a typical split level.

Fig. 4–15. Seemingly typical Chicago split level

As you begin your 360-degree size-up, you see critical facts about this nontypical split level. The lower personnel door tells you that there likely was a garage in that location that has been converted into some type of home or business office or second apartment (fig. 4–16). Note the location and direction the car in the driveway is facing. If this were a single-family home, it would most likely be pointing toward the house, as if the occupant had just pulled into the driveway. The driveway has been redesigned to allow cars to pull in forward, then back out and turn around before going back on the street. This appears to be a parking area for

customers coming and going to the lower office. The living space was expanded into a large addition toward the rear, with garages underneath.

Fig. 4–16. Key features that should alert you to the fact that this is not a typical split level

Additional Size-Up Practice Scenarios

Fig. 4–17. Additional practice scenario

You are dispatched to a mulch fire, 10 August, 0023 hours, 42 Tarawa Drive. While you are responding, the PD reports a working fire in the front of the house. Upon arriving, you notice what looks like a family huddled near the neighbor's yard. They have panicked looks on their faces.

What is your on-scene report? You confirm the address as 42 Tarawa Drive, report a working house fire in a two-and-a-half-story wood frame. One of your most important actions is to determine if everyone is out of the house. Are the people on the lawn the occupants of the home that is burning? You should ask the police officer if everyone is out of the house. Don't forget to ask the next most important question: "How do you know?" The police officer points to the family nearby. You ask them, "Is everyone out of the house?" If they respond without much consideration, ask again: "Are you sure?" As the parents look around at each other and the kids, a confident look comes over their faces. They repeat that yes, everyone is out. You just conducted a very important part of size-up: asking questions. Remember, the police officer was probably there before you and had time to sort out some key size-up factors for you.

From the front of the house you see flames from every window. The roof collapses before the first apparatus arrives. The fire building is completely involved and not salvageable. The exposures on both B and D sides are threatened. You see the vinyl siding on each melting vigorously. These observations and facts will help you determine where your first hoselines will be directed.

The 360-degree size-up was important at this fire as well. The family lost everything but their lives. The firefighter who did the 360 for you rescued the family pet while they were walking around the structure (fig. 4–18). The reunion in the neighbor's yard brought smiles to a devastated family.

Fig. 4–18. Dog rescue. Courtesy Tom Bierds.

Below is another scenario to help you practice your size-up skills (fig. 4–19).

Time of dispatch is 0232, Tuesday evening, 23 December, at 44 Saipan Drive, a good neighborhood. The fire was reported by a neighbor returning home after a Christmas party. Upon arrival the caller meets you and tells you there is a family consisting of a mother, father, two children, and an aging and ill grandparent who all live in the home. Put together everything you've learned from the chapter and develop a good size-up of this fire.

What is your on-scene report? I'm on the scene at 44 Saipan Dr. We have a two-and-a-half-story 20 × 40 ft balloon-frame home with fire from the A side, second floor, possibly extending to the attic. Neighbor reports possibility of persons trapped.

Fig. 4–19. Second additional scenario

What is your size-up of this fire? With only the front view, it appears that this is a one-room fire that has extended (as you would expect) into the attic. Was the aging grandparent smoking in bed? Since smoke is not showing from any other areas, it is likely that either the bedroom door is closed or this house is energy efficient and not leaking smoke from windows or through walls or siding. If the bedroom door is closed, the fire can be kept to the room and escape by occupants is possible. Entry, visibility, and mobility for search crews should be good if the bedroom door is closed. The porch roof is not a viable option for VEIS because the window is blocked by heavy fire, and with the addition of air from the VEIS the fire will quickly intensify. If possible it is always a good idea to look directly beneath the window, as salvageable victims may be found there.

The stairs to the second floor as a general rule are behind the front door. Since the ridgepole of the roof is at a 90° angle, it is reasonable to expect that the upstairs hallway has a similar angle to it.

Your 360-degree size-up reveals much more critical information. For the life hazard aspect, a car is parked in the rear driveway that you could not see from the front (fig. 4–20). Certainly not a reliable indicator that someone is home but it is a good clue. The house looks well kept, another clue that it is not a vacant structure. Your trip around to the C side verifies that occupants have not escaped the building through the rear door (if that was the case, a dangerous search may not be necessary, or at least not as important). Additionally, the second-floor windows are intact and a quick look at the shrubs and the base of the home shows that no one has jumped from upper floors. Your view of the windows and rear roof shows you both the horizontal and vertical ventilation possibilities.

Fig. 4–20. Notice the car parked in the driveway.

Regarding the whole fire, it now looks like there may be a basement fire in this balloon-frame home that has spread to the second floor and attic. You will not be certain of this until members get inside, but if the fire is in the basement, the integrity of the first floor will be an issue to consider. The fire was intense enough to fail a basement window in the rear so it is a good assumption that there is significant fire down there and members will have to operate above it.

The rear view of the home shows an opportunity for VEIS on the bedroom window on the right side of the picture. The window in the middle might be the landing at or near the top of the stairs. The window on the left is the fire room. The rear door provides access to a small porch and the basement clamshell doors are just to the right of it in this picture (B-C corner). The clamshell doors provide an opportunity to get a line down on a level with the fire as opposed to climbing down the interior stairs.

References

Knapp, Jerry, and George Zayas. 2011. "The Dangers of Illegally Converted Private Dwellings." *Fire Engineering* 164 (6). http://www.fireengineering.com/articles/print/volume-164/issue-6/features/the-dangers-of-illegally-converted-private-dwellings.html.

Mittendorf, John. 1988. *Ventilation Methods and Techniques.* El Toro, CA: Fire Technology Services.

Moran, Lee. 2015. "Ark. Deputies Find Meth while Trying to Save Cats after Fire." *New York Daily News*, May 6. http://www.nydailynews.com/news/crime/ark-cops-find-meth-rushing-house-fire-save-cats-article-1.2212025.

National Institute for Occupational Safety and Health. 2007. *Career Engineer Dies and Fire Fighter Injured after Falling through Floor while Conducting Primary Search at a Residential Structure Fire—Wisconsin.* NIOSH Report No. F2006-26. https://www.cdc.gov/niosh/fire/reports/face200626.html.

Reeder, Forest. 2014. "Understanding the New SLICERS Acronym." *Fire Rescue Magazine* 9 (2). http://www.firerescuemagazine.com/articles/print/volume-9/issue-2/training-0/understanding-the-new-slicers-acronym.html.

Smith, James. 2002. *Strategic and Tactical Considerations on the Fireground.* Upper Saddle River, NJ: Prentice Hall.

5

Command, Control, and Fire Attack Strategies

Summary

This chapter describes command and control strategies and lays out and provides examples of the four basic strategies for rescue and fire attack operations: offensive, defensive, offensive-defensive, and defensive-offensive. (This chapter assumes you already have a working knowledge of the NIMS/ICS principles.)

Introduction

This chapter will discuss both general and specific command, control, communication, and fundamental fire attack strategies to provide new firefighters with an overall understanding of the strategic and tactical objectives they and other department members will be executing. For experienced firefighters, line officers, and commanders, this chapter provides key elements that can be used to evaluate and possibly improve fireground command skills for house fires. For the less experienced, it provides an overview of command, control, and communication at house fires.

Scenario

Fig. 5–1. House fire scenario

Situation: It is 0530 hours, Tuesday morning, 3 July. A hysterical neighbor tells you that she just saw the husband go to work about an hour ago, but the wife and kids are home.

Resource report: Your first engine has a good hydrant and has laid a supply line into the front of the building, leaving room for the truck that pulled up immediately behind. Your inside team has forced the front door and is making progress down the hallway toward the bedrooms.

Radio report: "Inside team to command, the interior door to the fire room, A-D corner of the building is just starting to show fire."

"Outside team to command, no fire showing out the rear. We have made entry into the rear bedroom off a 24 ft extension ladder and found one child and are bringing her down the ladder now. Where is EMS?"

Questions

- What is the top priority at this fire at this point?
- Is the building safe for members to operate in?
- What search operations can members perform from the outside if necessary?
- Is this fire going to an offensive or defensive operation?
- How do you want your first hose to protect the members inside?
- What other strategy could you employ and why?
- What would happen if you went defensive then offensive?
- How does the first hoseline get into place and where does it go?

Firefighter Safety

As we have seen in the previous chapters, and surely based on your own fireground experience, house fires are very, very dangerous alarms. The American fire service's nationwide emphasis on firefighter safety over the past few years has resulted in a dramatic reduction in our injuries and deaths.

To further reduce fireground injuries, commanders, line officers, and firefighters must use firefighter safety as the cornerstone against which we judge and balance all fireground decisions and actions. However, we have always taken and will continue to take calculated risks, especially when there is much to gain by working in that particular risky situation. We will continue to risk our lives during search operations when it is known or strongly suspected that there are occupants in danger who can be saved. Thanks to 2016 UL research projects, we now operate with an understanding of just how quickly the ventilation-limited house fire will go into firefighter-killing flashover.

We will continue to ventilate the roofs of fire buildings to minimize hazards for interior crews. We will continue to push hoselines into house fires to protect search teams when there are minimal dangers of collapse or other threats so that we can execute our second prime directive: saving property. The goal of every firefighter and fire officer is to take manageable risks when there is something to be saved and never indulge in senseless risk-taking for any reason. In essence, we risk a lot to save a lot, always measuring the level of danger to ourselves and our members. As a chief or a firefighter always ask yourself: "Is what I am doing worth the risk?" Another way to phrase this question is, "What am I risking? What am I saving?"

Financial Future

House fires are some of the most devastating for our customers. Their home contains all they hold precious; their lives and family's lives, photos, memories, and pets. Their home is often the culmination of their life's work. We should take all reasonable steps to save property, especially at a house fire. But in the end, sometimes what we are able to save is not up to us.

We train hard as firefighters to be aggressive; to conduct SAR operations under zero-visibility conditions, push hoselines into raging fires, and ventilate roofs to make conditions better for our firefighters inside. It only makes sense and is human nature that we want to apply these skills to the challenges of the fire building and the conditions it presents. There is nothing in the firefighting world like the feeling of making a great stop. We must, however, consider the realistic future of the building. If there is heavy fire, smoke, and water damage, what will be the fate of that home? The ultimate judgment will come down to money: if the cost of repairs is close to or exceeds the cost of the building, it will probably be torn down and replaced. Surely you can recall more than one great stop where you risked your life and maybe your members' lives only to see the excavator and wrecking ball tear it down the following week. As firefighters we like to think of our profession as one of life and property saving. More often than not, it is a matter of dollars and sense: knowing when there is nothing, lives or property, for us to save.

Strategic and Tactical Discussions

Some firefighters may argue that VEIS is the safest and most productive way to search at house fires. Others may counter that if a firefighter does this in a one-story ranch house, the engine company attacking through the front door is likely pushing heat, steam, and products of combustion toward the firefighters engaged in search. VEIS is simply a tool or option, and frankly we can give you the tool, but not tell you how or when to employ it. Firefighting must not be done with a cookbook mentality: do this, do that, and the end result will be perfect. All tools have strengths and weaknesses, advantages and disadvantages. Think of these acronyms and suggested procedures as tools. These summaries must be used by firefighters and incident commanders along with thorough and continuous size-up and always with firefighter safety in mind. Today's fires are far too dangerous to not be evaluating what you and your team are doing at every step. A carpenter does not use a hammer to put in a screw. Tool selection, like strategy and tactic selection, requires evaluation and thought.

One may further argue that VES should be VEIS: vent, enter, isolate, and search. Isolate of course refers to shutting the door to the room you just entered to isolate you and protect you from the fire. While this is certainly a practical and effective tactic, the point here is not to split hairs, but rather to put forth some strategic and tactical considerations for you to use wisely and appropriately at your next residential fire.

While reading this chapter it is imperative to keep in mind five important considerations:

1. The strategic and tactical options presented here must be applied with a primary focus on firefighter safety and based on a sound understanding of fire development and firefighting operations.

2. We have assumed that the reader has basic firefighter training and experience. Absent that, the reader must consider what is presented in this book as an overall plan that must be applied with great discretion.

3. If a simple cookbook approach—when you see this, do that, and everything will be fine—could be successfully applied to firefighting, firefighters would not be getting injured or killed anywhere near the current rates.

4. There is simply no substitute for a substantial amount of training and experience to produce excellent results.

5. You must constantly be thinking at house fires.

No matter how you arrange, add, or delete the letters of an acronym, the priorities are still the same:

1. Firefighter safety
2. Civilian safety
3. Property conservation

The only thing that differs is how *you* personally package the tool (acronym) to help *you* recall critical functions and execute them safely at your next fire.

Command Principles

The command principles and plans presented in this and other chapters are summarized with acronyms. Acronyms (by necessity and design) are freeze-dried, shrink-wrapped versions of complex concepts. Some of the most popular "acronymized" command plans are CRAVE, CAN, RECEO VS, and SLICERS. As we have seen so dramatically, the house fire battlefield has changed, but we must still fulfill our age-old duties—saving life and property—with the careful eye first and foremost on firefighter safety: saving and protecting *our* lives. It is absolutely critical to remember that the house fire battlefield is not what it once was, for the many reasons discussed in chapters 2 and 3. We must continue to search for viable victims in house fires, but bear in mind the ground-breaking research done by UL (see ch. 3), which shows us how quickly a one-room, modern house fire will flash over if water is not applied. Clearly we should place a high priority on search operations (life safety is our first customer priority) even though these are high-risk operations, but we should also support and protect those searching firefighters with a hoseline to control the fire. The new battlefield of house fires demands this in order to fulfill our absolute first priority: firefighter safety. To maximize firefighter safety, we must consider the entire scenario: building, fire location, flow paths, anticipated fire growth, and so on.

Command and Control

These are the days of formal incident command systems. It is more than just pulling up, throwing a green light on the roof, announcing that you are command, and activating a communication system to talk to the world. The incident command system (ICS) is a standardized, on-scene, all-hazards incident management approach that

- Allows the integration of facilities, equipment, personnel, procedures, and communications operating within a common organizational structure;

- Enables a coordinated response among various jurisdictions and functional agencies, both public and private; and

- Establishes common processes for planning and managing resources.

ICS is flexible and can be used for incidents of any type, scope, and complexity. It allows its users to adopt an integrated organizational structure to match the complexities and demands of single or multiple incidents.

Establishing command announces that you are in charge of this event from that point forward. You take responsibility for all decisions and outcomes of that event legally, morally, and ethically. Yes, it is an enormous undertaking. You are responsible for the lives of everyone on your fire scene. You must have the resources to protect them.

As the hazards of modern house fires increase, so do the means to convey them. Placards for hazmat locations and signage indicating truss roof constructions are replaced with CAD information and mobile data terminals. Three-ring binders are replaced with electronic command boards. However, the function of processing information for the incident commander remains the same.

The level of detail that is often available to modern ICs in the form of floor plans, diagrams, and digital imagery can lead to information overload. Therefore, the ICs at today's house fires must be multitaskers, able to retrieve information from various electronic and digital sources, and must also have the experience required to determine the interior conditions from what they can see and the reports from interior teams.

Establishing a command post

Location, location, location: The location of a command post (CP) needs to be in area away from the noise and diesel exhaust of the apparatus. The noise and exhaust not only are unhealthy but can be distracting for the decision-makers. It should provide a full view of the whole fire scene. It should be in an area that is easily identified by incoming units and outside support agencies while maintaining security to prevent intrusion by the media or spectators.

The location needs to be able to support the number of agency representatives present. The fire agencies may include subject-matter experts (SMEs) from other branches, fire investigators, and public information officers (PIOs). Law enforcement may have officers tasked with crime scene preservation and security of the property who will have a liaison at the CP. Emergency medical services may have members at the CP for victims as well as response to a trapped or injured firefighter.

Utility crews will often be necessary to isolate the building from public utilities. Public officials often respond to the scene of fire to demonstrate support for the community and offer assistance. Depending on the office the public official holds, this could add to the group dynamic at the command post. Code enforcement officials are often part of the group

of public officials. You should also never underestimate the number of nongovernmental organizations (NGOs; Red Cross, Salvation Army, church groups, etc.) that may appear following a house fire. They can be instrumental in providing aid that the public sector cannot provide to displaced persons. They will need to consult with someone in command and the general staff at some point. If the incident has escalated to the point where these positions need to be staffed, identify these liaisons quickly to allow the IC to focus on the emergency.

Get a good view of what you can't see

Once the CP location has been established and you exit the vehicle, it is important to gather as much information as possible as quickly as possible. You need to see what you don't see! Doing a 360 does not mean running around the house (unless that is easily accomplished). The intent of the 360 is to make the IC aware of what they can't see from the CP.

If you approach the structure from the front and pull past the building, you will get a view of three sides. You only need to go around one side to see the final side. The area in the photo from this chapter's scenario would be a good place to locate a CP, since it provides a good view of the two sides with the active fire situation. The CP may need to be moved later based on the placement of apparatus but that can be determined as apparatus arrive. When you do get to look at the rear, you might find an addition with a roof area that provides a platform to enter windows and search for victims in need of rescue. This may change the priorities in your fire attack plan.

CAN reports

For an IC to manage a fire they need accurate, complete, and timely information. This comes in the form of a conditions, actions, and needs (CAN) or progress report. It is how an IC assesses the scene. Bill Gustin, a longtime friend of the authors, has put his own twist on CAN reports and gave me (Chris) some advice when I was promoted to lieutenant in the FDNY: "To be a good company officer all you have to do is be able to answer three questions when the chief asks." He said the chief will ask, "What do you have, what are you doing, and what do you need?" Let's look at a few examples of how CAN reports are transmitted.

Example 1. "Chief, we have a carbon monoxide detector activated and we're getting elevated readings on our meters. We're searching for the source. When the medics get in, can you have them check the family in the front yard? They're complaining of headaches."

Conditions: CO alarm, elevated readings, and symptoms

Actions: Metering, searching for source

Needs: EMS response for victims

Example 2. "Chief, we have an unattended pot on the stove. We cooled it in the sink, and there is no extension to the walls or cabinets. Can we get a PPV fan set up at the front door? There's a light smoke condition in here."

Conditions: Pot on the stove, no extension, and light smoke condition

Actions: Moved pot to sink, beginning ventilation for nuisance smoke

Needs: A fan at the front door

Example 3. "Chief, we are moving down the hall extinguishing visible fire, but it's not cooling down; we may have extension to the attic. Can we get someone to check the attic and start a second line? My crew will need relief soon."

Conditions: High heat in areas not being cooled by the hoseline, possible extension to the attic

Actions: Extinguishing visible fire, making progress

Needs: Backup line for protection, confirm fire in attic, plan for relief of members

These examples demonstrate how a short, concise report can provide actionable information to the IC as they move through the incident. Whatever the language and jargon of your department, it is important that the person you are transmitting the information to understands the conditions, actions, and needs.

I'm sure every department has a story of a miscommunication where a second alarm was transmitted because the aide (or someone) told the chief, "The dispatcher wants to know what we're using," and the chief replied, "Give me a second," prompting the aide to request a second-alarm response when the chief was just looking for some time to formulate an answer. This becomes humorous when the second-alarm units check in at the command post and the chief wants to know why all these people showed up. On the other hand, if the fire situation had been expanding and a miscommunication caused a delay in the request for additional resources, the situation could have been deadly.

Transfer of command

If you are in a position where you will transfer command to a higher-ranking officer, you need to be prepared to provide a current situation report that is complete and accurate. When the relieving officer feels they have sufficient information to assume command, they will accept command and announce over the radio that they have done so.

For house fires, this process will be a lot more intuitive and "on the fly" than the formal process outlined in the official FEMA protocol in the sidebar. The process is very similar but much more streamlined. Recall that the FEMA process is often used for very large operations, such as huge wildland fires, flood responses, or other large-scale events. House fires, while still requiring a smooth transfer of command to ensure the highest degree of firefighter safety and department efficiency, are much smaller.

Using our last CAN report example as the base, transfer of command may sound like this: "Chief, upon arrival we had heavy fire in all three bedrooms on the B side extending into the hallway. The occupants were on the lawn and were certain that everyone was out, so we didn't conduct a primary search. Engine 1 is moving down the hall extinguishing visible fire but it is not cooling down, so we may have extension to the attic. I requested a second alarm for one additional truck to open the roof and one engine company to back up the operating company. There is no fire below the hoseline and our preplans show that the roof doesn't contain trusses or engineered lumber. The additional units are arriving on the scene now. Unless you have any questions, I'm going inside to check on my crews and see if we have fire in the attic. Also, the first-due crews will need relief soon."

This short brief answers all the questions and builds naturally on the CAN report and the basic questions: What do you have? What are you doing? What do you need? The incoming IC now has a clear picture of the situation and the current strategy and may choose to continue with this plan or radically change it. Their initial and ongoing size-up (How much fire is in the attic, if any? What progress is being made with the fire suppression?) will determine the forward course. Command of the incident was transferred smoothly and without detriment to the safety and effectiveness of operation forces.

Accountability

As we have mentioned, the technology of incident command provides many levels of electronic information. Though systems exist that can monitor air capacity in individual SCBAs, pinpoint firefighters' locations, and even track the ambient temperature firefighters are working in, these are no substitute for the incident commander's knowing where their people are and what they are doing.

We have seen accountability systems consisting of a pad and pencil or "cow tags" used just as effectively as electronic ones for monitoring firefighters on the fireground. The benefit of electronic monitoring is the telemetry features that allow firefighters' movements to be tracked from a remote area during the chaos of the active fire scene. They provide a level of overarching monitoring and accountability data that can be recovered and examined following a catastrophic event at the incident.

The Personnel Accountability Report (PAR) allows the IC to check on the status of the members with whom they may or may not have contact. It could be a simple roll call of units operating to ensure that everyone is OK. Other systems have been developed to allow an incident commander to conduct a PAR via the firefighter's portable radio. The electronic PAR reports allow for status updates without interfering with the voice communication during operations. Electronic reports are generated through an Electronic Firefighter Accountability System (EFAS) and are viewed on command consoles at the CP. These are excellent ways to determine that your members are safe, but may require additional support staff to properly utilize. When staffing is often the critical factor in fireground operations, think carefully about relying on systems your department may not be able to fully support.

Mayday, Mayday, Mayday—How prepared are you?

Knowing where people are and what they are doing is a priority for the IC, as we have said. What happens when the catastrophic happens: You hear the radio transmission, "Mayday, Mayday, Mayday!" Believe it or not this is a better transmission to hear than the emergency alert tone after a low rumbling sound from inside the building. Neither is one the IC wants the hear, but in the case of the Mayday, the firefighter is able to communicate on some level instead of only being able to activate the emergency alert. The worst case then becomes the radio silence after a collapse preceding the automatic activation of the PASS devices on the SCBA. This means that the member is trapped or injured so severely they cannot communicate and the PASS is the last resort to locate them.

FAST and RIT. Do you have a plan in place to provide assistance to trapped or injured firefighters? For those answering that you have a FAST or RIT, that's good, but are you sending your members in before these teams arrive? If so, what is your plan? The initial stages of an incident, when staffing is at the lowest, is when you need the most things done in the shortest amount of time, but this is not the time to overlook safety for the sake of expediency. The NFPA is regarded as the national expert in the field, and their standards have become the industry's best practices and the performance measures by which we are judged. In 2015, they wrote NFPA 1407, a standard dealing specifically with training rapid intervention crews (RICs; similar to FASTs or RITs) in order to help them meet the NFPA 1710 (*Standard for the Organization and Deployment of Fire Suppression Operations, Emergency Medical Operations, and Special Operations to the Public by Career Fire Departments*) requirements.

Remember that NFPA standards are nationally recognized recommendations. You should work toward these standards since OSHA likes to adopt recognized standards, so early adoption may also save you some work down the road. Regardless, don't you want to work from the highest standard available? Don't you want to do all that you can to protect your firefighters?

2 in, 2 out. OSHA's provisions for interior structural firefighting (which include the basics of 2 in, 2 out) are: "At least two employees enter the IDLH [immediately dangerous to life or health] atmosphere and remain in visual or voice contact with one another at all times; At least two employees are located outside the IDLH atmosphere; and All employees engaged in interior structural firefighting use SCBAs." There are also two notes related to these provisions: "One of the two individuals located outside the IDLH atmosphere may be assigned to an additional role,...so long as this individual is able to perform assistance or rescue activities without jeopardizing the safety or health of any firefighter working at the incident," and the second, which states: "Nothing in this section is meant to preclude firefighters from performing emergency rescue activities before an entire team has assembled" (OSHA, n.d.).

Other OSHA standards. The Public Employees Safety and Health Act of 1980 (PESH) gives OSHA the right to make regulations for numerous agencies, including fire departments. The fire brigades standard (OSHA standard number 1910.156) and the respiratory protection standard (OSHA standard number 1910.134, where 2 in, 2 out comes from) are some of the firefighting-related laws that OSHA enforces. The following information on firefighter safety protocols comes from OSHA section 1910.134(g)(3) (procedures for IDLH atmospheres):

> One employee or, when needed, more than one employee is located outside the IDLH atmosphere;
>
> Visual, voice, or signal line communication is maintained between the employee(s) in the IDLH atmosphere and the employee(s) located outside the IDLH atmosphere;
>
> The employee(s) located outside the IDLH atmosphere are trained and equipped to provide effective emergency rescue;
>
> The employer or designee is notified before the employee(s) located outside the IDLH atmosphere enter the IDLH atmosphere to provide emergency rescue....
>
> Employee(s) located outside the IDLH atmospheres are equipped with:
>
> > Pressure demand or other positive pressure SCBAs, or a pressure demand or other pos-

itive pressure supplied–air respirator with auxiliary SCBA; and either

> Appropriate retrieval equipment for removing the employee(s) who enter(s) these hazardous atmospheres where retrieval equipment would contribute to the rescue of the employee(s) and would not increase the overall risk resulting from entry; or

> Equivalent means for rescue where retrieval equipment is not required. (OSHA, n.d.)

The IC must ensure that they have adequate personnel on scene to meet these requirements *before* beginning interior operations.

Some departments have "determined that the Pump Operator [and truck chauffeurs] will jeopardize the safety of others if he or she abandons his or her duties [in response to a call for assistance], making the Pump Operator ineligible to be one of the standby persons [safety team members]" (Port Gibson Fire Department 2014). If pump operators and truck chauffeurs have been identified, who else should be excluded? The incident commander? Safety officers? Accountability officers?

Documentation

A short note is better than a long memory, but the IC should not be filling out ICS forms while trying to command a developing fire situation; this is just not practical. Don't overlook this task, as it will be essential in report writing because it's easier to prove what you did than what you didn't do, but this is a task to hand off to subordinates. A chief's aide or scribe can capture information and record times, notifications, and contact information from agency representatives on the scene. In growing incidents, these forms are a framework for providing situation reports to incoming units, information for liaisons and PIOs, and an additional level of accountability.

Gathering complete data is essential, particularly if there is ever any litigation. All notes, reports, and documents are public record and must be secured as such. Most states do not allow fire reports to ever be destroyed; they must be archived forever. Injury and exposure reports must be kept 30 years past the member's retirement date as part of their permanent record. Once a report is completed, you will likely have requests for copies from outside agencies or insurance companies. Have a policy in place for the release of this information. Do you provide the electronic report with all the coding from the National Fire Incident Reporting System (NFIRS), or do you provide some other simplified document? As we said, a fire record is a public document and subject to Freedom of Information Act (FOIA) requests. It is important that the person responsible for your records understands what they can release to whom and under what conditions.

Members' photos, social media, can also be subpoenaed as evidence. Have a policy informing members of their rights to privacy (and the limits of these rights) and HIPAA laws. Instant messaging chats, texts, and tweets where members shared information about an emergency have caused problems for departments.

None of this means you should command a fire as though the threat of a lawsuit constantly hangs over your head, but you need to know how to protect yourself and your members from the possibility. The hazards of the modern house fire do not end when the flames go out.

Fire Attack Strategies for House Fires

In this section, we focus on the four basic fire attack strategies. Firefighters sometimes get bogged down in grand, over-arching discussions of strategies, tactics, and the topics of the day, which can include anything from bailout bags to nozzle selection. Since this chapter precedes one on SAR, it is vitally important to take a short diversion in the interest of our safety and emphasize the importance of the first hoseline to any attack strategy and the safety of our members. Chapter 7 covers hoseline facts in more depth, but for now, let's look at some hardcore actions that must be taken on the fireground to support the SAR and briefly revisit the critical importance of the first hoseline at a house fire.

The first hoseline

We have a moral obligation to get the first line, the heart of our fire attack operation (and actually every line), stretched and into play, flowing decisive amounts of water quickly and effectively every time we pull up to a fire. Every time. Not sometimes, not on a good day, but every single time we're called to a fire.

At house fires, this is even more important. Why? We have intentionally put firefighters' lives—the SAR teams—in imminent danger! We have put these members in this position under a critical assumption: that one or more hoselines are going to quickly follow them or be in the right place to protect them from the fire.

In general, at house fires the first line protects life: the means of egress for occupants or firefighters. This usually means the line goes through the front door to protect the stairwell. As we saw in chapter 2, most homes have open and unenclosed stairwells, which lead to rapid vertical fire spread. For single-story homes where there are no stairs, getting the first line into the front door and into the hallway can cut the fire off from the sleeping areas and protect the egress route. For homes with more than one story, there are several tactical advantages to having the first line protect the stairs:

1. Protects the search crew trying to reach the upstairs bedrooms

2. Protects the means of egress for occupants who might be able to self-evacuate

3. Divides the fire area into smaller rooms of fire to eliminate the synergistic effect of fire controlling the entire first floor

Stretching, operating, and advancing an aggressive interior attack line is our most basic play. It is also our most important task. We are firefighters and this is what we get paid for in one way or another. More on the critical factors for this line in the next chapter, but for now, realize how important it is to get the proper lines stretched and operating, quickly, with minimum manpower, to apply decisive amounts of water on the fire. The lives of *our* firefighters depend on it.

The first line and the first priority for the engine company is to protect the firefighters who are conducting the search. The situation will dictate how best to accomplish this but never forget that this is the first priority for the engine company when a SAR operation is in progress. There are two fundamental rules regarding initial hoseline placement when a search is in progress or is necessary:

1. The first line goes between the fire and the occupants.

2. The first line protects the means of egress.

Embed these two rules in your head and draw on them at your next house fire. They will never fail you or your members.

The majority of your fires will probably not have a significant SAR component to them. The job of the engine company (and the entire operation) is much less complicated when the lifesaving issue is removed and the focus is on saving property. However, keep in mind that though the occupants who live in the house may be out, there are still lives in the building that need to be protected—*your* firefighters—or as we refer to them, the most recent occupants.

Additional hoselines

Additional lines mean additional manpower, which may be scarce on your first alarm, but if you are stretching one line inside the building, there must be immediate consideration for additional lines. Fire attack and hoseline positioning will be covered in greater detail in chapter 7.

There is no cookbook answer for the second and third hoselines. There are, however, a few considerations. The second line can back up the first if manpower permits. In concept, this is the safest bet because it increases the ability of the first line to accomplish its task. If the first line bursts, the first crew is injured, or anything else happens to make the first line less effective, the backup line can finish the suppression.

Another option for the second line is to attack the fire in coordination with the first. In a fully involved ranch house, one line goes left through the front door and one goes right, and just like that there is a two-pronged fire attack in progress. It is important to remember that handlines flowing in the 180 gpm range will control the fire quickly. This is a house, not a warehouse with tons of burning material that the stream cannot reach. Of course, if the fire has made it into the concealed spaces of the home there will be a lot of opening up to do to get water on the fire, so fire control will not be as rapid as we would like.

The second line could also go to the floor above the fire. Fire burns upward quickly, but if we can get ahead of the fire and cut it off, we generally end up with a positive outcome.

A general rule of thumb that has been reinforced repeatedly at house fires is that one line is good for about two rooms of fire. This takes into account the crew's fatigue after stretching and operating in one room then moving on to the second room, firefighter injuries, short stretches, and a myriad of other fireground realities. Since this is a generalization, it does not always hold true, but it is a good estimate when assessing manpower and equipment needs.

If manpower permits, a third line is a great idea. This line can either support the first two lines if necessary or go to an area that has not had water applied yet, such as concealed spaces that members are just opening up. Another reason to stretch a third line is for the training value. Sure, we all brag about how good we are at stretching, but when the hose meets the road, we may not be as good as we think. The third line, often standing fast at the front door, is usually used either because the first two are not suppressing the fire or it is needed during the overhaul process.

Overall Strategy

Retired Deputy Chief Anthony Avillo summarizes the overall primary factors that need to be considered for any rescue and fire attack with the acronym CRAVE: command, rescue, attack, ventilation, and extension prevention (2015, 258). It is important to note that this particular summarized version of priorities is what worked for Chief Avillo, in his part of the country with the resources he had to work with at the buildings he and his members typically fought fires in.

When we present, both at major fire conferences like FDIC or at the smallest rural firehouse training sessions, we say that there is only one best way to do any fireground operation: the way that works for you in your department with your members and your resources in your buildings. It is the same with command and control; what works at our department in New York City may not work in another fire department. Parts of successful command and control procedures, strategies, and tactics are extremely useful, transferrable, and applicable, but rarely can you or should you use someone else's entire system. The bottom line here is this: you have to research, think, train, and practice with any tool or system to see how it will work for you, then apply it to your situation and evaluate the outcomes. It is also critical to recall that these are general and broad priorities and critical tasks on the fireground. Important tasks, assumptions, and priorities (like firefighter safety) are built into many of these summaries.

RECEO VS

When in command on the fireground, there are a multitude of things to remember, observe, evaluate, and execute. Selection of a basic strategy for your operation will be dependent on a few very important factors. One of the best acronyms we have come across for developing an overarching strategic plan for house fires is RECEO VS. It is not new. We used this in the size-up chapter to guide you in your information collection, and it is equally useful for framing your strategic goals for the entire house fire operation. It does not cover every little detail but does provide an excellent overall framework for most house fire operations. We are not sure of the origins or whose idea it was originally. What we do know is that it is really effective.

There are a number of ways to display this acronym but the display in figure 5–2 seems to be the most practical. While ventalation is off to the side, ventilation operations, including door control, limiting ventilation, and ventilation coordination with fire suppression, support and are key to the success of all other tasks, missions, and objectives on the fireground, so they are often assumed. Ventilation is thoroughly covered in chapter 8.

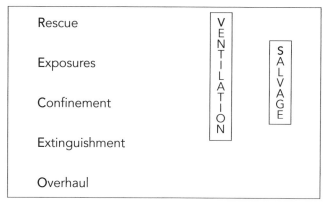

Fig. 5–2. RECEO VS

As you should know by now, there is no panacea for all your command problems. As Bill Herman, the first Rockland County (NY) fire coordinator often said, "You have to think to be a firefighter." These memory aids are very useful and practical on the fireground but can never replace your best tool: your brain.

Let's look at RECEO VS in the context of command, control, and strategies on the fireground at house fires.

Rescue: Are victims in need of immediate rescue or are there possibly others in the building?

Exposures: Is the fire spreading to other rooms, floors, or buildings?

Confinement: Can you confine the fire with the resources available?

Extinguishment: Can you extinguish the fire quickly, minimizing all your other problems?

Overhaul: Final step before you leave the scene to prevent rekindling and ensure that the building is safe. This is not an issue for strategy selection.

RECEO VS is good for helping officers prioritize resources on the scene of a working house fire. It tells them what to do and it is easy to remember. It also forces firefighters to consider their next tasks after they complete their initial size-up so they can anticipate and plan their next actions on the fireground.

If we use this acronym as a general guideline, tempered by firefighter safety, experience, and good judgment, it is a valuable tool for fireground strategy planning at house fires. Let's look at the steps in a bit more detail.

Rescue

This is the first priority for obvious reasons and will likely be the deciding factor in whether there should be an offensive operation. We usually cannot wait for the fire to be extinguished before rescuing the victims: if we put the fire out then conduct our SAR operation, it often results in a body recovery, not a rescue. However, there are instances where extinguishing or controlling the fire with a hoseline may be the best strategy for conducting or supporting rescue operations. As the late Andy Fredericks was fond of saying, "More lives are saved with properly placed hoselines than any other tactic."

Recall that during a SAR operation, all resources support the SAR as the situation dictates. For example, if you don't have enough of something (water, people, hoselines, etc.) to fully support the SAR effort *and* protect exposures, protecting exposures will have to wait: the strategic priority is rescue.

Exposures

Let's say there is a house fully involved and the houses on either side of it are starting to light up. If we direct our efforts on the original fire building we will lose the houses on either side and maybe the houses on either side of those! Logically, after the rescue priority is addressed as the situation dictates, the next priority is to keep the fire from getting any larger, involving new areas or buildings. The exposure can be an exterior exposure or it can be another area inside of the building (interior exposure). In the scenario at the beginning of this chapter, the attic is an exposure, as are the rooms adjacent to the fire room.

Confinement

After we have taken care of the SAR component and protected the remainder of the home or, on a larger scale, protected the neighborhood from a conflagration, confining the fire in preparation for extinguishment is the next step. Confining the fire to a room is essentially protecting the exposure (other rooms, floor above, or hallway) and extinguishing the fire outside of the room. In the scenario at the beginning of the chapter, confining the fire to the room of origin can be accomplished by placing a hoseline at the door of the fire room. If it was a garage fire, the fire could be confined with a hoseline at the interior door between the garage and the basement. It is important to keep this door closed because it is likely a rated fire door, probably the only one in the house! More on this later but for now, consider cutting the fire spread route off and extinguishing the fire with a second hoseline.

Tactically, it is important to be conscious of the reach of your hose stream. For example, a line can be operated with just the stream going down the hallway, especially in a home where the halls are generally short, and drive the fire back into the room, simultaneously allowing searching firefighters to advance and making their search and rescue unencumbered by the line and other firefighters in the same small hallway.

Extinguishment

In a perfect world we move seamlessly from exposure protection through confinement to extinguishment. We may not be staffed to conduct these operations simultaneously; life is not perfect, so our fire attack will also not be perfect. Generally speaking though, after rescues are complete, exposures are protected, and you have limited the fire spread by confining the fire, extinguishment is the phase where you have to aggressively gain access to areas where the fire has spread. In our scenario at the front of this chapter, judging by the smoke pouring from the soffits of this home, it is quite likely that the fire has extended into the attic. It is also possible that it has extended to the attic space on the exposure 2 side, the lower attic space over the living room. This needs to be determined in order to confine the fire and not cut off the means of egress. The heavy involvement of the fire room probably caused the sheetrock to fail, allowing the fire to extend into the attic. Extinguishing this fire will take more manpower, hoselines, and equipment.

Proper ventilation will make the difference between success and failure for this important phase of your operation. See chapter 8 for a thorough discussion of both the strategic and tactical importance of ventilation.

Overhaul

This is the last priority before we clear the scene. Certainly not part of our rescue strategy but a vital piece of our second mission: property conservation. Command concerns during overhaul are many but none supersede the safety of our members. There is a tendency to think that the hazards have been minimized because the fire is out and we are just overhauling, but command must consider the following:

- Fatigue of members—Have they been through multiple bottles of air, physically exhausting them? Is it time to pull them out and use fresh firefighters to do the labor-intensive task of overhaul?

- Emotional exhaustion—Members may be emotionally spent and not cognizant of common hazards, particularly if this fire involved single or multiple fatalities, children, or the serious injury or death of a fire service member.

- Building stability—If fire has weakened the building or structural hazards exist such as unsupported chimneys, walls, floors, or roofs, overhaul may not be possible or practical.

- Cause and origin determination—In many municipalities the task of determining cause and origin is the responsibility of the ranking fire department officer on scene. When the cause cannot be determined or all accidental causes are ruled out, it is then considered suspicious and possibly a crime scene that requires additional investigation.

- Crime scene—If you have called arson investigators to the scene or if command or police have determined that the incident requires an investigation, do not overhaul the scene any more than is necessary to stop the spread of fire. Valuable evidence may be destroyed during a total overhaul. We assist arson investigators by not performing unnecessary overhaul until they arrive and collect evidence or take documentary photos or videos.

The first part of overhaul is to be sure there is no hidden fire that will develop after we have left the scene. The second is salvage. Salvage is often overlooked, but we should teach firefighters to act as if it were their home. After a significant fire or even a one-room contents fire, the occupants' property has been significantly damaged and they may have lost large amounts of their personal property. If it was your house, you would want to save and protect as much as possible, so if manpower and the situation permits, start your salvage operation—saving what you can—as early on as possible, especially before the first hoseline starts operating. If the fire is on the second floor and you have enough manpower, see what you can save from destruction on the floor below. Are there family pictures on the walls or movable furniture? Can you get a salvage tarp over the furniture before it gets destroyed by water coming through the ceiling? Property conservation is our second mission, though we tend to forget it sometimes.

The final phase is to hand the building back over to its owner or occupants and advise them on appropriate next steps, or make the building safe based on the fire or structural conditions at the time. This of course may include preventing access to the building until modifications are made or the building is evaluated by fire and building inspectors to ensure safety.

Four Fire Attack Strategies

Aggressive or offensive interior fire attack is what firefighters live for. It is our ultimate challenge. It is what we train for and when we use this strategy, we are saving the most lives and property. But it is also the most dangerous strategy for us. It is important to remember that sometimes we cannot employ our favorite tool or strategy for a variety of reasons. Let's look at the four basic strategies and see how and when each should be applied to house fires.

Offensive

There is nothing like a good offensive fire attack. Forcing the door, conducting a rapid SAR operation, and possibly making a rescue. Or pushing a hoseline into a house, fighting your way down the hall, and directly attacking the fire, confining it to the room of origin. OK, that is the over-romanticized version of what we do powered by way too much gung-ho energy. Let's take a look at the actual details and what we must think about during or when considering offensive operations. This may be our favorite strategy, but like everything we do, an offensive fire attack operation at a house fire should only be applied under certain conditions:

1. There is something worth risking firefighters' lives to save. This may be occupants or the building itself or its contents when there is significant salvageable property and a manageable risk to firefighters. Although we don't trade firefighters' lives for civilian lives, we will take manageable risks to attempt to save a life.

   Conversely, a fully involved house has little value. Occupants cannot survive a fully involved structure and are therefore considered unsalvageable. Contents involved in fire are worthless, particularly compared to a firefighter's life or limb. The building itself may also be worthless because of the fire damage it sustains; for the owner or insurance company, it might be cheaper to demolish and rebuild new. Vacant homes should never result in firefighter injuries or deaths. You should consider the future of the building in terms of economic value and from a community planning perspective. If the home or building is old, in bad repair, or on the site of a future shopping center, the house has less value and we should risk less. How many times have you seen a new and much improved building rise from the ashes of a fire building you fought and risked so much to protect? It happens a lot and strictly for financial reasons. Which is cheaper—to renovate, repair, and modify to meet new codes, or to build new? It is almost always the cheaper and better option for the owner to build new.

2. The building is structurally safe for firefighters to enter. In chapter 3, you became familiar with the rapid collapse potential of lightweight construction. If no life is involved we must carefully weigh the risk versus the gain for fighting fire in a home where we are the only occupants. A fully involved wood-frame home is likely not safe for firefighters to enter to make an offensive fire attack, especially a home built of lightweight structural materials. Extinguish the fire from outside, enter carefully when visibility is good and maybe after you have had an opportunity to examine and evaluate the structural integrity of the building (Is it going to collapse under the load of a firefighter standing on the floor? Is it going to

collapse onto our firefighters?). Consider also that even if a collapse occurs after the fire is extinguished, members are not falling into a fatal situation in a burning basement.

3. There is a reasonable potential for or confirmation of salvageable human life. In the scenario at the head of this chapter, there is no salvageable human life in the fire room. However, there may be salvageable human life in other parts of the house; often a closed door is all that separates an occupant from life or death. In this case, an offensive strategy is correct and the following actions should proceed.

    a. SAR teams aggressively enter the home using the VEIS method and the two-team approach.

    b. Ventilation is accomplished ahead of the hoselines so as to not endanger the SAR crews.

    c. Hoselines are stretched to protect the means of egress and SAR teams (to protect them from flashover) and to the seat of the fire for rapid extinguishment.

Defensive

A defensive strategy is appropriate when you are defending the neighborhood from a well-involved to fully involved house fire. There is nothing left to save—no salvageable people, things, or even the fire building. A primary search will not be conducted because there is no salvageable human life in the building, it may be unsafe to enter, and water is being directed into the building from outside (fig. 5–3).

Fig. 5–3. This house was fully involved upon arrival, so a tower ladder was used as a defensive strategy.

A defensive operation should be considered when the house has conditions such as collapse or potential collapse, hazardous materials, uncontrolled fire, or severe wind or weather conditions blowing fire uncontrollably. It should also be considered after a high-risk operation is completed, rescues have been made or attempted, or the risk to occupants

has been cleared and decision made to decrease risk for firefighters.

This is a "surround and drown" operation. A defensive operation does not commit firefighters to the interior of the fire building and keeps them out of the collapse zone. Hoselines are applied to surrounding exposures to keep them from igniting from radiant or convected heat or, in extreme cases, flying brands.

Here are some conditions that should help you decide to use a defensive strategy:

1. The building is not safe for firefighters to enter. The building may not be safe for a variety of reasons, including full fire involvement, structural instability, or lightweight construction.

2. It is reasonably likely or confirmed that no one is in the building or there is no chance (as in fully involved fires) of salvageable human life or property remaining after the fire has been extinguished.

3. The building is essentially worthless, or at least the value is low enough that it is not worth risking firefighters' lives.

4. Defensive operations may be necessary if firefighters cannot approach the building due to wires down, chemicals or explosives involved, or other severe hazards.

At a fire where a defensive strategy has been chosen,

- SAR teams do not enter, but an exterior search should be conducted to see if victims jumped or are hidden by shrubbery;

- Ventilation is not necessary or possible, and the fire may have self-vented;

- Hoselines are stretched first to protect exposures and limit fire spread to the building of origin;

- Large-caliber streams are used from safe areas outside the building; and

- A collapse zone is established for firefighter safety.

Offensive-defensive

This is a strategy that we don't use enough. Once we commit firefighters to the building, we tend to want to keep them there until the entire operation is completed. It is also tough to pull overzealous but well-intentioned firefighters out of a building they think they can save (fig. 5–4).

Fig. 5–4. It is important to remove members from the building before starting a defensive operation.

The scenario at the beginning of this chapter is a great example of a situation where an offensive-defensive strategy would be a good idea. Clearly there is a need and justifiable risk for firefighters in an offensive mode inside the home to attempt to save this family. But there is also the risk of the truss roof collapsing and trapping or killing firefighters inside. Another hazard to firefighters in this scenario is fire from the well-developed attic fire extending down onto unsuspecting firefighters on the top floor or cutting off the escape route.

In this scenario, offensive operations should attempt to make the rescues while protecting the SAR team (and means of egress) with the hoseline and possibly extinguishing the fire. If the attic trusses become exposed to moderate-to-heavy fire or there are other signs of potential structural weakness, the offensive strategy needs to be terminated and changed to defensive strategy.

The scenario may go like this: The SAR teams complete searches and primaries show everyone is out of the building (this can be confirmed by rescued victims or neighbors). Interior crews have knocked the fire down in the bedrooms, but the fire has now taken control of the attic, weakening the trusses. Because of the potential for collapse, interior crews are ordered out of the building. The attic fire is knocked down either from portable ladders with hose streams operated through the gable vents or an aerial device that has been employed to control the attic fire in a defensive manner.

This strategy is good as it allows an appropriate risk to firefighters when there is an appropriate gain or reward, and minimizes risk to firefighters when the reward potential changes to low or nonexistent.

The fire in figures 5–5 and 5–6 started off as an offensive operation. As fire spread into the walls of this large balloon-frame home and took control of the attic, it became clear that it was going to transition to a defensive operation. Tower ladders will soon start pouring thousands of gallons per minute onto a building that is already structurally compromised by fire. Firefighters on the porch roof endanger themselves for little gain. When the operation switches to a defensive operation, pull out all firefighters from the building to prevent them from being injured or killed by collapse, misdirected master streams, or other hazards. There is nothing left to save at this building and firefighters should risk nothing.

Fig. 5–5. Offensive-defensive operation in a large balloon-frame home

Fig. 5–6. The offensive fire attack operation has now gone to a defensive or exterior mode.

Defensive-offensive

After the determination has been made to commit to a defensive operation, there is generally little left worth risking firefighters for. This combination is not often used at house fires, but as with everything we do, there are several exceptions.

The most common instance is when a special hazard delays access by firefighters to the house. Downed electric lines or hazardous conditions such as flooding, vicious dogs, or even an armed perpetrator prevent you from attacking the fire, and lines should be positioned in defensive positions to protect exposures as the situation allows. When the hazard is cleared, an offensive operation can be initiated if practical and safe.

Another notable exception is when an offensive attack is warranted by the fire condition or other circumstances, but you cannot get into the home because of high-security locks, doors, or bars on the windows.

SUMMARY OF ATTACK STRATEGIES

Offensive: This is the aggressive interior attack mode, performed at the fires firefighters train for. Firefighters will force entry, advance hoselines, and meet the fire face-to-face to do battle with the red devil. Other firefighters will be searching for victims, making rescues, and venting to support the hoseline advance in a coordinated operation. All of the roles are equally important and support a successful outcome. Purely offensive operations provide the greatest opportunity for successful outcomes.

Defensive: This may be the most frustrating for firefighters. These operations happen when units arrive after the fire has already developed to the point where no lives or property can be saved; we are now focused on protecting the neighborhood. The focus of this operation is to prevent the fire from extending to adjacent structures, contain it, and then extinguish it. This tactic is used for advanced fires where the risk does not match the reward; there is nothing left to save in the fire building or the structure itself. This is usually an extended operation that is labor intensive. Not-so-affectionately referred to as surround and drown.

Offensive-Defensive: This strategy is used when an attack is necessary but there are limits to safety. The fire situation is tenuous and if knockdown is not accomplished quickly, forces must be withdrawn. At house fires, this strategy means that hoselines are stretched to support a rapid search and remain in place until the search is complete or the fire situation overwhelms the available resources, at which point the hoselines and firefighters are pulled back and the operation becomes defensive. The key to success in these operations is interior forces and commanders knowing that the time is limited and that when the time comes, they must back out.

Defensive-Offensive: This strategy is deployed when several priorities need to be addressed simultaneously and sufficient staffing, equipment, or water is not available to satisfy all the priorities. The aggressive fire attack operation must be put on hold. This may require that a line be stretched to contain the fire to a room or protect one or more exposures and not extinguish it while salvageable victims are being removed. Containment may be to a room, a floor, or an area of the house, garage, or even the house of origin if no salvageable victims remain. When victims are out of harm's way or additional resources arrive on scene, then the firefight can transition to an offensive mode.

It can be argued that using a hoseline for containment to support SAR is an offensive tactic, because you are controlling the fire. We feel that it is more of a defensive maneuver because you are protecting the SAR crew and not advancing on the fire. What is important is that the mission of the hose team be clearly understood and coordinated with other fireground activities.

Scenario Discussion

Situation: It is 0530 hours, Tuesday morning, 3 July. A hysterical neighbor tells you that she just saw the husband go to work about an hour ago, but that the wife and kids are home.

Analysis of Info: This sounds like a reliable report. It looks like the house is well kept and would probably be occupied based on time of day and this report. On 3 July the kids would not be in school, though they could be away, as could the family except for Dad, who was working or perhaps leaving to join them on vacation. A good idea is to verify that what she is telling you is true. Ask her how she knows. Maybe she babysits for the family, is a relative, or is very close to the family and knows their whereabouts.

Resource report: Your first engine has a good hydrant and has laid a supply line into the front of the building, leaving room for the truck that pulled up immediately behind. Your inside team has forced the front door and is making progress down the hallway toward the bedrooms.

Radio report: "Inside team to command, the interior door to the fire room, A-D corner of the building is just starting to show fire."

"Outside team to command, no fire showing out the rear. We have made entry into the rear bedroom off a 24 ft extension ladder and found one child and are bringing her down the ladder now. Where is EMS?"

Questions

1. What is the top priority at this fire at this point?

   The safety of firefighters is always a top priority. Rescuing occupants (if there are salvageable human lives inside) is the operational priority. It is important to attempt to confirm the life hazard (how many people could be in the house). In this case (based on time of day), it is likely that any occupants would be in the bedrooms. Your outside team has already made one rescue, and the interior team will probably make one or more successful searches and rescues.

2. Is the building safe for members to operate in?

   Your building information system (preincident intelligence) shows that the building has a truss roof system with stick-built floors and walls. This information should guide your strategy. The windows have completely burned out of the fire room, indicating that the fire has been on those windows for some time. The soffit is probably covered with a plastic vent covering or at best ¼ in. plywood that likely has burned through. This will allow fire to rapidly communicate to the attic.

   Once in the attic, fire will rapidly burn up the underside of the roof sheathing that is being preheated as more fire comes through and lights up the sheathing near the soffit. In this case, it looks like fire is already in the attic (collapse potential) because of the smoke showing so heavily at the soffit vents. Consider the fuel load in the attic from the homeowners' stuff—if the attic is a truss system, the web members of the truss may limit the storage and that could be a good thing. If not, can the truss carry the weight of that load on the bottom cord? If it is conventionally framed, has the attic floor been covered with a layer of plywood to facilitate storage? This can cause a heavy fire to develop in the attic and create challenges in applying water.

3. What search operations can firefighters perform from the outside if necessary?

   If this is going to be a defensive operation, firefighters can gain access to a second floor window via a porch roof or ground ladder and search just inside the exterior wall, under the window visually, or with a pike pole or tool. Another good method is to get to a door and look under the smoke to see if you can find a victim who was trying to get out but did not make it all the way.

4. Is this fire going to an offensive or defensive operation?

   Right now it looks like it will be an offensive operation for a short time—until the primary search is completed—then go to a defensive operation because of the truss roof collapse potential and the probable fire in the attic. The time to switch from offensive to defensive is a critical judgment call but life safety of occupants and firefighters will be key elements in your decision. The risk to interior firefighters is the truss roof collapsing and trapping them (or rapid fire extension downward from the attic); the gain is multiple rescues.

   One option is to risk firefighters with an interior attack and attempt a rapid extinguishment of the fire to support the SAR teams. The one-room fire will knock down quickly. Depending on how much fire is in the attic, it may take time to pull the ceilings, work around stuff stored in the attic, and extinguish that fire. You will need a second line and more manpower. The first-due crews are spent and in need of rehab by now because of the intensity of the SAR operation.

   The second option is to conduct the offensive primary search, supported by a hoseline, assuming you have enough manpower to do this. The line will protect the means of egress and if luck is with you this morning (and your size-up was correct and your members well trained to deliver decisive amounts of water), that line will extinguish the bedroom fire quickly. In the interest of firefighter safety, pull firefighters out of the building before the truss roof comes down, trapping them and raining burning debris on them. If you have rescued all occupants of the building, the only "occupants" left are those wearing fire gear. If the attic is a full truss, there may be little or no storage in the attic, which means that the attic fire could be knocked down with hoselines operated off ground ladders in the gable vents of the house from the outside.

   Another consideration would be to knock the fire down rapidly with an exterior line. The disadvantage to this is that smoke, heat, and steam may be driven onto search crews if they are already inside. This tactic may reduce the fire hazard to interior members (assuming they are held outside until the stream is shut down) but is not a preferred tactic, especially if you are working under the premise that this is an aggressive interior operation. Recent research has shown that a line operated in a window from outside will not drive heat and smoke onto occupants, will actually increase the chances of survival of occupants, reduce the temperature, and delay or kill the flashover. (See ch. 10 for a complete discussion.)

5. How do you want your first hose to protect the members inside?

It must protect the means of egress as a first priority. In this case, the hallway from and between the bedrooms is that point and will protect the interior teams. If there are no surprises, this line should be able to extinguish the fire in the room of origin.

6. Why would you consider going offensive at this fire?

There appears to be a high life hazard with salvageable victims and there is a reasonable chance that we can quickly knock the fire down. The hoseline can protect the means of egress. The risk is reasonable and manageable to firefighters.

a. What other strategy could you employ and why? You could consider going strictly offensive. This option does not consider the danger from potential collapse of the truss roof system.

b. What would happen if you went defensive then offensive?
Sadly there are several examples of this caught on video. One famous video shows a department extinguishing a bedroom fire in a split ranch home from the outside using a hose stream through the window. It was a clear day, plenty of manpower, a reliable water supply, all the necessary ingredients for an aggressive interior attack. While this line is pouring water through the window, another fully manned and charged hoseline takes root on the front lawn. After the fire is extinguished, the chief enters the home through the front door and retrieves a lifeless older man. Resuscitation efforts on the lawn were too little, too late.

7. How does the first hoseline get into place and where does it go?

If you have determined that this will be an offensive fire attack, the first line goes through the front door to protect the search teams and possibly attack the seat of the fire. The first line goes between the victims and the fire or to protect the means of egress. Remember, your hose stream will control the fire quickly if there are no special hazards in the house. Protecting the means of egress is a priority to ensure the safety of your members.

Additional Practice Scenario

Fig. 5–7. Scenario

Friday, 4 January, 1323 hours. Dispatch reports a fully involved house fire at 223 Betio Street. PD confirms fully involved.

Size-up: The initial dispatch should have you thinking about defensive operations. Although not always correct, dispatch info was verified by the PD. Upon arrival, your size-up reveals a fully involved house fire with no chance for salvageable human life. Heavy fire is showing out all the windows and doors.

Rear view confirms your decision to go defensive and minimize the risk to your members (fig. 5–8). Fire is out all the rear windows. There is little to save here; no human life and no property. Clearly, nothing worth risking firefighters' lives for. A quick look around the back shows no victims at windows or signs that they have jumped. No one is on the rear porch. Fire has complete control of the building.

Fig. 5–8. Rear view

On-scene report: "I'm on the scene at 223 Betio Street, at a fully involved two-and-a-half-story wood frame, 30 × 30 ft. Exposure problems on both B and D sides. This will be a defensive operation."

Discussion: Based on your observation of this fire, you could argue that there might be salvageable human life in the garage and that might be true. This is a modern house with parallel cord trusses supporting both the first and second floors. It is unclear how much fire is in the basement or the garage. You may choose to have members open the garage door and do a visual search from the outside by looking under the smoke. Crossing the threshold of the garage door puts members at significant risk of being caught in the collapse of the first floor. If additional reliable information is available, a quick search of the garage can be made, but always keep in mind the fire load that was, is, or will be weakening the truss connections, resulting in collapse.

References

Avillo, Anthony. 2015. *Fireground Strategies*. 3rd ed. Tulsa: PennWell Corporation.

Occupational Safety and Health Administration. "N.d. 1910: Occupational Safety and Health Standards, subpart I: Personal Protective Equipment, standard no. 1910.134: Respiratory Protection." https://www.osha.gov/pls/oshaweb/owadisp.show_document?p_table=STANDARDS&p_id=12716#1910.134(g)(4).

Port Gibson Fire Department. 2014. "The Two In, Two Out Rule." Last modified March 23. http://www.portgibsonfd.com/uploads/1/9/9/7/19973347/two_in_two_out.pdf.

6

Search and Rescue Operations

Summary

This chapter provides both strategic and tactical considerations for search-and-rescue (SAR) operations at house fires. Considerations, plans, and techniques provided in this chapter must be applied with consideration given to the specific conditions, personnel, types of apparatus, types of homes, and response times. The specific conditions in which you work will dictate which of the specific strategies and techniques are useful and applicable to you. What works in the big city may not work in your situation. Search information is provided for you as skills and techniques to apply based on conditions you face.

Introduction

Firefighters are taught from day one of recruit school that life is always our first priority. It is the same for house fires: saving occupant's lives while minimizing risks to firefighters is our first priority. The best way to achieve this goal is to provide well-researched, well-planned, and well-orchestrated SAR, ventilation, and fire attack operations. This chapter will focus on specific techniques, strategies, and tactics for SAR operations at house fires.

For the purposes of this chapter, search and rescue is being dealt with separately from size-up, ventilation, and fire attack operations. Obviously they all work cooperatively together for the same goal but are difficult to examine in detail together as they are all complex, multifaceted, and dangerous operations. Scenarios will utilize ventilation, search and rescue, and fire attack as mutually supporting and cohesively applied in chapters 10–12.

Scenario

Fig. 6–1. Scenario

Situation: It is Wednesday morning, 0223 hours, and you are tapped out to a fire at 2 Trophy Point Drive. Caller reported seeing the house across the street on fire. There is no additional information. Upon your arrival, the police officer reports that she heard a dog barking inside.

Resource Report: You are on the scene and your first-due piece of apparatus is approximately 2 minutes from the scene. There is a good hydrant three houses away from the fire building.

Radio Report: Dispatcher reports that the 911 caller from across the street said there are four kids in the family and they are all home.

360-survey has no additional fire showing in rear or in rear windows, no victims visible, nothing unusual, back door closed, all windows closed.

Questions

Based on figure 6–1, answer the following questions and use this scenario as the basis for the topics examined in this chapter.

1. To start the SAR evaluation and planning process, you need to obtain the following information and complete the following tasks:
   a. Construction type
   b. Occupied or vacant?
   c. What is the floor plan of the house?
   d. Where are the bedrooms? Stairs?

2. What is your occupant survival size-up?
   a. Salvageable life—Is there any?
   b. Where do you think the victims are most likely located?
   c. What is the condition of the means of egress?

3. What is your size-up of the fire situation?
   a. One room?
   b. Multiple rooms?
   c. Nearest exposure?
   d. Fully involved?
   e. Will this be an offensive or defensive attack?

4. Give your on-scene report.

5. Deploy your first assignment of both personnel and apparatus. Be specific with tasks and assign the number of firefighters you think those tasks will require.
   a. Begin a search?
   b. Stretch a hoseline?
   c. Personnel for both?
   d. Ventilation?

6. List additional resources you will need in the next 30 minutes.

Protecting Life

Saving lives on the fireground is our absolute first priority. It is often referred to as search or rescue or search and rescue. It may be much more appropriate and more modern to describe it as "protecting life." If we think about search and rescue in those terms, firefighter safety—protecting lives, ours included—is the prerequisite of any fireground action or priority. The house may not be occupied before our arrival, but it certainly is a few moments after the air brakes go on if we are too aggressive and try to go places we should not be. We need to have just as much concern for our own safety as we have for the safety of the occupants. We need to always weigh the risks versus the potential gains. As we go through this chapter and the SAR tactics and strategies, tools and techniques, and alternatives available to you, firefighter safety is an assumed and unalterable priority for everyone on the fireground from the newest firefighter on their first day to the chief officer planning their retirement.

Safe search and rescue can be accomplished a number of ways:

- Not committing firefighters to the building before the fire is suppressed
- Suppressing the fire concurrent with search team's entry
- Taking manageable risks to attempt to save the lives of trapped civilians

Search and rescue often conjures up, in the minds of inexperienced firefighters, dashing through the flames and snatching a person (why do we always think of a baby?) from the flames of hell. To firefighters who have been trained and have executed search operations, it may mean going left or right in a room during the most emotionally and physically draining and dangerous operation they may ever do in their career. With the knowledge from this chapter, search will take on a whole new meaning as you begin to understand the complexities and options available to you. These options and decisions will be much more than a left or right search pattern.

To company and chief officers, search requires gathering critical size-up information and making decisions on just how much or how little risk to take to save a person trapped by fire. These are hard life-and-death decisions based on the best information available at the time. Decisions on whether to search or not, how best to do it, and how much priority to place on search are judgments that must be made quickly, often with incomplete information. They are, however, always made with firefighter safety in mind as a first priority.

General considerations

There is no fireground operation more important than search and rescue (SAR) at a house fire. SAR is such a priority mission that we will risk firefighters' lives to accomplish it. Note that we said *risk* firefighters lives, meaning manageable, well-thought-out risks. We will not *trade* firefighter lives for civilian lives. We should take manageable risks when there is the possibility of salvageable human life inside the fire building.

From Firefighter I through your last day as a chief officer, saving and protecting human life is your top priority. It does no one any good to save the building (put the fire out) then pull out the occupants, now very dead victims; that is a body recovery, not a successful SAR operation. So for the purposes of this chapter we will assume that the search-and-rescue operation will be the first priority (after size-up, locating the fire, and considering fire suppression before, during, or after the search starts) and will be done before the first lines are stretched and operated or at least concurrently with the fire suppression operation. As in all things, there are always exceptions, contradictions, and modifications, and fire conditions may prevent a search from commencing or knockdown of the fire may be the best way to protect the victims. The old adage "A well-placed hoseline saves more lives than any search operation" is true, now more than ever. Water on the fire, quickly and in decisive amounts is a great tactic to protect life, ours included. The guiding principle for search must be: all buildings must be searched within the limits of firefighter safety, tempered with the probability of the presence of viable victims inside.

Integral to the search-and-rescue equation and fireground decision-making is the recent research regarding fire development in residential buildings. Chapter 8 contains a thorough discussion of this vital research. In summary and related to search operations (in broad general terms), live burn testing showed that for single-family-home fires we can expect flashover between 100 and 200 seconds after the search team forces the front door if the fire is not suppressed or extinguished (Kerber et al. 2012, 4). Obviously, protecting the search team(s) by putting water on the fire is of paramount importance.

Generally, if you have arrived on a first-due apparatus to a working house fire, the mission of the first SAR crew is actually threefold: first to search for fire victims, second to rescue those victims, and third to specifically locate the fire and radio the location and size of the fire to command. Note that finding the fire is not a specific mission but something to be done while conducting the search operation, when you locate the fire. If the first search team proceeds directly to the proximity of the fire, they are actually searching for those in the most danger first, those closest to or adjacent to the main body of fire. From here, the team can move to other areas likely to contain victims.

Removal of victims is a very physical operation. If you find victim(s), you will likely need help removing them from the fire building. Consider an average person (175–250 lb of dead weight) and how hard they will be to move. If the victims are burned, their skin will be very slippery, not allowing you to get a good grip. They may have their clothes burned off or be scantily clad in sleepwear or less. Less clothing means fewer good handles for moving the dead weight of a victim. When you find an unconscious victim, call for help; you will need it.

Search

Most of us learned SAR during our recruit training with only the basic standard search process taught and practiced. Often, there was little thought toward overall firefighter safety or victim survivability. The mentality was, "Get your facepiece on quick, get in there quick, make a big decision (go left or right), and conduct the primary search as quickly and thoroughly as possible."

During training, we are also taught to draw a mental picture of the room as we search it in zero-visibility conditions, remembering all the turns, furniture, and other landmarks along the way to ensure a safe egress route from the home if things go bad. Sounds great in theory, but in reality, there is so much going through your head during an intense search operation that there is little mental ability and memory space left to really accomplish this. The firefighter's mind is occupied with a number of critical issues to determine and monitor during the search: What is the fire condition? What is the structural stability of the home? Is the hoseline really attacking the fire? What kind of vent ops are in progress? Where do I think the victims are? Where is the fire? What is my radio telling me? How intense is the heat? Will I be caught in a flashover? What room am I in? How is my air supply? What is the best way out? Are there windows nearby? Did firefighters ladder them? And the most important of all: Can I get myself, my partner, and a victim out in time?

Searching safer

John "Skip" Coleman, retired assistant chief of the Toledo Fire Department (OH), defines four types of search operations (2011, 8–9):

1. Standard search—The one you probably learned as a recruit.

2. Team search—Used in large buildings; uses a rope for orientation and to ensure we can find the exit.

3. Vent, enter, and search (VES)—"The original vent/enter/search is defined as an emergency means of searching second-floor bedrooms off a porch roof for probable victims, using a roof ladder for access

to the second floor from the exterior. The original method is extremely dangerous. However, when two firefighters use this method combined with the oriented method of search, it becomes a very effective and safe form of search."

4. Oriented search—Divides responsibility into two specific areas: The officer is responsible for the safety of the crew and developing a search plan. The crew's responsibility is to search effectively.

Oriented search. For a firefighter to remain oriented (know where they are, be oriented to the area of refuge or escape route, and monitor conditions, radio transmissions, etc.) while conducting a search is nearly impossible. The oriented search method divides critical tasks and uses the officer to ensure these critical tasks are accomplished for his search crew. Consider just how important it is to know the best way out, after you find a victim. According to Chief Coleman,

> The oriented method of search is defined as using an officer or "oriented position," and one or more searchers conducting a one-firefighter search evolution. The key to this method is that the officer is not actually searching—that is, not sweeping under beds or in corners or closets. The oriented position has a specific list of duties to focus on while the searchers are conducting one-firefighter search patterns. This allows focus to be split into two separate and distinct places: the safety of the crew and the actual search. (Coleman 2011, 46)

The advantages of the oriented search method are:

- **"The safety of the crew is maintained in all times"** (Coleman 2011, 46). The primary duty for firefighter safety is with the officer, who monitors three critical factors: fire conditions in the search area, structural stability, and advancement of fire toward the search crew. Fire advancement on the search crew is particularly important at house fires since a fire downstairs can quickly move up the open and unenclosed interior stairs. Hot, flammable gases can cause a rapid flashover, trapping searching firefighters.

- **"Searchers are allowed to focus on conducting search patterns in their assigned areas"** (47). Searchers can focus all their mental and physical energy on search because they know someone is monitoring key safety issues for them.

- **"Searches are conducted faster"** (47). This conclusion is based on several fire department studies.

- **"Search continuity is maintained"** (47). The oriented officer keeps track of what areas have been searched and which areas have not and can pass that information on to the oncoming crew, especially if a victim has been found and removed by the first SAR crew.

Clearly, the oriented search seems to be the most effective search method. Presented here were just the highlights of the oriented search plan. For a complete background and thorough explanation, refer to Chief Coleman's book, which we believe should be required reading for all firefighters and fire officers, current and future.

DEFINITIONS

Primary Search: A rapid search, early in the operation, of a house fire focused on areas that are normal areas of occupancy or egress for the occupants. This includes bedrooms, hallways, stairs, and other typically occupied living areas. Primary search teams should penetrate to the seat of the fire where the victims are in the most immediate danger. Simultaneously, this unit can find the seat of the fire, relaying this vital information to the IC and the engine company. The second search unit should search areas adjoining and above the most exposed areas.

Secondary Searches: Slower, more methodical searches, generally after the fire has been controlled or extreme danger to firefighters has been lessened. This is a very thorough search of all areas of the home, both typically occupied areas and nontypical areas such as attics and basements. Debris piles may need to be moved to search and ensure a victim was not covered with drapes, sheetrock, or other rubble. This search guarantees that no victims are in the home. It is a good practice to conduct the secondary with a company that did not participate in the primary to be sure this search is accomplished thoroughly with a fresh set of eyes.

Critical considerations

To provide a basis for this all-important fireground action, we need to look at a few critical considerations. These are strategic-level issues that will impact whether you search, how you search, and how effective your operation will be. We will get much more specific later on in the chapter but you must first understand the overarching concepts that affect every search-and-rescue plan or operation.

Primary search always? Many departments conduct a primary search at all fires. There is a changing trend across the nation to abandon this mindset to improve firefighter safety. Incidents like the six firefighters killed searching an abandoned warehouse in Worcester (MA) helped bring about this change. The "we always do a primary search, no matter what" mentality has its roots in too much macho firefighter spirit untempered by rational thinking. It is easy having one simple rule rather than needing to think, evaluate, and make judgments at dynamic fire scenes. A way to view this is, as previously mentioned, to recognize that firefighters are occupants too. If the IC puts four firefighters in for a search in dangerous conditions, we know four lives are 100% in harm's way.

The traditional, immediate primary search strategy is often begun before the hoselines are in place: obviously a very dangerous operation. The rationale for the risk to firefighters is to provide a high degree of certainty that there is no civilian life hazard. Another reason to conduct a primary search is that occupants, in the panic of a fire in their home, may forget that someone is still inside; in other words, everyone may not have escaped. Although this sounds like a noncredible statement, there are cases where people have had guests staying with them and not remembered they were inside. This exact scenario occurred at a house fire in Rockland County (NY) in 2007 and resulted in the death of a 15-year-old house guest.

Search operations conducted after the fire has been controlled or extinguished are called secondary searches. Secondaries often reveal burned bodies, which are hard to distinguish from other fire debris. The purpose of a secondary search is to determine with 100% certainty that there are no human lives unaccounted for in or around the building. Check shrubs, porches, garages, and other surrounding areas to be sure there are no victims there. Secondaries inside the house must examine any and all areas where a victim might be. Some of these locations include under beds, under cabinets, in drawers of furniture, in closets, bathrooms, beside beds, and under piles of debris or clothing and toys. After your secondary, you should know with certainty you have accounted for all the victims.

Rescue profiling. Firefighter Michael Bricault from the Albuquerque FD (NM) breaks down the possibility of the home being occupied into what he calls "rescue profiles":

> A Rescue Profile is a method of categorizing the way firefighters look at a particular fire with regard to the potential or known life hazards and subsequent rescue actions that are required. A Rescue Profile in the scene size-up provides in common language an indication to all responding units the status of the life hazard and the actions being taken by companies operating on-scene.

There are four levels or categories to a Rescue Profile for a residential fire:

> Low rescue profile
>
> Moderate rescue profile
>
> High rescue profile
>
> Urgent rescue profile

Notice that there is not a " No Rescue Profile" category. Why? Because there is no such thing as an empty building until we, the fire department, have searched and determined the occupancy to be "All Clear." At the very least, a residential occupancy cannot be considered devoid of human life until firefighters have completed a primary search. (Bricault 2006)

Bricault goes on to define the four profiles. Low means "you suspect or believe the occupancy to be empty; it appears to be an abandoned house or apartment building that no one should be living in." Moderate is when "you are unsure if people currently are inside the structure," but it obviously is an occupied building." High is when there are clear signs that the building is occupied and "you have every reason to believe that someone is still inside" (Bricault 2006). Urgent is when people are hanging out of windows or you have a very reliable report someone is inside.

The rescue profile for figure 6–1 is high for the following reasons:

1. The house appears to be occupied and the time of day (0232 hours) indicates that any occupants would likely be sleeping.

2. Holiday—in this case, Halloween—decorations are carefully and intentionally arranged outside the home.

3. The property looks well kept.

4. Trash cans outside the home give an indication of recent use.

Clearly, there is no one-size-fits-all answer. Different scenarios require different strategies and tactics. However, there are three constant tactical and strategic considerations for search-and-rescue operations: 1. Only conduct SARs for salvageable human lives, 2. Firefighters are occupants too, and 3. Everything on the fireground must support the SAR.

Salvageable human life. Command must determine (or make an assumption based on available facts and size-up information) whether there is or is not salvageable human life inside. A fully involved house most likely contains no salvageable human life; thus, firefighters' lives should not be put at risk in an SAR operation. Smoke or fire showing out one window may mean the fire is contained to a room and lives in other parts of the house may be salvageable; this represents a worthwhile and manageable risk to firefighters. Obviously, if the size-up indicates the possibility for salvageable human life, the SAR operation must start as soon as possible and in accordance with your department's SOGs.

Determining if there is salvageable human life inside a burning home can be a difficult task. Ideally, we will err on the side of the possible victims while at the same time exercising care not to endanger firefighters unnecessarily.

An article in *Fire Engineering* by Captain Stephen Marsar, FDNY, makes a strong case that fire victims may not be able to survive as long as we think, meaning that we are risking firefighters to rescue corpses:

> Survivability profiling asks—if people are suspected or known to be trapped—is there a reasonable assumption that they may still be alive? If not, we should slow down and attack the fire first and complete the searches when it is relatively safe for our operating forces to do so. Some will argue that using survivability profiling will kill people. No, fires and smoke kill people (many times before we even arrive on the scene). Survivability profiling will save firefighters' lives. (Marsar 2010)

Marsar cites several medical and statistical studies of fire victim survivability that describe the effects of exposure to extreme heat; inhalation of carbon monoxide, cyanide, and other products of combustion; and hypoxia on the fire victim. The sum is death in a very short time, so the key question before beginning SAR operations must be, Can we execute the SAR in time or are we risking firefighters for nothing? How long have occupants been exposed to dangerous conditions of toxic gases and heat? How intense were these conditions?

The best we can do is conduct a thorough size-up and make strategic decisions based on the fire conditions, all on a case-by-case, scene-by-scene basis. The outcome—successful rescue versus body removal and injured firefighters—will be the ultimate evaluation of the quality of these judgments. This is the "one-third experience" portion that depends on the "one-third art" portion of firefighting. As a very wise person once said, "Experience is the best teacher, but it is also the most expensive."

Firefighters are occupants. Consider this: You think there *might* be someone inside, so you send in a team of two to four firefighters for an SAR. Now there are *definitely* occupants in the building. You have increased the life hazard and for what gain? That is the ever-present question, especially for house fires.

Support for the search and rescue members. Everything on the fireground must support the SAR operation. Forcible entry teams must quickly open up normal exit ways to allow swift entry for the search and hose teams. Most people try to exit their home the "normal" way and are often felled by smoke and toxic gases along the way.

Hoselines must be positioned to either quickly control the fire or protect the means of egress of both occupants and firefighters. Ventilation must be done appropriately to enhance the SAR operation and possibly the fire attack. (Chapter 8 contains a thorough discussion of ventilation considerations.)

It is the incident commander's choice how to deploy their (likely limited) personnel. All the first-due staff will be assigned to SAR if, for example, people are at the windows in extreme danger. First-due personnel may be immediately consumed with placing ladders to the outside of the building. In other situations, it may be more effective to dedicate some personnel to an aggressive interior fire attack, protecting the means of interior egress or containing the fire and thereby limiting all other problems. This action can enhance the SAR operation.

The fire attack operation must in no way slow down, impede, or stop the SAR operation, unless your plan is to suppress the fire first. The ventilation must not further endanger victims or firefighters and the forcible entry effort must be swift to allow truck company members into the home to start the search. If the outside ventilation firefighter takes windows before the line is in position and the rooms light up on the search team, the SAR is over because the search team is itself in survival mode. Another common example of fire suppression impeding the SAR operation is the first hoselines (using fog streams) pushing smoke, steam, or superheated gases onto the SAR teams, making their job harder and making them slower and less effective. No other operations can be allowed to hamper the SAR operation. This is easy to say and hard to do on the fireground due to the complexities of each incident, but it should always be a goal of the IC.

Nozzle personnel must also be aware of the potential of their line to push smoke, steam, and heat onto the search teams. We are talking about temperatures of 2,000°F near the ceiling for a simple living room or contents fire. Conditions become untenable for firefighters at around 300°F when the air is moist, and our skin starts to blister when it reaches 124°F. Pushing smoke, heat, and steam onto your SAR teams is essentially killing them with friendly fire. Fog streams are especially notorious for moving products of combustion due to the huge amount of air entrained in the stream. (See ch. 7 for details on superheated air movement caused by hose streams.) Recent research by Steve Kerber and his Underwriters Laboratories teams has shown pushing fire itself is very unlikely unless a fog stream is employed (Kerber 2010, 302).

It is often an effective strategy to knock down the fire with a stream from the outside to allow entry for search or to slow the fire extension (fig. 6–2) (Kerber 2010, 4). This is especially effective if there is a delay in forcing the doors or windows, as security bars or highly secured doors may drastically slow the search operation. Letting the fire burn unchecked during this time is often not practical. Directing hose streams in windows from the outside is certainly not the first choice, but in some cases it may be the only choice. Straight or solid streams entrain very little air and can be used to suppress the fire without pushing huge amounts of steam and smoke into the home and onto search crews. An outside stream may also be practical to knock down fire extension, such as fire on vinyl siding that is extending into a window or into soffits covered only with more vinyl or thin plywood.

Fig. 6–2. This house was fully involved on arrival. The fire was knocked down with large-caliber streams before entry was made because it was determined that any occupants in the house would not have survived these heavy smoke and fire conditions. Knocking down the fire minimizes dangers to firefighters when they do finally enter the home.

Is the building safe to enter? This must be determined as a key component of the SAR size-up and subsequent strategy. As we saw in chapter 2, lightweight and modular construction present almost immediate collapse hazards. We must have these building hazards identified on preplans to protect our members. As an important reminder, NIST experimental fire test data showed complete collapse of lightweight floor joists (I-joists and trusses) in 6 minutes!

Consider the common and uncommon hazards identified in chapter 3. Always be on the lookout for both classes of threats to members when planning, developing, and conducting an SAR operation.

Platform-frame homes without lightweight components and older stick-built homes will resist the attack of fire generally long enough for at least a quick interior operation. Balloon frame homes present the hazard of rapid fire spread throughout the building. As we saw in the case histories in chapter 3, lightweight construction will be on the fringe of failure upon our arrival. These buildings will continue to kill firefighters until we execute the cultural change of not making an aggressive interior fire attack or searching in these disposable homes.

Your department's SAR plan. You must have a solid SAR plan and your members must know it long before you are toned out. It should be part of probie training and part of recurring in-service training. Yes, it is that important. SAR is, after all, one of our most dangerous missions and certainly our most important mission. Your department's SAR plan is one of the most important considerations on SAR for your department. Responding to a call is not the time to build, test, evaluate, and train your members to effectively execute the plan. If you try to set up an SAR plan on the fly at a working house fire with the massive stress of people trapped, you and your department will likely not be successful.

WHO ARE WE SEARCHING FOR AND HOW DID THEY GET INTO THIS SITUATION?

MAKING OF A FIRE VICTIM

"A civilian fire death occurs every 143 minutes in the United States. Fire fatality rates in the United States roughly equate to three jumbo jets crashing every month for an entire year. Fire deaths due to toxic gases and/or oxygen deprivation from smoke inhalation outnumbered fire deaths resulting from burns" (Schnepp 2009, 3). When we think of fire fatalities in this sense, about 300 per month, the massive scale of this loss becomes real. It is not usually presented this dramatically, because Americans die from fires in small numbers, one or two (rarely more than two) at a time. The one or two deaths per fire don't make national news so we aren't aware of them. Small numbers directly affect the local community, but do not attract the national attention they deserve.

As we saw from the statistics in chapter 1, over 75% of the Americans who die by fire die in their own residence. Let's take a look at several factors that often combine to transform a person in a house fire into a fire victim, a person trapped and killed by fire in their own home.

LIKELY FIRE VICTIMS

In a 2014 NFPA report by Marty Ahrens titled *Characteristics of Home Fire Victims*, a short summary of the report paints a clear picture of who are likely fire victims and who we are historically searching for:

During 2007–2011, home structure fires killed an estimated average of 2,570 people and caused an average of 13,210 reported civilian injuries per year.

Only 13% of the US population is 65 or older, but 30% of fatal home fire victims were at least 65 years of age.

The percentage of fatal home fire victims under five years of age fell from 18% in 1980 to 6% in 2011, while the percentage of victims 65 or older increased from 19% to 31% over the same period.

Compared to their share of the population, African Americans were roughly twice as likely to be killed or injured in a home fire in 2007–2011 as the overall population. The difference was even greater for children and older adults.

Males were more likely to be killed or injured in home fires compared to females (56% of the deaths and 53% of the injuries).

Who are the most likely fire victims? According to the NFPA, one age group is especially vulnerable: the elderly (older than 65; Ahrens 2016, 6–7). These Americans likely suffer a disproportionate number of deaths for the following reasons:

They are less able to escape, often due to physical or cognitive impairments.

They have a lower tolerance for toxic gases.

They may not have been trained on how to react properly.

CAUSES OF THE TRANSFORMATION

Rapid fire development or flashover can cause almost instantaneous incapacitation or death. We know that it occurs without warning, suddenly and silently. A civilian has no concept of rapid fire development and can be caught in this deadly trap very easily. Inhalation of fire or superheated gases generated by flashover can burn the trachea and dry out the lungs in a single breath. (See chapter 3 for more details on flashover.) However, more commonly, transformation from occupant to victim is a result of smoke and toxic gas inhalation.

Smoke or products of combustion are made up of

Soot—unburned or carbonized particles or liquid balloons of whatever was burning,

Heat of combustion (heated or superheated air and radiant heat),

Toxic gases, and reduced oxygen.

Soot can accumulate in nasal and respiratory passages, physically clogging the airways' ability to move air. My company rescued an intoxicated young man from a house fire early one morning. His nose was almost completely obstructed with soot and mucous and his teeth were etched with black soot. In addition to being a physical obstruction, soot carries with it toxic chemicals that eventually paralyze the cilia on the lower airway passages because they simply cannot clean or eject the massive amount of material that is being breathed in. The toxic chemicals in both the soot and the smoke can cause actual sloughing of the lining of the trachea. This, in addition to massive mucous production and the edema caused by the chemicals, can quickly reduce the oxygen available to the occupant, who is quickly becoming the victim. Reduced oxygen to the brain causes people to do irrational things (Crapo and Nellis 1981).

Rob Schnepp explains other toxic products of combustion: "Rarely acknowledged are compounds such as ammonia, hydrogen chloride, sulfur dioxide, hydrogen sulfide, carbon dioxide, the oxides of nitrogen and soot—a known human carcinogen" (2009, 3). These chemicals, most of which are strong respiratory irritants, make escaping from a house fire more difficult for the untrained occupant. Ammonia burns the eyes, making seeing more difficult. As the body tries to rid itself of these toxins, violent coughing leading to vomiting further inhibits swift egress from the building. With limited sight, difficulty breathing, panic, and inhibited mental capacity from lack of oxygen and carbon monoxide inhalation, even the most rational person will quickly become a fire victim.

But there is yet another major contributor: "Even less frequently named are cyanide compounds, particularly hydrogen cyanide, recently identified as a major factor in smoke inhalation fatalities" (Schnepp 2009, 3). The role of cyanide poisoning at residential fires is just now being understood by firefighters, doctors, and paramedics. Donald Walsh, assistant deputy fire commissioner of the Chicago FD, explains the role of cyanide poisoning in the tragic Rhode Island night-club fire in February 2003 that claimed 100 lives: "Results of the NIST experiments are consistent with the possibility that an elevated level of hydrogen cyanide was among the causes of incapacitation and death in the Rhode Island Nightclub Fire. Hydrogen cyanide is a highly toxic combustion product that is formed during combustion of any material containing nitrogen—that is to say, during combustion of almost any material found or used in the construction of human dwellings" (Walsh 2007, 6).

In this same article, Walsh demonstrates the pervasive effects of cyanide poisoning among firefighters:

> A recent case of cyanide poisoning of a firefighter occurred in early 2006 during a structural house fire in Providence, Rhode Island. During the fire, a firefighter collapsed into cardiac arrest while operating the engine pumps outside the fire building. The firefighter of Engine 6 was resuscitated at Rhode Island Hospital and was found to have hydrogen cyanide in his blood. The firefighter had no memory of the fire.
>
> Following the fire, the fire chief had the poisoning investigated. Ninety-one of the department's firefighters responded to three fires in March of 2006. Of those 91, 28 sought medical treatment, and 27 had their blood tested for cyanide. Eight of the firefighters, including the firefighter from Engine 6, had high levels of cyanide in their blood. (Walsh 2007, 4–5).

Although the biochemistry is complicated, the lethal effect of cyanide on the body can be explained easily. Hydrogen cyanide is inhaled with the smoke and enters the lungs then the blood stream and finally the cells of the body. When individual cells in the body try to use the oxygen the red blood cells have delivered to it, the cyanide chemically blocks the uptake of oxygen and the cell eventually dies. As multiple cells become incapacitated and die, the organ they are a part of loses efficiency and also eventually succumbs. Consider the brain: as it loses efficiency, rational thought is first to go, eventually leading to unconsciousness and inability to escape the fire scenario.

As the victim remains incapacitated but still breathing, they inhale more and more carbon monoxide. This toxic gas is produced in vast quantities and also prevents the red blood cells from uptake and delivery of oxygen to the other cells, resulting in more cellular death and subsequent organ death.

An Effective SAR Plan

You can utilize facts and experiences specific to house fires to plan your department's SAR operations. First, however, let's define the SAR plan.

A search-and-rescue plan is a predetermined plan that defines how you will deploy your limited personnel and equipment to gain access to the fire building and conduct an effective, efficient, and swift search for and removal of salvageable human life. Note that it is much more than "You go to the right, I'll go to the left," which is just a tactical decision far below company or department levels. The SAR plan is a strategy-level plan. This plan is a place to start; you can always deviate from the plan during your operations. As General Eisenhower said, "The plan is nothing but planning is everything." However, if you don't have a plan to deviate from, your fireground will be ineffective chaos and wildly unsafe for your members.

An SAR plan that works well (at least in the eastern US) is the one used by the FDNY. It is specific to house fires. This plan utilizes limited first-due personnel and equipment for maximum benefit and effectiveness while still providing a reasonable degree of safety for members. It works well for the FDNY, but may not be applicable to your particular department. It is presented here as an example of an SAR plan. You should only apply all of it or parts of it to your department if it will work for you. Before you try to put the proverbial square peg in the round hole, you must consider any potential differences your department may have from the FDNY: personnel, response times, water supply, varying house design, construction, occupancy, and any special considerations you may have such as unique occupancies (college town, migrant worker population, etc.). Unquestioningly cutting and pasting concepts and SOPs that work in other parts of the country, in other departments, and in other conditions may result in more dangerous operations than your department is currently using.

Internalizing these concepts can result in a great increase in efficiency and safety for you and your members. If you want to use any of the skills, tactics, or strategies you read here, remember that they must be woven into the fabric (SOPs) of your daily operations and learned and honed during training.

Force multiplier

One of the main benefits of this plan is that it is a force multiplier: it gives the best results with the least personnel. It cohesively synchronizes the most important fireground tasks—size-up, forcible entry, ventilation, search, and fire attack. It enhances interior fire attack (if that is what is chosen by the IC) while simultaneously supporting ventilation. It has proven effective over time and is based on a large amount of fireground experience.

Victim profiles

This plan, specific to SAR operations at house fires, concentrates primary search efforts on the bedrooms and means of egress. According to the NFPA, "Only one in five (20%) of the reported home fires occurred between 11:00 p.m. and 7:00 a.m. but half (52%) of the home fire deaths resulted from incidents reported between these hours" (Ahrens 2016, 5). This is the time when most folks are sleeping in the bedrooms of their homes.

What are people usually doing when they are transformed from occupant to fire victim? According to another NFPA report that analyzes data from fatal house fires from 2007 to 2011, 34% were sleeping and 36% were trying to escape (Ahrens 2014, 47). These data clearly show where 70% of home fire victims are and what they are doing at the time of their death; therefore, if your department's SAR plan concentrates on both the bedrooms and the means of egress, you stand an excellent chance of encountering a fire victim and saving their life.

This plan is based on a two-team approach, with both teams focused on gaining access to the house, searching for, and finding occupants as quickly as possible. Employment of this plan assumes there is light to moderate fire in the home and the house is safe to enter. That is, there is no lightweight construction, no special hazards, and the fire does not have full possession of the house and victims, yet.

This plan is also based on the fact that fatal fires in homes often start on the first floor or basement level of the home (where heating equipment tends to be located). According to the NFPA, heating equipment is ranked second in causes of home fire deaths overall (tied with cooking equipment; Ahrens 2016, 8), but is the leading cause of fire deaths in one- and two-family homes (10).

FDNY Deputy Chief John Norman (retired), in his excellent book, *Fire Officer's Handbook of Tactics*, summarizes the benefits of this plan:

> In developing this plan of action, FDNY took many factors into consideration; based primarily on a great deal of past fire experience, and they established a list of priorities that are reflected in the members' assignments. The interior team is the largest team, and it has the experience of the officer directly available. This is logical, since the primary duty of officers is to protect their people. The firefighters on the inside are in the greatest danger, trying to get close to the fire as well as above it via the stairs. Under conditions of high heat and dense smoke, the task of the interior team, to conduct search and rescue, is quite formidable. For this reason, three members are initially given this task.
>
> A high priority is placed on search of the means of egress and the bedrooms, since the area of highest life hazard—regardless of the time of day—is the bedroom. Statistics nationwide, as well as New York City, bear this out, and it is easily explained. Alert human beings will flee a fire if they are physically able. Persons who are asleep aren't alert and thus cannot flee. This includes many people who work nights and sleep during the day, as well as many older folks who nap each day. Young children in cribs, the temporarily disabled, and older bedridden folks will all be found in bedrooms the great majority of the time. The odds are very high that, if you are going to have to rescue a victim from a dwelling fire, it will be from the bedroom. The plan also recognizes that fires most frequently start on lower floors, severely exposing the interior stairway and rendering it useless as a means of access until the fire has been darkened down. By having the three remaining members perform VES, access into the critical bedrooms is made immediately. By working from the ladder, roof, or fire escape, making their way into a window, the members aren't exposed to the severe interior conditions as long. They are also in less danger of having their escape route blocked by fire. If conditions worsen, they are only a matter of feet from their escape route. (Norman 2012, 170–71)

This plan contributes to the overall success of SAR operations in many different departments across the country for three important reasons:

The first is that an interior team, two members minimum, must begin the search and assist the attack team in the interior (fig. 6–3). The second point is that roof venting isn't initially required, but horizontal venting, which greatly speeds the advancement of the hoseline, is required. The third is that, by properly selecting a window for entry, you have a very good chance of entering the location of any missing occupants. Instead of six members, this tactic can be used with almost any number. (Norman 2012, 172)

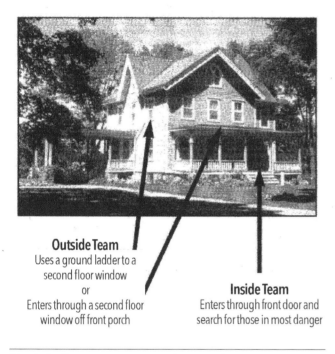

Outside Team
Uses a ground ladder to a second floor window
or
Enters through a second floor window off front porch

Inside Team
Enters through front door and search for those in most danger

Fig. 6–3. The two-team approach to SAR has many advantages for the overall operation.

This SAR plan is statistically proven by the following fireground facts from a USFA report based on data from 2013 to 2015:

- 50.7% of civilian fatalities are in their bedrooms (National Fire Data Center [NFDC] 2017, 9).

- 51% of civilian deaths in residential buildings occur between 11 p.m. and 7 a.m. This time period also accounts for 47% of fatal residential fires (2).

- 37.4% of civilian fire fatalities were trying to escape and 31.4% were sleeping (7).

The inside team

The interior team is made up of two to three members. This could be the officer and one or two firefighters from the first-due apparatus. The officer has typical coordination duties—size-up, on-scene report, positioning the truck, and

so on—as you would expect. One firefighter has forcible entry tools and the second has a 6 ft hook and a pressurized water extinguisher. This part of the plan can be executed by two firefighters if necessary; if you have enough personnel to staff this team with three, even better. This team will be in the most danger so it should be composed of your most experienced and capable members.

The goal of this team is to enter through the front door and search for victims on the first floor or closest to and/or above the fire. (Ch. 8 covers the ventilation effect on the fire of chocking the front door open after forcing it, but, in essence, it might be much better for both victims and firefighters to close the front door as much as possible to limit fire growth.) The front door is usually the fastest way into a house. The door is accessible from the street, front lawn, or driveway and provides access to the stairs, which are often located behind the door. Unless you are in a high-crime area, forcible entry is not usually difficult. Wooden doors are easily forced and often glass panels on either side or in the door make access swift, since all you have to do is take out the glass and reach in and unlock the door. This of course after you "try before you pry."

This team's mission is to search the first floor area and report the location of the fire. When this is complete, they can move to other areas based on the officer's direction or the specific circumstances of the fire, such as location of the fire, to assist other search teams in likely areas to find or help remove victims (location in upstairs bedrooms, cellars, and attics).

The inside team has a pivotal role in the entire outcome of both the SAR and the fire attack operations. By opening the front door they accomplish the following three goals:

1. They access the most likely escape route for trapped victims. Victims may be lying on the floor near the door or on the stairs just inside. Even if heavy fire conditions exist, open the door and look under the smoke for victims from the front door. The inside team can then move directly to the fire area or room. This brings them to the victims who are in the most danger first and allows a search on the path away from the fire. Conducting a left- or right-oriented search pattern provides a greater sense of orientation and is the preferred method for areas adjoining or above the fire (fig. 6–4).

2. They pave the way for the hoseline advance. You are not going to put out the fire with a Halligan or a hook no matter how hard you try. Controlling or extinguishing the fire minimizes all other life safety

problems for both civilians and firefighters. The sooner the line gets water on the fire the better. The line can protect the stairway and help improve your safety and the safety of the exterior team operating upstairs.

Fig. 6–4. Despite heavy fire and smoke conditions, a firefighter can look under the smoke and visually search for victims on the floor using a hand light.

3. They also play a critical role in overall safety. They are inside and readily available to assist the outside team in case of a Mayday or other emergency. Since they are already inside, they may assist the outside team bringing a victim down the interior stairs.

There are four important factors that drive a successful search by the inside team. These are best described as quotes from Ladders 4 (Fire Department of the City of New York [FDNY] 1997, 8):

1. Aggressive leadership by the officer is the most important factor in conducting the inside attack.

2. The dominant consideration of the "inside team" is search and rescue of those occupants who have a chance of survival if immediately removed. Therefore, if officer and members of the inside team are met at the front door by fire they will not wait to advance in behind the line. Rather, they will seek another means of access into those rooms not yet involved in fire. In this event one man will be designated to remain with the Engine to perform any required Truck work as needed.

3. The time of the incident plays a key role in search operations. During sleeping hours a heavy emphasis

must be placed on bedroom search. Limited time is spent on Living Room, Dining Room and Kitchen areas. Concentration must be on the first floor bedrooms (if any).

4. When search of the first floor is completed, the inside team shall make a rapid advance to the upper floor bedrooms via interior stair.

Tool selection for the interior team is obvious. Since forcible entry is a major duty, a flathead axe and Halligan are good choices for one firefighter. Commonly called the "can man," the second firefighter in this team carries the can and a 6 ft hook. The extinguisher can control minor fires, such as a mattress or couch, quickly. The extinguisher can be used to hold back a room-and-contents fire long enough for members to get by it to make a rescue. The hook can be used to pull the door to the fire room shut, either to get by to continue the search or to contain the fire until the engine crew can get water on it.

After the line is in place and has controlled the fire, the hook (pike pole) can be used to open up and examine for fire extension. This allows the member to remain at the point of operation without having to leave the area to retrieve the necessary tools (fig. 6–5).

Fig. 6–5. 6 ft hook

Rear doors. In the interactive scenario portion of the house fire seminars we have taught for the past 20 years, firefighters often want to assign the inside team to enter the home through the rear door. Certainly, if the front door is blocked by fire or other hazards, the rear door is a good second option.

However, using the rear door has several disadvantages. First, the engine company now does not have access to the building, because the door has not been forced. The front door is usually near the stairs to the second floor, so the engine crew cannot protect the means of egress and vertical fire spread as quickly as they could if the front door had been opened. Additionally, the engine now has a much longer stretch around the rear of the home, maybe through a fenced-in yard, around cars, toys, swing sets, and other unforeseen obstacles. For departments using preconnected hoselines, your 200 footer will likely not even be long enough. An important skill that we have somewhat lost is extending a short stretch. One easy way to do this is to take 50 or 100 ft off your bumper line to extend the attack line. The trash or bumper line is easily accessed, probably folded so you can carry it, and you can quickly choose how many lengths you need.

Second, victims will likely be found using their primary means of egress, which is usually the front door. Victims who have been found upstairs (closer to the front door) or en route to the front door now have to be taken out the rear of the house and brought around to the front of the home, which is another time- and personnel-consuming action. As a general rule, you can expect to spend four rescuers on each victim, especially if you have to move unconscious victims any distance.

For house fires, don't ignore the rear (your 360-degree survey should let you know what is going on there), but focus on the front; it is where your resources (engines, trucks, EMS, and so on) are, and it's a good place to start your operation.

The outside team

The outside team's job is to ladder the second-floor windows, enter the bedrooms and conduct a search, and report the location of the fire if they find it. Again, let's look at the FDNY plan and how you can adopt it for your specific buildings, staffing level, and responding apparatus. For FDNY, this team is made up of the chauffeur, roof firefighter, and the outside vent firefighter. Your department can easily use this plan with two firefighters for the outside team. You may choose to put several outside teams to work to enhance your SAR if you have the personnel on your first-alarm apparatus or as additional units roll onto the scene. Although the absolute first priority for engine company members is to establish a water supply and stretch an attack line, members conducting the outside operation can be from an engine company not needed for fire attack using portable ladders carried on your rigs, again if staffing in your specific situation permits.

The outside team consists of the following members:

- The roof firefighter, who ladders the front porch roof and gains access to the bedrooms using a roof ladder and 6 ft hook;

- The chauffeur, who raises an extension ladder to one of the upper floor windows and gains access to the bedroom for SAR operations; and

- The outside ventilation firefighter, who teams up with the chauffeur for this operation and has an axe or Halligan and a 6 ft hook.

There are several options available for utilizing these three members to maximize the speed and efficiency of the SAR and overall fire attack.

Fig. 6–6. Flat-roof porch with access to multiple bedrooms

Porch roof. The porch roof provides a stable and easily accessible platform to work from to gain access to one or more bedrooms. (See figs. 6–7 through 6–10 for visual examples.) The porch roof is a great tactic for the following reasons:

1. It is easily accessed by one firefighter using a roof ladder, because this ladder is lightweight, easy for one firefighter to carry, and long enough to do the job.

2. It is a safe platform to work from. You are working outside the smoke and heat. You can clean the window of all glass, screens, shades, blinds, curtains, and whatever else is blocking it, making a door out of a window. This obviously provides you a nearby and rapid means of escape after you enter the room.

3. Should you find a victim, you can bring them out to the roof—an area of safety—quickly. This does create a problem since then you have to move the possibly unconscious person down a ground ladder, but the good news is that it is a short ladder and by

this time the second-due apparatus and personnel have arrived to assist you. An additional advantage is if the victim is conscious; you have brought them out of the toxic smoke and they might be able to get themselves down the ladder while you continue to assist them or search for other victims.

4. The roof provides some protection from the fire venting below.

5. The roof usually provides access to more than one bedroom.

6. Unless there is an air conditioner in the window, access through the window is easy; often the window is unlocked and can be opened for an immediate quick look before taking the glass, sashes, and other window treatments.

7. The progress of the VEIS team can be easily monitored by others such as the chauffeur or officers in the street.

Fig. 6–7. Many homes have porch roofs that are an excellent platform for us to operate from safely.

Fig. 6–8. This wraparound porch has access to five windows on the front (A side) alone and additional windows on the B side.

Fig. 6–9. Even this small porch roof has access to three bedrooms.

Fig. 6–10. This porch roof not only has windows to multiple bedrooms, there is a door that leads out to the porch roof! Don't expect to see this at your next fire; I have not seen one since snapping this picture while on vacation at a seaside community.

The bedroom window via ground ladder. This is a two-person operation due to the weight of the 24 ft extension ladder. Generally it will take this entire ladder length to reach the second-floor windows and it will take two firefighters to move and deploy it. This is another team whose mission it is to go directly to the location of the victims.

This team has the same duties as the firefighter on the porch or garage roof: take out the window, enter the room, and search for salvageable human life. Depending upon the situation, if a person or object is available to buttress the ladder, the second firefighter can climb the ladder and assist the firefighter entering the window. This firefighter can keep verbal contact with the firefighter inside.

Successful rescue by outside team. It was 0530 hours on a very hot and muggy July Saturday morning when we pulled up to a single-story ranch with fire showing out the front and side doors. Our first-due engine had four firefighters plus a driver and officer. The officer split the crew: two for the line, two for search. It was a good neighborhood, lots of young families with kids.

Fig. 6–11. Upon arrival, the front door was blocked by fire.

The fire had complete possession of the kitchen and was extending out the front door and the living room, which had a large picture window that was beginning to fail. This would soon light up the entire living room, further exposing bedrooms in the rear of the house (fig. 6–11).

The side door had fire showing out it about halfway down (fig. 6–12). The police officer was in the front yard and said to me, "The victim is in the back bedroom." I asked how he knew and he said he saw him at the still-intact window and heard him hit the hardwood floor when he passed out due to smoke inhalation. Hearing this, the search team made their way to the rear of the house, did a VEIS, and found the victim just were the police officer had said he was (fig. 6–13).

Fig. 6–12. The side door was blocked by fire.

We stretched the line off the engine, quickly knocking down the fire extending into the living room, then made quick work of the kitchen fire. As we made our way toward the back bedroom, the firefighter who had made the rescue called a loud verbal Mayday that we heard through his mask. He was physically and mentally exhausted from the excellent rescue and we were able to help him out of the building.

Fig. 6–13. The outside team placed a roof ladder against the window sill and a firefighter climbed, vented the window, entered, and rescued the male victim unconscious on the floor.

This is an excellent example of a rescue made by the outside team. The outside firefighters bypassed the fire, used a roof ladder to gain access to the rear bedroom window, searched, and found and handed the victim out to waiting firefighters. Despite severe smoke inhalation, the victim made a full recovery.

The cause of the fire was unattended cooking. The two brothers who lived in the house were out on the town Friday night. They returned home late Saturday morning, and one started cooking and fell asleep. Unattended cooking, as we saw in the statistics in chapter 1, is the cause of numerous deadly fires. Upon noticing the fire, the other occupant drove his car to the police station to report the fire, leaving his brother trapped in the rear bedroom.

There are five main lessons to be learned from this case study:

1. Always anticipate a different grade in the rear of the house (a value of the 360-degree size-up).

2. The outside team, going directly for the victims, can often make a swift rescue.

3. Always position the lighter roof ladder on the outside of the heavier 24 ft extension ladder on your pumper. One firefighter can easily take off the roof ladder and go with it, but it typically takes two firefighters to remove the 24 ft ladder, even if you have a drop-down ladder rack.

4. The police usually arrive on scene before we do; they have had a minute or two to ascertain important information from occupants who have escaped, bystanders, or their own observations. Ask the officer on scene what they know about the incident if they do not volunteer this critical information.

5. The two-team approach increases firefighter safety by limiting exposure of the outside team and by allowing the inside team to assist the interior team, as they may have their own Mayday or need assistance removing a victim.

Vent, enter, isolate, search. At first glance, it seems unsafe to put a lone firefighter up on a porch roof or up an extension ladder and then have them enter a burning building. In fact, this is one of the safer methods of conducting SAR operations at house fires. Are there hazards? Sure, but consider that these firefighters are not trying to get by the fire if it is on the first floor and can isolate themselves by closing the bedroom door. Some also argue that this is a good risk particularly during nighttime fires, where it is very likely that salvageable victims are in the bedroom(s).

The firefighter is in relative safety until they enter the room. They have made a large escape opening by taking the window. Most bedrooms are relatively small, so the firefighter is not far from the window and complete safety. Once inside the room, they can reach the bedroom door and close it, keeping smoke, heat, and fire out of the room while they complete a quick primary search. Essentially, this is a search in an area of refuge free from advancing fire, at least for a short time.

To further increase safety in bad conditions, the firefighter can use the 6 ft hook (hooked on the window) as a clear route back to the entry point. A 6 ft hook combined with the average 5.5 ft reach of the firefighter's arms will enable a rapid, safe primary search in the worst conditions (fig. 6–14).

Compare this against the team entering the front door of the house. This team is slowed to some degree by forcible entry, then possibly by fire. They will likely operate near the fire or adjacent to it or, worse yet, above it. If the first floor search is negative, they then proceed quickly to the second floor.

Fig. 6–14. Use the 6 ft hook as an anchor point back to your means of egress.

Tactics. Let's look at some critical details of the outside team's actions during an aggressive VEIS operation. It is 2330 hours on a Tuesday night and you are dispatched to a possible house fire, 20 Continental Drive. A passerby reported via cell phone what looked like smoke coming from a house. Upon arrival there is heavy smoke pouring from most of the downstairs windows and fire out a few windows. There seems to be a good body of fire on the first floor but it is not showing due to the heavy smoke. Clearly it is a ventilation-limited fire. You are assigned to the outside team so you take the 14 ft roof ladder and ladder the porch roof (fig. 6–15).

Fig. 6–15. If porch roof is present, the outside team can gain access to several bedrooms without passing the fire.

You clean out the window, sound the floor, and make entry, then make a push for the bedroom door, which you immediately close. You search the room—under and on top of beds, in the closet, in the hope chest and toy box—and find no victims. You are at a decision point. Your choices are

1. Exit the room immediately via the window you entered;

2. Open the door and exit via the window you entered;

3. Open the door, cross the hallway, and search another room; or

4. Open the door and exit down the stairs.

Exiting immediately has the following advantages: It is the safest option available. Leaving the door closed almost 100% ensures that you will make it back safely to your window. You have limited the spread of fire into that room. By keeping the door closed you have not possibly created the first or an additional flow path for the fire. If you acted alone, you have done a great job searching a portion of the house that was likely to contain victims and have done this in relative safety. However, leaving the door closed does not provide any ventilation for the engine company pushing a hoseline in on the fire. You also did not search any other areas.

Opening the door before you exit provides an excellent vertical ventilation opening to lighten up conditions for the engine crew attacking the fire either on the first floor or the floor you are on. The decision to leave the door shut or open should depend on how well the fire attack is progressing. If the line is not operating on the fire for some reason, or has stopped due to a burst length, it is probably best to keep the door closed and exit. You don't want to create or improve a vertical chimney or flow path if the fire has already established it. If the hoseline is operating and flowing water, it might be good to leave the door open. Lifting high-heat conditions will make the push by the engine easier, quicker, and safer if the fire has been controlled. The importance of this timing cannot be overstated. Too early and you create a flow path, providing the oxygen this ventilation-limited fire needed to progress quickly to flashover. This endangers everyone in the house, but ventilation after the fire has been controlled reduces heat and increases visibility.

Prior to the UL and NIST studies, the American fire service believed in and routinely practiced leaving the door to the bedroom you just performed a VEIS into open. The theory was that since there was no life hazard in that room, if you drew fire up into it through the open door and window there was no harm done. We reasoned that life was our absolute priority, and if you inadvertently draw fire up from downstairs into this room, oh well. You'd already confirmed that no one was in there, so you have lost little of value. Sure, the fire gets worse, but the advantage is that it is not drawn up into another room that may be occupied or not searched yet, or occupied by a firefighter searching. We now know, courtesy of Steve Kerber and the UL, that this is the wrong action to take because it creates a flow path for the fire. More on this in chapters 8–11.

In regard to opening the door, crossing the hallway, and searching another room: If there is a high life hazard, a good option is to open the door a crack and look out into the hallway, then make your decision. What do you see? If you see heavy fire roaring up the stairs, it is most likely a good option to keep the door closed and exit. If you look out and conditions are not too bad, you can decide whether you want to cross the hall and search another room. The obvious advantage of crossing the hall is that you will immediately be searching another likely area for salvageable victims. This vastly increases the efficiency of your search. It is much quicker than retreating down the ladder, taking the ladder down, moving it, raising it to another window, taking the window, and so on.

But this method also puts you in more danger since you don't have a confirmed path out of the next room or a confirmed area of refuge if fire is advancing toward you. You also know nothing about the room or its windows: Are there bars on the windows? Is it a three-story drop to the ground? Remember that it is important to report your aggressive actions to the IC for everyone's safety. If the IC knows you are going across the hall, they may be able to position ground ladders to provide you a secondary means of egress should you need it. However, the ladder may prevent the use of your bailout rope, which will be a high-risk evolution for you.

What was the situation on the fireground? Was there a report of several children trapped or are you conducting a primary search as a standard operating procedure? Obviously, a reliable report of people trapped may warrant a higher risk. The key factor here again is whether there is water on the fire. If there is, going across the hall may be a good option.

The final option—exiting the room you are in and going down the stairs—can be useful if you find a victim, because then you don't have to take the victim out the way you came in, which is often much more complicated, particularly if the victim is unconscious. This is a good option if the fire has been controlled and there are no more areas to search, or if you need to exit the building for an emergency (if you are hurt or become ill). Since the interior team has already searched the downstairs, this option adds little to the search effort.

We can't tell you the right answer; that will depend on the conditions during that call, your assessment of the conditions, and the progress of the hoseline. The important point here is that you see it as an important decision point, an important decision point for your safety.

Chief Anthony Avillo (retired) makes several very important comments regarding ventilation and the VEIS process:

Venting for life in these situations will most likely be accomplished by personnel accessing upstairs bedrooms via ground ladders or low roofs. During this operation, every attempt should be made to confine the fire to as small an area as possible. This may be accomplished by closing doors. This will not only tend to confine the fire, but buy some extra time for the search team operating above the fire. As a matter of routine in any type occupancy, personnel entering from elevated positions should immediately and as safely as possible make a beeline for the door and get it closed, both negating the effect of a ventilation-created draft on the fire and buying them some time to accomplish the search.

The question will also arise as to whether the VES firefighters should enter the room at all, or search for a victim by performing a sweep beneath the window. This is a tough call that should be made by an experienced crew after considering several factors. First and foremost is the location of the fire. Remember that by breaking the window, a vent path will be established. If it has no other way to go, the fire will come that way. The race to see if the fire gets to the entrance door to the room before the firefighter can get inside and close it is not a gamble that should be taken by unseasoned personnel. Other factors to consider include if water is being applied to the fire, if the stairs are open, and what the wind direction is. If water is not yet being applied, if the stairs are open and in close proximity to the room door (which they usually are in private dwellings), and if the wind is blowing toward the venting crew, the risk may not be worth the gain. (Avillo 2008, 248)

The reality of search operations is that they are emotionally charged, high-stakes operations with lives—ours and civilians'—hanging in the balance. Often these intense emotions and split-second decisions drive our other fireground decisions. All levels of firefighters, from probies to chief officers, must use the risk versus gain measure to determine the most correct course of action during search operations and, as much as humanly possible, eliminate the intense emotions of the moment.

Engine Company Search Procedures

Life is of course our first priority. For the engine company, putting out the fire or protecting the means of egress (for both civilians and firefighters) is their top priority during a search operation associated with an aggressive interior fire attack.

However, the engine company can and should conduct their own search operation. As the line is advanced, the officer can rapidly search rooms and areas off the side of the line and where the fire has been knocked down. Typically this is done with a hand light on the floor under the smoke but can be a full search of adjoining rooms if necessary. A search is done by default as the engine crew moves into areas of the home.

If the hoseline is the first group of firefighters into the home, their efforts at a concurrent search are very important. There is of course a balance between spending time and effort on search and time and effort on controlling the fire. If the engine crew spends all of its time and effort searching, who is extinguishing the fire? The answer is obviously no one, and the situation for everyone in the home is going to turn bad in the very near future as the fire grows and possibly flashes over.

The important point here is this: should your engine crew find a victim, then call for help, as you may have to use some of your crew to assist the victim out of the building. If possible and practical, leave at least one firefighter on the line to control the fire or maintain the means of egress until the victim is removed from the building. Good judgment on the scene based on circumstances you are facing should always prevail.

Search Rope

As a volunteer firefighter I don't often get the opportunity to develop a solid experience-based skill set for conducting searches. One of the factors that holds us back is the threat of getting lost, disoriented, and possibly caught by flashover and killed. After the experience described above, I developed what I call a search rope (fig. 6–16). Thirty years ago a search rope was a new tool that was being experimented with in some departments across the country. As a result of a lot of testing in our live burn building at the Rockland County Fire Training Center in Pomona (NY), I carry 35 ft of tubular Kevlar rope in a small bag. If I have to search under difficult conditions, I hook one end of the rope with a small carabiner to a substantial object in an area outside the building or fire area and put the bag in my tool-free hand and pay out rope as I go. It lies on the floor, assuring my partner and myself that we have a quick way out. It is not a bailout rope, not a life rope, just a trail of bread crumbs to get us out quickly if necessary. On the floor, it remains out of everyone's way. We conducted exhaustive testing and experimented with many different types of configurations of ropes—long, short,

Early in my career I (Jerry) got off the first-due ladder truck at an apartment fire in a garden apartment complex. It was a Saturday around 1030 hours so there were lots of civilians exiting the building and gathered outside by the time we arrived. As I made my way into the building I saw the engine company crew at the apartment door waiting for water. The door had been forced and heavy, nasty black smoke was pouring out into the hallway. Assuming water was coming soon I waited a few seconds then asked, "The primary is done right?" Their answer was no, they were waiting for water. Testosterone overcame good sense and I climbed over them and did a primary search of the bedrooms in the back of the apartment.

However, the fire was growing quickly and I became convinced it was time to make a hasty exit. The only problem was that I was completely disoriented. Inexperienced and so focused on the search, I had lost situational awareness. Another veteran firefighter must have heard me crashing around and repeatedly slapped his big gloved hand and pounded on the floor near the door with his Halligan while telling me to come to the noise. I did and was really glad to see him. The engine crew was extinguishing the fire while I was considering the mistakes I had made.

attached to two firefighters, retractable dog leashes—and found that this simple system seemed to work best. I have since used it on a number of searches and it has been very successful.

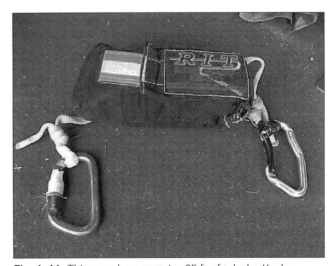

Fig. 6–16. This rope bag contains 35 ft of tubular Kevlar-based webbing. Carabiners are attached to each end to speed attachment outside the fire area. If not needed for a search, it can be used as a utility rope as well. It is lightweight and relatively small.

It is important to note that this is not a bailout rope. It is a search or utility rope for hauling tools to upper levels. It should not see the light of day in emergency bailout situations.

There a few methods for using this rope, but I find it works best when you don't take it with you into every room but leave it in the hallway, search the room, then pick up the bag on the way to the next room. It is not a tool that you have to use at every fire but is a good one to have in your tool box.

There are some other benefits to the search rope, especially in the limited-personnel situations that face both career and volunteer departments today. If firefighters come in behind me and the rope is on the floor, they know not to start their search in that room. If my partner and I need assistance, either with a Mayday or a victim removal, the rope is a direct, quick path to us.

Some may think the rope is a foot tangle hazard but experience has shown that it actually is generally not a problem. It lies flat on the floor and it is difficult for large boots to catch on it. Use of the search rope is shown in figures 6–17 through 6–20.

Fig. 6–17. The search rope bag is first connected to a substantial object outside the fire area.

Fig. 6–18. As you start your search, hold the bag in one hand and pull it along with you, paying out rope as you go.

Fig. 6–19. The search rope continues to pay out and remains on the floor, out of the way of other firefighters.

Fig. 6–20. If the search team member needs to make a hasty exit, they can follow the search rope back to a safe area. The search rope gives great confidence to the search team that they can make their way out quickly if needed. This is a great tool, especially for firefighters who are not regularly called on to conduct primary searches and may lack confidence to be as aggressive as necessary.

This rope can be a lifesaver if you have to search by yourself. Although we will all swear this never happens, we know it really does, especially in many departments with low staffing on first alarms. Searching alone is certainly not a recommended practice (unless you are doing the VEIS described above from a position of safety). Unfortunately, in the harsh reality of today's understaffed fire departments, single-firefighter searches are sometimes the only option. The rope creates an increased level of safety at very little cost in both dollars and personnel.

The advantage of using the search rope in a private dwelling is that it provides a known way out for you. The disadvantage is that it can be just another danger to get tangled up in.

Moving Furniture

Certainly you need to search over, under, and on all furniture, but throwing the furniture around in the room is a bad idea. Furniture is almost always located so that the room has adequate ingress and egress routes (fig. 6–21). Often in homes these routes can be narrow and even cluttered with toys, video games, magazine racks, coffee tables, and other hazards. If you go in and start tossing over chairs and tables, knocking lamps off end tables, you are

- most likely blocking your own means of egress: the couch was against the wall when you entered, and you just tossed it into the middle of the room (fig. 6–22); and

- disorienting yourself and other team members by rearranging all your landmarks.

Fig. 6–21. In this case, the front door is behind the recliner with steps to the second floor just behind the door. There is a large path to the left of the recliner to move into and out of the room.

Fig. 6–22. If firefighters toss furniture around during the search, normal exit paths can be blocked, resulting in disorientation and delays in removing victims and impeding rapid exit from the fire building. Just moving this one piece of furniture (rolling this recliner onto its side) blocks the exit from the first floor and impedes firefighters coming down the stairs and moving toward the exit. Especially in cases of rapid fire development, a hasty exit may mean the difference between life and death for our members. If other furniture was tossed around, escape may be impossible.

Bottom line here is this: search thoroughly but move as little as possible, especially toward the center of the room or toward doors and windows that you may have to use for a hasty exit.

BAD INFORMATION

I (Jerry) have done nine search operations as a first-due firefighter so far in my career. Of these, seven were based on completely incorrect information that there were people trapped. More than two-thirds of the times that I have risked my life entering a well-involved house before suppression had begun was for no possible gain. If you read and study the case histories (such as the one described in fig. 6–23) of firefighters getting killed or injured conducting SAR operations at house fires when there were no occupants inside, often they are risking their lives based on incorrect information from bystanders. These people usually don't mean any harm, but I believe they get caught up in the action at the scene and want to help so badly they say things that have no substance.

Fig. 6–23. A fire in the basement raced to the attic of this balloon-frame home. While walking down the street to this fire with my camera in hand, an occupant of a nearby house informed me that the attic was occupied and the resident was home. This turned out to be completely incorrect information. The attic was not occupied and the nonexistent occupant was, of course, not there!

Fire service history is full of examples like this. The Worcester Cold Storage and Warehouse Co. (MA) fire is another classic example of firefighters searching for victims who had long since left the building. 3 December 2017 marks the 18th anniversary of the fire that resulted in the line-of-duty deaths of six brother firefighters, the Worcester Six:

Paul Brotherton, Rescue 1

Jeremiah Lucey, Rescue 1

Lieutenant Thomas Spencer, Ladder 2

Timothy Jackson, Ladder 2

James Lyons, Engine 3

Joseph McGuirk, Engine 3

In December 2011, Worcester FD suffered another devastating loss of one of their own at a house fire based on unreliable information. Jon Davies of Rescue 1 was killed when the triple-decker-style home partially collapsed, trapping him and fellow firefighter Brian Carroll. Davies and his crew had reentered the home to conduct a second search when neighbors reported that an occupant was missing. The occupant was not at home (National Institute for Occupational Safety and Health [NIOSH] 2012).

Information from neighbors should always be suspect. In my experience it has been totally incorrect much more often than it has been correct. Although time may be limited, you should further question the validity of the information before wantonly risking firefighters' lives in a very dangerous operation. Simple quick questions such as: "When was the last time you saw the 'trapped person?'" and the always pertinent, "How do you know?" I have found this question to be invaluable in my career as a paramedic, hazmat chief, and firefighter. More often than not, this simple but effective question garners answers like: "Well, I thought he was in there" or "I think he is home but his car is not here" or "I assumed he was home...." Answers to this question often show the initial statement is incorrect. It is important to be a bit of a police detective here for our own survival! Watch people's reactions, what they say to further questioning. You will be surprised at how often this question—how do you know?—results in a total change of the initial statement. Commanders must judge the value of information from neighbors and decide if it is worth risking firefighters' lives based on that information. What appears to be a reliable statement (especially in the psyche of firefighters) may be totally unreliable with tragic consequences for us.

GRANDMA AND THE PINK SLIPPERS

The West Haverstraw FD (NY) responded one morning around 0530 to a group of old saltbox homes. They were close together and built around 1900. The house was well maintained and appeared to be occupied. As I got off the rig and walked toward the front door I saw an older woman on the porch next door, fuzzy pink slippers and all. I asked her if anyone was home. No answer. "Ma'am, are they home?" No answer. Finally my frustration peaked and I yelled, "Hey lady, is the family home?" She responded that they were in Florida on vacation. I asked, "Are you sure?" She said yup, she was definitely sure.

Clearly you are not going to make life-and-death decisions based on this kind of information but it may intensify or diminish your search efforts and the risks you are willing to take. Another important point here is that you may have to press people hard for information. Ask the question, "Is everybody out?" and immediately follow up with "How do you know?" Think of this questioning more like an interrogation than an interview. You don't have a lot of time and you need hard, clear facts. Be firm and to the point.

Search Safety

Conducting a search operation above the fire or before the fire is controlled is very dangerous. Chief Vincent Dunn (retired) offers these thoughts to help protect you and your members during a search above the fire:

> The degree of danger or threat of being trapped above a fire is greatly influenced by the construction of the burning building. Of the five basic types of building construction (fire resistive, noncombustible, ordinary or brick-and-joist, heavy timber, and wood frame), the greatest threat to a firefighter who must search above a fire is posed by the wood frame building. Vertical fire spread is more rapid in this type of structure. In addition to the three common avenues of vertical fire spread—the interior stairway, windows (autoexposure), and concealed spaces—flames can trap a firefighter above a fire in wood frame buildings by spreading up its combustible exterior. The combustible materials often used for interior wall coverings, halls, and stairs in a wood frame building can also rapidly spread fire upward to the floor above. (Dunn 2015, 98)

Chief Dunn offers the following specific recommendations to improve your safety during a search above the fire (2015, 100):

1. Notify your officer when you go above a fire. Even if your assignment has been preplanned, confirm the action by portable radio. This information is a form of fireground accountability and control that increases firefighter safety. A company officer should know where all of his or her assigned firefighters are operating during a fire.

2. Let the officer in charge of the attack hoseline know that a firefighter is moving above. When the officer is crouched down in the hallway about to advance the hoseline into the burning room, a tap on a shoulder and a finger pointed upstairs can convey the assignment.

3. When firefighters are available and fire conditions warrant it, leave a firefighter at the foot of the stairs to warn of deteriorating conditions on the fire floor.

4. If numbers permit, one or two firefighters should be assigned to assist an undermanned or inexperienced attack hose team on the fire floor.

5. When there is a danger of flashover in the hallway above a serious fire and a difficult forcible entry operation is required, locate an area of refuge that is not above the fire, such as an adjacent apartment. If conditions suddenly grow worse in the hallway, this space may save lives.

6. Firefighters going above a fire must be equipped with a portable radio and full protective fire gear. They should know that if they become trapped above they should give a Mayday report, activate their PASS alarms, and notify the officer in command of their condition and location.

7. When entering the floor above a fire, first locate a second exit such as a window leading to a fire escape or a portable ladder before starting to search for victims.

8. When climbing or descending a stairway between the fire floor and the floor above, stay close to the wall and face it. Heat, smoke, and flame flowing up a stairway will rise vertically near the stairwell's center, around the banister.

QUICK REVIEW OF THERMAL IMAGERS

GENERAL RULES

Thermal imaging does not require visible light.

Thermal imaging measures infrared (IR) waves.

Thermal imaging is not affected by smoke.

Microbolometer technology converts IR waves to viewable images.

LIMITATIONS

Does not see through glass or plastic

Shiny, waxed, or polished surfaces will reflect image

Will *not* see through water

TEMPERATURE SENSITIVITY

Ability of camera to distinguish differences in temperature

Displayed as shades of gray or color

QUICK TEMP INDICATOR

Indicates temperature in "crosshairs"

Not atmospheric (ambient) temperature

Can be used to compare objects

THERMAL CONTRAST

Differences between the object and the environment

Low contrast: cold scene; objects close to ambient temp

High contrast: warm scene; large difference in temp

High thermal energy in room

Sharpest image in camera

Camera works best when looking from cool to warm

Low temp: "HIGH sensitivity mode"

Objects appear similar color / featureless

Temps above 392°F appear red

High temp: "LOW sensitivity mode"

"EI" or "L"

Temps over 1,112°F appear red

WHITE OUT

Sensor at thermal saturation point

Move away or cover lens to reduce input

OVER TEMP WARNING

Thermometer in triangle flashes

Electronics in camera overheating

CONVECTED HEAT

Appears as white waves or smoke in camera

Direction and velocity may indicate location of fire

MASKED IMAGES

Materials that cover or block the heat signature of an object

Blankets/bedding

Layers of flooring or drop ceilings

DISTORTION

Objects may not have their familiar shape.

FOGGING

Steam or water spray *will* affect the image.

Facepiece, lens, and screen—wipe all three.

TACTICAL CONSIDERATIONS

The camera is *not* a life-safety device.

The camera does *not* replace standard SAR practice.

Operate as if you don't have a camera.

RECOMMENDATIONS FOR USE

View all six sides of the room: four walls, floor, and ceiling.

Turn around and look at your way in—it is your way out!

Check the ceiling first.

Find the thermal layer, pick a landmark, and monitor for heat banking down.

Survey the room and note the layout.

Check everything you *can't* see (e.g., under covers on a bed).

Scan from the doorway *before* you enter the room.

Scan slowly—recognize and interpret what you see.

Look for holes (heat) in the floor.

Communicate what you see to others.

Let them look.

JUDGING DISTANCE

Focal length will distort the view. Plot your course: pick an object and move toward it.

Navigate from point to point.

Use landmarks.

SEARCHING

2 types—Led / Directed

Led—follow the leader with the camera

Directed—member with camera directs searchers

Camera at known point

Better accountability

Members searching must be able to hear directions

"Move to your left," "Go toward 11 o'clock"

Not "Over here"

MSA EVOLUTION 4000

Displays in 26 shades of gray.

54° field of view—like looking through a paper towel roll.

Focal length 3 ft—closer objects may be distorted.

Objects may be closer than they appear.

Operates for 3 hours on two batteries.

BULLARD T-3

Displays in gray and colors (yellow and red).

Yellow and red are the hottest surfaces.

300°F–500°F is the low-temp mode.

At or above 1,100°F is the high-temp mode.

Moving Victims: Quick Webbing Harness

Moving victims from the danger area in the home to the outside can be a difficult task. The use of webbing can vastly increase your effectiveness in this operation as described in this article originally published in *WNYF Magazine*. For basic victim movement techniques we recommend you see *Fire Engineering's Handbook for Firefighter I and II* (Corbett 2013). This book and associated training videos provides a comprehensive review of these critical techniques.

Many departments provide each member with a 20 ft section of 1 in. tubular webbing as part of their personal protective equipment (PPE) to assist in the removal of a trapped or injured firefighter. However, this is not its only use. The webbing can also be used to remove a civilian. Often we must remove a civilian who is not dressed in heavy clothing, so dragging and handling them safely becomes difficult. With our webbing we can add "handles" to the victim to assist us in their removal.

You will probably encounter incapacitated civilians in a head-first and face-down position. There are several reasons for this. If the victim was conscious when the fire started and attempted to escape, they probably were moving toward the door you came in. If the heat and smoke forced them to crawl or when they were overcome, they would have collapsed face down with their head facing in the direction of travel (fig. 6–24). Firefighters encountering a civilian victim should note the position of the body, as this will provide clues to the chiefs and fire marshals in their investigation, especially if they are not found in the usual manner.

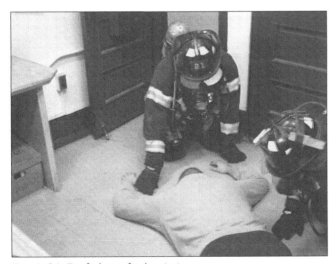

Fig. 6–24. Firefighters find a victim.

If you find the victim in the position described above, here is how to quickly deploy the webbing (figs. 6–25 through 6–28).

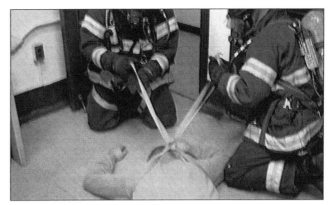

Fig. 6–25. Remove the webbing from your pocket and unroll it. Slide the webbing under the victim and bring both ends up under the arms and behind the shoulders. Bring both ends together and tie an overhand knot between the victim's shoulders.

This webbing is an essential part of our fire gear. Members should also practice working with the webbing with gloves on and simulate conditions that may be encountered in an actual situation.

Fig. 6–26. Roll the victim over and the webbing provides handles to drag the victim while supporting the head and shoulders.

Fig. 6–27. If two firefighters are necessary to drag the victim down a long hallway, tie the knot closer to one end.

Fig. 6–28. Making the loops of the webbing uneven allows two members to drag the victim down even an narrow hallway with relative ease.

Search and Rescue at Converted Private Dwellings

As we saw in the size-up chapter, converted private dwellings (CPDs) can contain many illegal modifications that can become killers when operating at these dangerous buildings. The hidden hazards presented by these deceiving houses are a problem for every fire department in America. We must alter our standard house fire procedures for operations at CPDs. They look like houses from the outside, but require very different strategies and tactics.

Tactical considerations for search

The interior team that entered the front door and made the second or third floor of a CPD does not have an area of refuge should the fire rapidly develop while they are in the hallways. Locks (often secure locks) on single-room occupancy doors may cause firefighters to become trapped and burn in the hallway. Forcing a door in conditions growing untenable is difficult. In a typical single-family dwelling (SFD), these doors would be unlocked and have at most a flimsy privacy door and lock and so would be accessible as an area of refuge.

The risk to members searching the second and third floors must not be underestimated. This is especially critical if a hoseline is not in place, is delayed, or has difficulty controlling the fire. Narrow hallways, locked doors, flammable wall coverings, and articles (bicycles, toys, etc.) stored in hallways can all add up to a fatal scenario for these well-meaning firefighters.

Consideration should be given to altering your search procedures. Instead of using the front door as the primary means of access for upper-floor search, use the vent-enter-search (VES) technique by way of the windows on the upper floors. Search teams enter upper floors using ladders placed at windows. The window is cleared; floor swept for victims, and then the floor is sounded. Controlling the door is top priority; thankfully, in an SRO [single-room occupancy] the door is usually closed immediately, putting the firefighter in an area of refuge and without impact on the ventilation profile.

This technique is not applicable to all fire situations, but it is especially important for the safety of members if the fire below is not controlled or is extending upward. (Knapp and Zayas 2011)

As a result of CPDs' often-illegal nature, the owner and occupants sometimes try to conceal neighbors' views inside. Normal escape routes for firefighters such as doors and windows can be intentionally blocked (fig. 6–29).

Fig. 6–29. Another viable means of escape for firefighters was not available to unknowing firefighters at this CPD. Note that the new window is intact behind a layer of sheetrock. The window was sheetrocked over, probably to prevent neighbors from recognizing the SRO on the second floor. VEIS will identify and resolve this safety issue quickly.

Room size. Another hazard to searching firefighters is the amount of furniture and other things (clothing, toys, etc.) that may be jammed into a small room. Prior to converting the home to apartments or SROs, the rooms were generally large. Furniture was well placed and provided relatively spacious accommodations. After the conversion, the house has been modified to contain more people in the same square footage, abnormally compressing any open floor space. This results in cramped conditions throughout the structure. Narrow hallways become even smaller to create rooms. Small rooms may seem advantageous for search; however, when overcrowded with furniture, small rooms can be death traps for firefighters (fig. 6–30).

Fig. 6–30. This small bath was too narrow for us to turn around in with our SCBA and tools. Hooking your foot on the doorframe to perform this search will solve this problem and ensure that you don't overcommit.

Blocked windows and doors. As a result of so much furniture and belongings in such a small room, windows and doors may be unusually blocked. This is an obvious serious hazard for firefighters searching if conditions become untenable (fig. 6–31).

Fig. 6–31. Bunk beds block a first-floor door and a window (to the left of this picture). It would be impossible for searching firefighters to force this inward-swinging door from the outside due to the weight of the beds. This bedroom was squeezed between the kitchen and the apartment's microscopic bathroom. Rapid egress from the area is impossible and entanglement hazards abound.

Occupied attics. We are finding these more and more often and they are a nightmare to search. Frequently they are occupied not by a tenant but by an adult or teenage member of the family or the owner who lives in the home. It provides a degree of privacy and independence from the crowded conditions on lower floors. As you might expect, access to and egress from the normally unoccupied area designed for storage is difficult at best. We responded to a fire one evening for a smoke detector sounding in the attic. Upon further investigation we found a living area similar to an apartment in the attic. Access was through a very narrow door in the bathroom, up a very narrow staircase. Because the stairs were near the outside wall, the pitch of the roof made access to the space possible only by crawling at the top of the stairs. The smoke detector was sounding from incense left burning (fig. 6–32).

Fig. 6–32. Limited ceiling height will cause smoke and heat to bank down very quickly in the attic, dramatically reducing visibility for firefighters searching above the fire. Flashover will happen also more rapidly in these smaller areas.

CPDs contain a variety of unusual hazards that, in combination with a few other seemingly insignificant events, can produce a failure chain severe enough to kill or injure firefighters. During our search operations, we must be aware of and report immediately the CPD conditions we find. Additionally, we must never forget or underestimate the dangers these building present to us.

Scenario Discussion

Fig. 6–33. Scenario

What obvious hazards or threats to your members do you see?

Construction type. As Frank Brannigan would say, this building did not just come from Mars last night and catch fire today! Your preincident intelligence should tell you any construction hazards.

Based on the look of the house and neighborhood, it is recent enough that it might be lightweight construction and pose a significant threat to your members operating inside. It could be platform construction with a stick-built roof, which will withstand a fire load for a short time. If it is a truss roof, collapse could be imminent, trapping your members inside. Your prefire planning information confirms that it was stick-built.

Fire. Looks like a one-room fire. From this vantage point it looks like the door to the bedroom may still be intact and containing the fire to that room. It is a lightweight interior door so time is running out fast.

Looks like the fire has extended into the attic, probably through the soffit or the interior ceiling or both. Is the black smoke from burning vinyl siding?

The intensity of the fire is high; it is well vented, and both windows have failed, probably due to the heat.

Flashover/backdraft conditions. The fire room is well vented and is past flashover conditions. The attic contains the warning signs of flashover or at least it is getting ready to light up. (The warning signs are high heat, dense smoke, free burning fire, and rollover.)

House layout. Unusual hazards—there is nothing obvious or outstanding at this point. This is a high ranch, open at the front door, stairs up to the living room and stairs down to the basement recreation room, garage, and boiler room. The downstairs may have a small kitchen, laundry, or bath area. Bedrooms are up the stairs and to the right. Probably two small bedrooms and one master bedroom. The stairs are directly behind the front door. As a general rule, stairs are usually found behind the front door. This design method minimizes unusable space in the home.

What is your occupant survival size-up?

There probably are salvageable lives inside (but not in the burning bedroom, because the doors are closed). House looks well kept and occupied. Halloween decorations on the front porch look fresh, the lawn is well kept, and there is a trashcan on the side of the house. Odds look good that the house is occupied and people are home. There isn't a car in the driveway, but it could be in the garage. Children's toys in and around the home are another good tip to who lives in the home.

Victim location. Statistics (based on nationwide fireground experience) show they are in the bedroom or trying to escape and are likely to be found in the hallways, stairs, or near doors or windows. They may be on the stairs right inside the front door.

Condition of the means of egress. Without an interior size-up or report, it looks like egress will be tenable for civilians if they stay low. This condition will likely deteriorate quickly, especially if the attic has a backdraft or rapid fire development and the ceiling in the hallways or nearby rooms fail.

Size-up. One room probably, extending to the attic

> Extension to multiple rooms—not yet but very soon
>
> Nearest exposure—the attic; fire appears to be there already
>
> Fully involved—one room

Will this be an offensive attack or defensive? Offensive

On-scene report. 360-degree survey—no additional fire showing in rear or in rear windows, no victims visible, nothing unusual, firefighter shuts off gas with spanner wrench at meter.

"I'm on the scene at 2 Veterans Street, confirmed fire in one bedroom of a high ranch with extension into the attic, autoexposure A-D side." This is a good on-scene report. It confirms the address, structure type, and fire location in the house. It paints the picture in incoming firefighters' minds so they can start to think about things like hoseline selection, length of ladders needed, the overall life and property situation, possible locations of victims, and other factors. The simple words "with extension into the attic" indicate that the fire is expanding in scope and may involve structural components leading to collapse (increased danger for firefighters) and also indicate what additional resources may be needed. The on-scene report is a key component to successful and safe operations.

Deployment. Truck 1—interior team, one firefighter, one officer to force the front door and search the means of egress and the first bedroom on the right, then the living room, dining room, and kitchen. Two firefighters take a 24 ft ladder and VEIS the rear bedroom through the window at the C-D side of the house.

Engine 1—lay a supply line from the hydrant on the corner and quickly advance a 1¾ in. line through the front door to the seat of the fire.

Engine 2—stretch a second line off Engine 1 as a backup and apply water to the involved attic if necessary.

These steps provide the following advantages:

- Forcing the front door paves the way for the engine crew. Water on the fire quickly reduces most of the hazards to both civilians and firefighters on the fireground.

- Allows rapid search of the means of egress and bedrooms with survivable conditions. We know from fireground statistics that occupants are trying to escape or are sleeping. There is a good chance you will find them there.

- VEIS provides another escape route for the interior team and allows for rapid and redundant search of likely victim locations with a minimum risk to firefighters.

- The hydrant establishes a reliable water supply.

- The line through the front door protects the search team, puts the line between the fire and any salvageable victims, and reduces other hazards.

- The backup line and crew can help extinguish the fire, continue the aggressive attack, and provide additional personnel inside as needed.

Truck 2—when the line is in place, take the windows in the middle of the A side and then the picture window in the front. Prepare a crew to cut the roof after you pull the gable vents.

Radio report. "Truck one interior to command. We have fire extending into the hallway from the fire room and one victim coming out the front door." This tells you these firefighters will likely be unable to access at least one of the bedrooms for SAR. A hoseline will be required to push the fire back into the room, making the hallway accessible and tenable for the search team.

Additional resources needed in the next 30 minutes. FAST, EMS, additional engine and truck company, utility company for an electrical fire cut, other resources depending upon your local SOPs.

Additional thoughts on this scenario

1. The rapidly developing fire needs to be addressed as soon as possible. As an alternative to sending in the search team first then the hoseline, the IC could send in a hoseline to quickly attack the fire. This crew could search as they advance with a high margin of safety. The search from this hoseline crew may not be as rapid as the truck crew but certainly is much safer.

2. Search via VEIS is a rapid and viable option for important areas that may contain victims, in this case the rear bedroom. If you are on this VEIS team, what important things should you consider?

3. As you carry out your assignment you hear radio reports of problems getting water on the fire. What are your thoughts now? Here are a few things: We vented the building upon arrival by opening the front door, so the fire has plenty of air now. Flammable gases in the smoke may be ready to drive this fire to flashover, especially if water is delayed. How quickly can you get out? Do you have a quick exit?

4. You saw children's toys in the yard on arrival. You have a reliable report of people, probably children, trapped. Aggressive fire pushes out the windows of one bedroom. You think about your search considerations: Kids hide under beds, in closets, in bathrooms, showers, under sinks. Infants may be in cribs in parents' rooms. What have you forgotten? Your own safety. It's natural, but you can't save them if you are hurt.

Drill Hint. Use photos of fire scenes from trade journals or other sources and ask a member to provide an on-scene report in accordance with your department's SOGs. Then show this photo to other members of the company or department and see if the report conveyed the critical parts of the scene in the photo.

References

Ahrens, Marty. 2014. *Characteristics of Home Fire Victims.* NFPA No. USS01. Quincy, MA: National Fire Protection Association. http://www.nfpa.org/news-and-research/fire-statistics-and-reports/fire-statistics/demographics-and-victim-patterns/characteristics-of-home-fire-victims.

———. 2016. *Home Structure Fires*. NFPA No. USS12G. Quincy, MA: National Fire Protection Association. http://www.nfpa.org/news-and-research/fire-statistics -and-reports/fire-statistics/fires-by-property-type/ residential/home-structure-fires.

Avillo, Anthony. 2008. *Fireground Strategies*. 2nd ed. Tulsa: PennWell Corporation.

Bricault, Michael. 2006. "Rescue Profiles for Residential Occupancies." *Fire Engineering* 159 (9). http://www. fireengineering.com/articles/print/volume-159/issue-9/ features/rescue-profiles-for-residential-occupancies. html.

Coleman, John "Skip." 2011. *Searching Smarter*. Tulsa: PennWell Corporation.

Corbett, Glenn, ed. 2013. *Fire Engineering's Handbook for Firefighter I and II*. Updated ed. Tulsa: PennWell Corporation.

Cote, Arthur, ed. 2008. *Fire Protection Handbook*. 2 vols. 20th ed. Quincy, MA: National Fire Protection Association.

Crapo, Robert, and Noel Nellis, eds. 1981. *Management of Smoke-Inhalation Injuries*. Salt Lake City: Intermountain Thoracic Society.

Dunn, Vincent. 2015. *Safety and Survival on the Fireground*. 2nd ed. Tulsa: PennWell Corporation.

Fire Department of the City of New York. 1997. *Ladder Company Operations: Private Dwellings*, vol. 3, book 4 of *Firefighting Procedures*. N.p.: Fire Department of the City of New York.

Grosshandler, William, Nelson Bryner, Daniel Madrzykowski, and Kenneth Kuntz. 2005. *Draft Report of the Technical Investigation of the Station Nightclub Fire*. NIST NCSTAR 2. Washington, DC: Government Printing Office. http://ws680.nist.gov/publication/get_pdf. cfm?pub_id=908938.

Kerber, Stephen. 2010. *Impact of Ventilation on Fire Behavior in Legacy and Contemporary Residential Construction*. Northbrook, IL: Underwriters Laboratories.

Kerber, Stephen, Daniel Madrzykowski, James Dalton, and Bob Backstrom. 2012. *Improving Fire Safety by Understanding the Fire Performance of Engineered Floor Systems and Providing the Fire Service with Information for Tactical Decision Making*. Northbrook, IL: Underwriters Laboratories.

Knapp, Jerry, and George Zayas. 2011. "The Dangers of Illegally Converted Private Dwellings." *Fire Engineering* 164 (6). http://www.fireengineering.com/articles/print/ volume-164/issue-6/features/the-dangers-of-illegally -converted-private-dwellings.html.

Marsar, Stephen. 2010. "Survivability Profiling: How Long Can Victims Survive in a Fire?" FireEngineering.com (blog), 1 July. http://www.fireengineering.com/ articles/2010/07/survivability-profiling-how-long -can-victims-survive-in-a-fire.html.

National Fire Data Center. 2017. "Civilian Fire Fatalities in Residential Buildings (2013–2015)." *Topical Fire Report* 18 (4). https://www.usfa.fema.gov/downloads/pdf/ statistics/v18i4.pdf.

National Institute for Occupational Safety and Health. 2012. *Career Fire Fighter Dies and Another Is Injured Following Structure Collapse at a Triple Decker Residential Fire—Massachusetts*. Fatality Assessment and Control Evaluation Investigation Report No. 2011-30. https://www.cdc.gov/niosh/fire/pdfs/ face201130.pdf.

Norman, John. 2012. *Fire Officer's Handbook of Tactics*. 4th ed. Tulsa: Fire Engineering/PennWell.

Schnepp, Rob. 2009. "Where There's Fire—There's Smoke!" In "Cyanide and Carbon Monoxide: The Toxic Twins of Smoke Inhalation," supplement, *Smoke* 2:3–8. http:// www.pfesi.org/publication_files/the-toxic-twins-of-smoke- inhalation.pdf.

Walsh, Donald. 2007. "Hydrogen Cyanide in Fire Smoke: An Unrecognized Threat to the American Firefighter." In "Perceptions, Myths, and Misunderstandings," supplement, *Smoke*, pp. 4–8.

7

Preparing for Successful Fire Attack

Summary

This chapter is all about your ability and preparation (of personnel and equipment) to put water on the fire. Extinguishing or controlling the fire is often the most effective way to reduce all other hazards to both civilians and firefighters. Recent research has provided irrefutable evidence showing how important it is to get decisive amounts of water on the fire quickly. Hoselines are needed to protect the means of egress in house fires, and the interior stairs are another critical area. Successful fire control depends on your ability to understand and execute the details of your fire attack system. This chapter will cover the details and mechanics of hoseline planning, training, and operation. You may consider it a back-to-basics review of our most important operation: controlling and extinguishing the fire. (Specific fire attack operations for basement, living area, attic, and garage fires will be presented in chs. 9–11.)

Introduction

This chapter covers preparing for an aggressive interior fire attack. We will drill down deeply into the critical concepts, mechanics, and the numerous factors that you must know and the skills you and your company must have for an effective fire attack at a house fire. The focus is strictly on putting decisive amounts of water on the fire as quickly and effectively as possible.

In our scenario for this chapter, we will assume there is no life hazard and our only objective is to extinguish the fire. This chapter also assumes you have established a reliable water supply from draft, tanker shuttle, or a good hydrant.

Aggressive interior fire attack is how and when we save the most people and things. It is our most common type of fire attack, the one we as firefighters prefer, but it is also the most dangerous. We must recognize when it is not appropriate to endanger firefighters. We do not risk firefighters' lives for little gain and we must recognize when to change from an offensive (interior) to defensive (exterior) mode. That sounds like a lawyer's disclaimer but it is an important point nonetheless. We don't always immediately launch blindly into an aggressive interior attack. There are times when an aggressive interior attack is appropriate and times when it is not, and size-up must be completed to determine this at every scene. It is important to recognize that there is a clear decision point before we undertake this dangerous task. Now, let's get started on our favorite play: aggressive interior fire attack.

Scenario

Fig. 7–1. Simulation. Courtesy of Brian Duddy.

Situation: It is 1830 hours, Saturday evening, 16 September. Mom and two small children are huddled together at the end of the driveway. The 5-year-old girl cries for Ruby, her pet Labrador retriever. Mom looks stunned. As you look up at the house, Dad comes stumbling out of the house, with Ruby right behind. You ask if everyone is out of the house and both Mom and Dad agree that everyone is out. The girl will not stop hugging Ruby.

In this case the life hazard is resolved, at least until we arrive on scene and enter the building! Firefighters are a definite life hazard. Effective fire attack is just as important because lives—firefighters' lives—still depend on it. Our job is to safely minimize the expanding loss of this family's personal property through an aggressive interior fire attack. The success of your fire attack will depend upon how well you prepared long before the bells came in for this fire.

Resource Report: Your first engine has a good hydrant and has laid a supply line into the front of the building, leaving room for the truck, which pulled up immediately behind. Your preplan for this neighborhood reveals no lightweight construction, no trusses or I-joists. The homes were stick-built.

Radio Reports: Lieutenant of Engine 1: "Engine 1 to command, a 360 shows fire out one rear window on the C-D corner. Looks like we have two rooms going good and it may be in the attic."

Dispatch: "Dispatch to command, the CAD indicates no trusses or lightweight construction at that address, home was built in 1960."

Questions

1. What type of fire attack strategy is appropriate and why?

2. Describe your fire attack system.

3. What is the target flow for attack lines in your department?

4. How and why did you establish your target flow?

5. How do you train and maintain the skills of your attack team?

6. What type of nozzle does your department use for this type of fire attack and why?

Fire Attack System

What kind of nozzle do you use? That's a common question in the fire service. We all have our favorite nozzle type, water pattern, and method of operating it during an aggressive interior fire attack. The nozzle is just a small part of the overall equation that this chapter covers: your fire attack system, which includes water supply, supply line, engine or pumper, pump operator, hose, nozzle, nozzle person, and engine crew. If one of these critical components or links in the chain fails, the fire attack will fail. Lives—often ours—will be needlessly placed in danger if you don't prepare properly. So let's take a very thorough look at the fire attack system.

If we were a sports team, making our fire attack system work would be our most basic play. It is our bread and butter;

we get paid to extinguish fires. It is critical for all of us to thoroughly understand the nuts and bolts, the mechanics and finest of details required to deliver decisive amounts of water at every alarm. This chapter will cover all those critical details that enable firefighters to kill the fire. As stated previously, at house fires, making your fire attack system work flawlessly, every time, is especially critical because we put other firefighters in harm's way during SAR and ventilation operations with the assumption that hoselines will be controlling the fire quickly.

Traditionally, the interior search team would force open the front door, to provide an access point both for themselves and the engine company with the hoseline. One of the startling results of the 2010 Underwriters Laboratories (UL) study based on instrumented, full-scale live burns of test homes is summarized in the executive summary:

> If you add air to the fire and don't apply water in the appropriate time frame the fire gets larger and safety decreases. Examining the times to untenability gives the best case scenario of how coordinated the attack needs to be. Taking the average time for every experiment from the time of ventilation to the time of the onset of firefighter untenability conditions yields 100 seconds for the one-story house and 200 seconds for the two-story house. In many of the experiments from the onset of firefighter untenability until flashover was less than 10 seconds. These times should be treated as being very conservative. If a vent location already exists because the homeowner left a window or door open then the fire is going to respond faster to additional ventilation opening because the temperatures in the house are going to be higher. Coordination of fire attack crew is essential for a positive outcome in today's fire environment. (Kerber 2010, 3–4)

What does all this mean? If your department sends search teams into the burning home and the hoseline does not get water on the fire quickly, especially if the search team chocks the front door open or there are other active sources of fresh air for the fire, expect rapid fire development that may injure or kill members of the search team.

After training firefighters across our country at various hands-on evolutions and seminars (especially at the FDIC) what we have found very interesting is that many, if not most, of us don't understand some of the factors that are critical to delivering the highly desired, decisive amounts of water with our own fire attack systems—the hose, nozzles, and engines we use every day! Therefore, before we discuss how to execute an aggressive interior fire attack, let's go through the critical factors, the nuts and bolts of how to make your

fire attack system work. The knowledge, skills, and abilities you read here will be useful for all types of fires but are especially important at house fires, where both civilian and firefighter lives are often in imminent danger. Without this vital understanding, your fire attack will be dangerous for your members and less than effective for your customer. As previously stated, our friend, the late Andy Fredricks, was fond of saying, "A properly placed hoseline has saved more lives than search-and-rescue operations."

Note that through this chapter we will use the phrase *engine company* to mean the group responsible for water delivery at the scene. Understand that some squirts, squads, quints, trucks, tower ladders, and other rigs may have hose and pumps and their crews may be stretching lines at a fire. The same fire attack system principles apply to these aerial-capable engines and other companies. For the purpose of this chapter, the engine company is the group of firefighters that is establishing a reliable water supply, operating the pump, and stretching and advancing hoselines to the seat of the fire.

Repeatable

Your fire attack system and target flow must be repeatable. In other words, the goal of your company and department should be able to execute the same effective fire attack (play) whenever it is called for. The bottom line is that your department should be able to deliver the same decisive amount of water at every house fire, especially from your first hoselines. With the ability to deliver a consistent and repeatable water volume, your nozzle teams, through experience, will be able to gauge how much fire their stream can extinguish. Without consistent flow, during the fire fight, you will never know if there is too much fire or you have too little water. You must do your homework, decide what that amount is (your target flow), then prepare yourself, your equipment, and your members to achieve it every time you roll out. *NFPA 1410: Standard on Training for Initial Emergency Scene Operations (*NFPA 2005*)* and *NFPA 1710: Standard for the Organization and Deployment of Fire Suppression Operations, Emergency Medical Operations, and Special Operations to the Public by Career Fire Departments* (NFPA 2016) recommend that the first two handlines placed in service flow a combined total of 300 gpm once all first-alarm resources have arrived on scene. Several of the standard initial attack evolutions cited by NFPA 1410 provide that the first two lines flow 150 gpm each.

Target Flow

Earlier chapters described the responsibility and mission of the engine company to flow "decisive amounts of water" to ensure the safety of members on the scene. Target flow is the amount, the decisive amount, of water (in gpm) your fire department has determined you should flow from your hoselines. Most fire departments in the United States have at most two sizes of handline: 1¾ in. and 2½ in. Some departments use a 2 in. handline exclusively for interior and exterior residential and commercial fires. Regardless of your choice of hose size and target flow, the starting point for perfecting your fire attack system is to answer the question: How much do we want to flow?

We have all been around long enough to know that perfect fire attacks only exist alongside unicorns and flying monkeys. There is no plan that will work 100% of the time. However, many fire department leaders and firefighters use this as an excuse to not plan at all; they are satisfied when they stretch a line and water comes out of the nozzle. This is the leprechaun method of fire attack: stretch your hose and hope your nozzle person is lucky and has adequate water to control the fire. Those who use the leprechaun plan for target flow say, "Aw shucks, we'll just crank up the pressure when the nozzle person needs more water." Sure, that works, but you know what size nozzle you have, what size hose, how long it is. Remember, it has to work for the nozzle person too! This is a safety issue. If your drivers did not get the numeric for the street address and answer you with something like, "We'll just go to the street and wander around until we find the fire" when you ask where you're headed, I don't think you would be satisfied. Hope is not a plan.

Consider this as well: When does the nozzle person need more water? When they are being overrun by fire and getting burned? It might be a bit difficult for them to get on the radio and request more water at this point. You must make your fire attack system work the first time, every time. If your nozzle person is not making progress on the fire, they need to know that it is because the fire is too big for the line, not because the line is flowing something less than your target flow. With some work and some training, it is easy; getting this right makes our job safer and makes fire attack more effective.

The bottom line is this: most fire departments are not flowing the amount of water they think they are. They are not achieving their target flow. We recently tested an engine that was supplied by a 4 in. supply line to a 1,750 gpm pump to a 2 in. handline with a 1 in. solid bore tip. This department thought they were getting flows of 200+ gpm. The day of the test that line flowed 125 gpm. That flow you could get out of an older style 1½ in. hose, which is a lot lighter and easier to move around than a 2 in. one. They missed their target by 75 gpm. Less water flowing than anticipated has been the rule rather than the exception in our experience testing fire attack systems.

Lack of water or adequate water has been the cause of many firefighter injuries and line-of-duty deaths (LODDs). Read the numerous National Institute for Occupational Safety and Health (NIOSH) LODD reports if you doubt this. There are various reasons for the lack of adequate water at the nozzle; what never varies is the result: dead or injured firefighters and at a minimum, needlessly lost property. Here are two excellent guiding rules to answer the question how much water will we need at a house fire:

1. The target flow for each interior attack line at a house fire should be 180 gpm.

2. You should have one line for every two rooms of fire in a house (plus backup lines and lines to floors above).

Why 180 gpm? You don't want your fire attack to be a fair fight. You want to overwhelm the fire quickly. A long, drawn-out firefight allows the fire to win, endangering firefighters and civilians and destroying more property. Are you worried about water damage? Captain Christopher Flatley, FDNY, has a famous saying about water flow: "How do you make a 180 gpm line flow 90 gpm? Flow it for 30 seconds!"

Another reason to use 180 gpm for your target flow is that there are a number of things that will reduce this flow. Kinks, elevation, and bends in the line are just a few of the most important. Also remember that theoretical friction loss numbers are for new hose, laid straight without kinks on a flat surface. If you shoot for 180 gpm and only get 150 gpm, you are still in the decisive-amount-of-water category. If you shoot for 150 and only get 120, you are in the fair-fight category. Anything below this and you are in leprechaun territory.

Theoretical required flows

There are several old and new quick methods for estimating required fire flows. The primary one that the authors suggest was developed by the National Fire Academy (NFA).

NFA formula. According to Chief James P. Smith (retired),

> The National Fire Academy (NFA) in Emmitsburg, MD, has developed a formula that allows for quick calculations. The formula was derived through a study of fire flows that were successful in controlling a large number of working fires, along with interviews of numerous experienced fire officers throughout the country regarding fire flows they have found to be effective in various fire situations. The NFA quick-calculation formula can be used as a tactical tool to provide a starting point for deciding the amount of water required at an incident scene. This will permit decisions to be made on the amount and

type of apparatus needed to deliver the water and the number of firefighters that will be needed to apply it.

The information developed by the NFA indicated that the relationship between the area involved in the fire and the approximate amount of water required to effectively extinguish the fire can be established by dividing the square footage of the area of fire involvement by a factor of 3. This formula is expressed as:

$$\text{Fire flow} = \text{length} \times \text{width} \div 3$$

....The formula can be applied to a single-family dwelling 60 feet long by 20 feet wide and one story high:

$$60 \times 20 \div 3 \times 1 = 400 \text{ gpm}$$

$$100\% \text{ involvement} = 400 \text{ gpm}$$

$$50\% \text{ involvement} = 200 \text{ gpm}$$

$$25\% \text{ involvement} = 100 \text{ gpm}....$$

In multistory buildings, if more than one floor in the building is involved in fire, the fire flow could be based on the area represented by the number of floors actually burning. For example, the fire flow for a two-story building of similar dimensions as the previous example would be:

$$60 \times 20 \div 3 \times 2 \, (\text{floors}) = 800 \text{ gpm if fully involved}$$

If other floors in a building are not yet involved, but are threatened by possible extension of fire, they should be considered as interior exposures, and 25 percent of the required fire flow of the fire floor should be added for exposure protection for each exposed floor above the fire floor to a maximum of five interior exposures. (Smith 2012, 17–18)

In essence, for house fires, a practical, street-wise, experienced-based estimate is that handlines flowing 150–180 gpm will control two rooms of fire. This takes into account flows, staffing levels, and other factors.

Inadequate flow

To demonstrate how very important target flows are, we'll examine the case history of a fire that resulted in the tragic death of a firefighter. In one of the most practical summaries of the importance of target flow ever researched and printed, Captain Jay Comella of the Oakland FD (CA) describes the importance of delivering your target flow:

The outcome of fireground operations depends on the outcome of the battle between the water the engine company delivers (gpm) and the heat (Btus) the fire generates. The flow at which the engine company can win the battle and kill the fire is defined as the critical flow rate. If the critical flow rate is not met, the battle will be lost. This dictates that the single most important characteristic of a hose and nozzle system is water flow capability. The water the engine company delivers must be sufficient to expediently kill the fire. Maneuverability of the hose and nozzle are important factors, but to sacrifice flow for ease of use has proved to be suicidal. (Comella 2003)

Comella highlights the need to apply appropriate amounts of water in this article because lack of adequate water is what caused the LODD he is describing. This is just one of many similar case histories. Fire service LODD reports are strewn with firefighters being burned to death by flashover or eventually overrun by fire because of inadequate water delivery. Comella describes the scene of this particular incident:

On arrival, the first-alarm companies encountered a heavy fire condition on the first floor with extension up the stairway. They made an aggressive interior attack using multiple 1½-inch handlines. Fire was not extinguished in time to prevent the loss of structural integrity. The resulting collapse of the second floor into the first floor killed one OFD [Oakland Fire Department] member and left two others with career-ending injuries. One of the three direct causes the Board of Inquiry report cited for the line-of-duty death was the inability of 1½-inch hose to flow sufficient water to extinguish the heavy volume of fire encountered. (Comella 2003)

Chief Alan Brunacini (retired) summarizes the firefighting battle this way: "The basic firefighting deal is pretty simple, always serious, and very severe. It's about who murders whom first. The fire shows up to kill everything and everybody (including us). We show up to kill the fire. This mutual murder contest creates a very dangerous situation." Simply, water is our weapon to kill the fire. Always bring enough—no, more than enough—to the fight and be able to deliver it in decisive quantities quickly (the quicker the better).

Test for the target flow

The only way to determine if you are actually achieving your target flow is to test it. Captain Christopher Flatley summarizes the test procedure in this diagram from an article he published in FE in October 2006 (fig. 7–2).

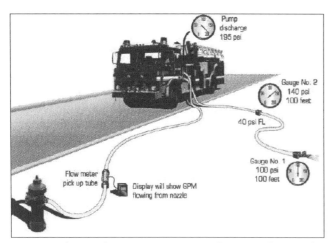

Fig. 7–2. Flow and pressure gauges used to test a fire attack system. In summary, a flow meter is put on the intake side of the pump.

Since water is not compressible, and assuming your engine is not losing water from some other source (open drain, water flowing into the tank or overflow), the volume of water coming in equals the volume of water going out through the fire attack system that you are testing. A gauge at the nozzle ensures that the correct pressure is reached at the nozzle, a gauge midway in the stretch allows you to measure friction loss for your hose, and a gauge on the engine allows you to calculate friction loss in the pump.

Items that reduce target flow

There are several factors that will affect your ability to achieve you target flow: friction loss in the pump, friction loss in the hose, setting the correct engine pressure, and choosing and using the correct nozzle to achieve your desired flow. This simple but effective test procedure measures the effect of the following on your target flow. Let's look in detail at these critical factors.

1. Friction loss in the pump. This is something we usually do not consider. The amount of friction loss in the pump might be zero and have no effect, or it might be as much as 35 psi, which will drastically effect the flow. We tested an engine that had a 2½ in. preconnected hoseline and a 1⅛ in. smoothbore nozzle. Theoretically, it flowed about 260 gpm at 50 psi nozzle pressure. Flow tests (numerous) revealed that it was actually flowing 180 gpm, because the heavy, bulky 2½ in. line really flowed the equivalent of a 1¾ in. line's water. When we looked at the piping, we found it to be only 2 in. in diameter.

We put a pressure gauge at the 1⅛ in. smoothbore nozzle and found that it was grossly underpressurized, resulting in the low flow. In this case, the piping and elbows reduced pressure and flow due to friction loss. Beyond the pressure sensor for the discharge is approximately 4 ft of 2 in. piping with two 45-degree elbows and one 90-degree elbow. The

friction loss in this case was 35 psi, which is very significant and is not considered by the pump panel pressure gauge reading or the hose friction-loss calculation. Flow testing is the only way to measure and account for these losses to ensure you are achieving your target flow (fig. 7–3).

Fig. 7–3. Pressure sensor and discharge valve leading to preconnected hoseline

2. Friction loss in the hose. Friction loss in hoses varies depending on several factors. The volume of the flow through the hose is an ever-present factor. More volume equals more friction loss; it's as simple as that. So when we talk about friction loss we must say that the friction loss is X psi at Y gpm. Example: 20 psi at 180 gpm. You must link an amount of friction loss with an amount of flow. For example, theoretical friction loss charts show a friction loss of 35 psi per 100 ft of 1¾ in. hose flowing 150 gpm. At 200 gpm, it is 62 psi. So what is the friction loss of 1¾ in. line? It varies with the flow! The gauge layout in figure 7–3 will measure the friction loss of your hose at your target flow. This system of gauges and flow meters is an excellent way to demonstrate the relationship between friction loss and flow. This procedure will also help determine if the gauges on your engine are not reading correctly and need calibration.

Another major factor in determining friction loss in the hose is the exact condition, type, and manufacturer of the hose. Assuming we are discussing the same diameter hose, the age and condition of the hose are key factors. Older hose is probably stretched out by the pressures it has experienced, causing the diameter to be greater. The greater the diameter of the hose, the less friction loss for all flows.

Conversely, hose that is older and has semipermanent kinks in the rubber lining or is rough from dry rot or damage will increase friction loss.

You would think that fire hose is fire hose; theoretically, 1¾ in. hose is 1¾ in. hose. As in life, nothing is that simple. There are many manufacturers and, similarly, many types of 1¾ in. hose. Simply stated, some stretch more than others, altering the diameter and friction loss. Intuitively, you would think this would be a minor factor but when combined with the different types of hose lining, friction losses can vary greatly.

It is important to remember that the rubber lining of all hose is not smooth. Some hose has in fact a rather rough inner surface. Mac McGarry and Bill Gustin have done extensive testing and measuring friction loss in various hose types and manufacturers and found huge variations in amounts of friction loss. For example, within just one manufacturer's various models of hose, friction loss varied by model of hose (all 1¾ in.) from 25 psi per 100 ft to 72 psi per 100 ft. These variations were flowing 185 gpm with a ¹⁵⁄₁₆ in. tip (personal conversation with McGarry and Gustin, June 2014). You need to test your system to determine the specific friction loss in the hose you are using.

3. Correct pump pressure. This is another factor that you would think would be a given for us, but it often is not. Remember, to calculate the correct pump pressure you need to add the following: friction loss in the pump + hose friction loss + nozzle pressure = pump pressure.

4. Correct nozzle pressure. Nozzles are designed to operate at specific pressures. Generally, smoothbores operate at 50 psi. Up until a few years ago when NFPA changed its standard, all combination nozzles were designed to operate at 100 psi. Now combination nozzles and automatics can operate at 50, 75, or 100 psi. You have to know what pressure your nozzle is designed to operate at so you can ensure that your target flow is possible and practical (fig. 7–4).

Kinks. Another common factor that will affect your nozzle pressure and reduce your target flow are kinks in the line (fig. 7–5). According to an article published in *Fire Engineering*, the effect or reduction in gallons per minute of your decisive amount of water depends on two interrelated factors: the severity of the kink and the type of nozzle on that line (Pillsworth et al. 2007).

Fig. 7–4. Left to right: A ¹⁵⁄₁₆ in. smoothbore, a low pressure (50 psi) combination, a 75 psi combination flowing 175 gpm, and an automatic nozzle operating at 100 psi. Each nozzle has its own advantages and disadvantages for an aggressive interior fire attack at a house fire.

Fig. 7–5. Kinks in the lines as they come off the rig

In these tests, an automatic nozzle (180 gpm at 100 psi nozzle pressure), a low pressure combination nozzle (150 gpm at 50 psi nozzle pressure), and a ¹⁵⁄₁₆ in. smoothbore (180 gpm at 50 psi nozzle pressure) were used. An ingenious system was used to maintain the same severity of kinks for each trial. To summarize the data, as the kink severity increased, target flow was reduced but with varying degrees of water loss (figs. 7–6 through 7–8).

Kinks	PDP	GPM	NP	GPM Reduction	Reach
No Kink	150	150	110	—	—
1–90°	150	120	115	20%	NSC*
1–135°	150	105	105	30%	NSC
1–180°	150	75	100	50%	Poor
2–90°	150	115	115	23%	NSC
2–135°	150	100	110	33%	NSC
2–180°	150	30	90	80%	Poor

*no significant change

Fig. 7–6. Automatic nozzle: flow reduction from 150 gpm to 30 gpm (80% reduction)

Kinks	PDP	GPM	NP	GPM Reduction	Reach
No Kink	90	150	55	—	—
1–90°	110	150	65	0%	NSC*
1–135°	115	140	60	7%	NSC
1–180°	120	120	45	20%	NSC
2–90°	110	140	60	7%	NSC
2–135°	130	95	25	37%	Poor
2-180°	125	105	35	30%	Poor

*no significant change

Fig. 7–7. Combination nozzle (low pressure): flow reduction from 150 to 105 gpm (30% reduction)

Kinks	PDP	GPM	NP	GPM Reduction	Reach
No Kink	120	180	54	—	—
1–90°	120	175	50	3%	NSC*
1–135°	125	150	40	17%	NSC
1-180°	125	135	25	25%	Poor
2-90°	120	155	40	14%	NSC
2–135°	135	105	20	42%	Poor
2-180°	130	115	20	36%	Poor

*no significant change

Fig. 7–8. Smoothbore nozzle: flow reduction from 180 to 115 gpm (36% reduction)

Obviously, the severity and number of kinks are the primary factors that determine how much of a flow restriction there will be. However, look at the losses of water (percentage from target flow) at each different kink for each different nozzle. The automatic gave up the most gallons per minute at each severity of kink but maintained good reach. The low pressure combination and the smoothbore kept significant amounts of flow but suffered some in reach.

Clearly, it is both the type of nozzle you use and the severity of the kink that will determine how much of your target flow you lose. The study used above helps support the rec-ommendation that you should pick 180 gpm as your decisive amount of water, your target flow. The reality of the fireground is that kinks and other obstacles to your target flow will occur and it will be reduced. To protect your members and be effective on the modern fireground, start high and expect it to be reduced by fireground realities. If you do this, you will have built in a margin of safety into your fire attack system for your next "mutual murder contest."

The Attack Team

Now that we have established how much water we want to deliver and all the variables that may affect it, let's turn to the human portion of your fire attack system: the attack team and their responsibilities. How well these firefighters can get the line in position, advance, and kill the fire will determine how effective the water in your target flow is. We acknowledge that not all departments will be staffed to have four firefighters stretching and operating their hoselines. What we would like you to take away from these duty descriptions are the key tasks that must be accomplished by someone (maybe someone doing double duty, filling both positions) to make it an effective stretch.

There are a thousand ways to pack, stretch, operate, and advance a hoseline for an aggressive interior attack. However you choose to do it, two factors are required. First, members must understand their specific responsibilities on the line, in both the stretch and the advance. Second, the attack operation must be a repeatable play or sequence of events.

Before we review some of the most common and critical responsibilities of the members of the attack team, a note of understanding and caution. Your department's fire attack SOPs and hose load may be vastly different from the plan that follows. Names of positions are not important. The tasks listed below must be accomplished by someone off the big red truck for a successful interior fire attack. Don't get hung up on the specifics; just focus on the common tasks for all loads and stretches. Whether you call it an attack team or engine company, the job is the same: put the fire out. As Andy Fredericks always said, "If you put out the fire, you don't have to jump out the window!" When asked what size bailout rope he preferred, Andy also liked to quip, "1¾ in., of course!"

Engine officer duties

Remember, we are discussing duties of the attack team with the assumption that size-up has been conducted by a chief on the scene (or has been done by the first-in officer) and other units have been assigned to SAR and ventilation. The engine officer's primary duty at this point is to get water on the fire quickly and use the first line to protect the members conducting the search (fig. 7–9). Other duties are listed below:

1. Receive orders from chief officer already on scene.

2. Locate fire and tell the crew the best route to it.

3. Be sure that the line and hose stream will not endanger the SAR crew or others.

4. Keep members out of the doorway flue.

5. Keep members at the floor level, below dangerous heat areas.

6. Coordinate timing of ventilation through command.

7. Be the eyes and ears of the attack team.

8. Supervise the nozzle person.

9. Call for relief when necessary.

10. Be responsible for the lives of the attack team.

Fig. 7–9. Engine officer's duties

Nozzle person's duties

It cannot be overemphasized that the attack team must know and be able to execute their duties without error, quickly and effectively. Especially at residential fires, other firefighters' lives depend on these skills. One of these critical skills is a simple one during the stretch. The nozzle person must be able to carry the nozzle and the first 50 ft of hose (working length). As they move away from the engine, they stop forward progress about 25 ft out and wait for the backup firefighter to shoulder their hose, then move toward the fire as a team. This simple skill is only developed through repeated and coached training-ground evolutions. The few seconds taken to wait for the backup firefighter will save minutes getting the line stretched and in operation, reducing the

danger to everyone on the scene (figs. 7–10 and 7–11). The full list of skills and duties are as follows:

1. Get direction from the officer regarding the size of the line, team's purpose, and best route to the fire.

2. Carry the nozzle and the first 50 ft of hoseline.

3. Stop and wait for the backup firefighter to shoulder their assigned hose.

4. Bleed air at door, check the flow, and don facepiece.

5. Communicate with officer and backup firefighter during stretch and advance.

Fig. 7–10. The nozzle person is responsible for getting the first length to the point of attack, where it will be charged and advanced.

Fig. 7–11. At the point of attack, the nozzle person must bleed the line of air. They must flow water to full capacity to ensure that they have enough pressure and volume. A squirt proves only that that little amount is available.

Fig. 7–13. After the stretch, the backup flakes out the line and does their most important job: relieving all the nozzle reaction for the nozzle person, allowing the line to advance smoothly.

Backup firefighter's duties

1. They are the second firefighter in the stretch (fig. 7–12).

2. They carry the second 50 ft of line and drop it at the appropriate time and place (fig. 7–13).

3. They communicate with the nozzle person.

4. They relieve the nozzle reaction for the nozzle person during the advance.

5. They watch the nozzle person's body and head positions during the advance.

Third firefighter or door firefighter duties

Many departments are understaffed and do not have a third firefighter in the attack team. This member may come off an additional piece of apparatus. Because of the many turns and twists the line can take once inside and the numerous obstacles that can stop the stretch (car tires, fences, hedges, steps, hand railings for outside steps, etc.) house fires in particular require this third firefighter for an efficient and timely stretch and advance (fig. 7–14). The third or door firefighter has six duties:

1. They are the third firefighter in the stretch.

2. Clears the preconnect bed or may estimate and disconnect the hose from the bed and connect to engine discharge.

3. Chases kinks after line is charged (fig. 7–15).

4. Feeds line from edge of fire area to nozzle team (figs. 7–16 through 7–18).

5. Can relieve nozzle or backup firefighter or officer, because they have conserved air by not being on air.

6. They are the safety link to outside; they know how far and where the team has advanced.

Fig. 7–12. During the stretch, the nozzle person must wait a few seconds for the backup to shoulder the hose and then advance as a team.

Fig. 7–14. Third firefighter's job is to estimate the amount of hose you will need to reach the seat of the fire, break it, and connect it to the pump or clear the bed if you are using preconnected lines.

Fig. 7–15. Though not glamorous, clearing kinks to ensure that the target flow is delivered to the nozzle person is everyone's job during a fire attack

Fig. 7–17. The door firefighter can simply lay a bend of hose on the floor to speed the nozzle advance to the seat of the fire.

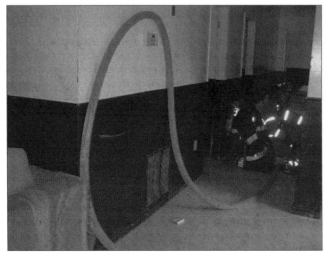

Fig. 7–16. Door firefighter can push hose up on a wall, making it easier for the nozzle and backup firefighters to advance.

Fig. 7–18. Even if the hose is obscured by smoke or darkness, a loop or S in the hose will allow the attack team to quickly advance without being pushed by the rigid hose.

Training your attack team

Training your attack team can seem like an overwhelming task. From our experiences with our engine company operations teaching program at the Rockland County Fire Training Center for the past 20 years, we offer the following suggestions:

1. Write a simple SOP describing the duties you expect each member to execute.

2. In the class room, review and explain these duties.

3. On the training ground, walk through the duties of each member in a slow-motion stretch and advance.

4. Advance to a half-speed evolution, coaching members as they go.

5. Continue to a full-speed evolution that includes donning facepieces at the door, calling for water, checking flow, and then advancing into a training building with proper skills.

We have found it critical to use this crawl, walk, run method both for retraining experienced firefighters and for training rookie firefighters. For veteran firefighters, it is a good review of procedures, and for rookies, it reinforces the team concept and how important it is for success and safety.

Another critical engine company training aspect when teaching hoseline advancement is not to use fire or smoke even if you have a live burn facility available to you. Although firefighters want to train and operate in smoke and heat conditions, it does not allow the trainer to focus on evaluating and improving the skills of advancing and operating the line through coaching. It is difficult to communicate under smoke conditions through a mask.

Nozzle Selection

Selecting the nozzle or stream pattern (straight stream or fog) has been a persistent hot topic in the fire service. The purpose of this section is to show you the advantages and disadvantages of some common nozzles and fire stream patterns so you can make an educated choice.

It is important to recall several important facts:

- We are dealing specifically with house fires here. Not industrial, not commercial, not a specific or high-hazard occupancy. With decisive flows, the fire should knock down fairly quickly.

- Live burn data show that ceiling temperatures can peak at 2,370°F+ for a well-ventilated living room fire. This is a high hazard to firefighters when we recall that our skin burns at around 124°F (Kerber 2010).

- The SAR teams might be operating in the building before or concurrently with the hoseline. It is especially important not to use a fog stream that will push fire or superheated fire gases on them.

- Ventilation (horizontal then vertical) will be accomplished as part of the fire attack.

Nozzle characteristics

For a house fire you want a line that is very maneuverable but at the same time delivers your target flow. Think about it: rooms are generally small, hallways tight, and 90-degree turns will be the norm rather than the exception. Even in the largest of homes, our handline streams will provide more than adequate reach and penetration. This is not a warehouse fire where you will stretch 300 ft straight from the door to get to the fire area. There will be lots of bends and, as you know, the fire may meet you at the door of the house.

Lower pressure nozzles and therefore lower pressure lines are easier to flex than higher pressure lines. Most automatic or combination nozzles operate at 100 psi. Lines pressurized to this level are very rigid. The advantage of this relatively high pressure is that many of the kinks tend to resolve themselves. The disadvantage is they are hard to bend around corners.

Nozzle reaction

Another disadvantage to a 100 psi nozzle is the nozzle reaction. Nozzle reaction is a function of both volume and pressure. A 100 psi combination nozzle flowing 180 gpm has a nozzle reaction of about 80 lb that the nozzle team must counter every second the nozzle is flowing. This high nozzle reaction obviously fatigues the nozzle person and the backup firefighter, which may ultimately lead to the attack's stalling when they are exhausted. In these days of understaffed engines, this is a serious concern. Relief for the initial attack team may be a long time coming.

It is difficult to judge nozzle reaction by holding a line in your hands. There are theoretical nozzle reaction numbers in many hose and flow charts. It is important to be able to accurately measure and demonstrate nozzle reaction to trainees and decision-makers alike. Trainees need to understand how difficult the backup firefighter's job (relieving all the reaction) is and decision-makers need to see the differences in nozzle reaction from one nozzle to the other before making unwise purchases and policies. Often it is the skill and muscle power of the backup firefighter that makes or breaks the fire attack operation.

Here is an excellent method we developed to measure nozzle reaction. A spring fish scale is attached to an anchor point using a piece of webbing (fig. 7–19). A choker is made

Fig. 7–19. A spring scale, an anchor, and two pieces of webbing can accurately measure nozzle reaction and help you choose a nozzle for your fire attack system.

of another piece of webbing and placed just behind the coupling at the nozzle. When the nozzle is opened, the scale will read very accurately the amount of nozzle reaction at this specific flow and pressure (figs. 7–20 and 7–21).

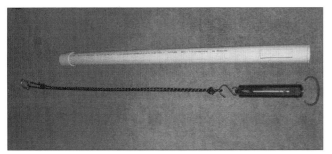

Fig. 7–20. This is the same spring scale, this time with a short piece of rope or webbing attached to the hook end. The rope is threaded through a piece of 2 in. plastic pipe that has a square hole cut out so the trainee can see the numbers on the scale.

Once you have a way to measure nozzle reaction, you need a way to translate that into a training method or device. I have found that it is critical to explain nozzle reaction and show and train new recruits how to be good backup firefighters and counter all the nozzle reaction.

A good alternative to the 100 psi combination nozzles is the low-pressure, 50 psi combination nozzle. The lower nozzle pressure results in a line that is very flexible and can easily make the multiple turns you will make as you push into the house to the seat of the fire. This softer line is more prone to kinks, but some firefighters think it is worth the trade-off. The reaction force for this nozzle is about 45 lb flowing 150 gpm. This is almost a 50% reduction over the 100 psi combination. This means less physical stress on the attack team and a more aggressive and, if needed, prolonged fire attack.

Fig. 7–21. The rope is attached to a solid anchor point with a piece of webbing and a carabiner as needed. The PVC pipe simulates a hoseline for the trainee. The trainee can now act as the backup firefighter, pushing the hose forward to relieve the previously measured nozzle reaction. The scale is visible through the cut out of the PVC, showing the trainee just how hard they have to work to counter the nozzle reaction.

This nozzle has another advantage. If you are using both smoothbores and low-pressure combinations, the pump discharge pressures will be the same since the nozzle pressure for both is 50 psi (figs. 7–22 and 7–23).

Fig. 7–22. A 50 psi combination nozzle that flows 150 gpm

Fig. 7–23. This low-pressure combination nozzle is a great addition to the smoothbore nozzles you may have on other lines. It operates at low pressure, low nozzle reaction, and relatively high flow.

The smoothbore is another good choice for an aggressive interior fire attack. It operates at low nozzle pressure (50 psi). A $^{15}\!/_{16}$ in. tip will flow 180 gpm with a nozzle reaction of 69 lb. This nozzle reaction is about the upper level of what one firefighter can handle alone. Teamed up with a backup firefighter, the attack team will be able to advance but will have to work together to ensure swift fire attack. The advantage of this nozzle is that it flows high volumes of water with acceptable nozzle reaction. It appears to be a good balance of high flow and low nozzle reaction. It is less prone to clogging from debris and scale that may make it through the hydrant, supply line, and pump.

If your target flow is less than 180 gpm, you can downsize to a $^{7}\!/_{8}$ in. tip that will flow 160 gpm with a 60 lb nozzle reaction. You should try out each of these tips in realistic fire attack scenarios to be sure your members can handle and advance the lines effectively. If you have access to a training building, make your members do everything they would during a real fire attack and watch them as they advance and evaluate their progress. Leave out the smoke and fire; if they operate the line properly the fire would go out anyway. If you build a training fire, you will not be able to observe and evaluate.

Air flow

An important characteristic of a nozzle (especially for fire attack at house fires) is the amount of air it moves into the fire area. During our careers when conducting aggressive interior attack we don't have time to observe the air movements in the area where we are working. Most firefighters have no idea how much their line will disturb the superheated air in the house. Additionally, many firefighters don't realize how much air their nozzle will move into the fire building. This air can push fire and superheated smoke onto the search teams, which will certainly end that critical operation.

A critical part of the Underwriters Laboratory study *Impact of Fire Attack Utilizing Interior and Exterior Streams on Firefighter Safety and Occupant Survival* was measuring the air flow from fire streams. Commonly called the air entrainment portion, it showed that air movement by solid streams from smoothbore nozzles and streams from combination nozzles were similar. If the nozzles were held in a fixed position they moved much less air into the fire area than if they were moved in typical fire service motions such as O, Z, or N while attacking the fire. This built on the nozzle testing and experimentation at the Rockland County Fire Training Center by Knapp, Pillsworth, and Flatley 2003.

UL data showed that when the smoothbore nozzle ($^{7}\!/_{8}$ in. tip, 150 gpm at 50 psi) was held still, about 1,000 ft^3 of air per minute (cfm) was moved. When the nozzle was moved in an O motion, air movement increased to 5,000 cfm. A straight stream from a combination nozzle went from approximately 1,500 cfm to 5,500 cfm. A narrow fog started out at over 8,000 cfm and went to 12,000 cfm (Stakes 2017).

It appears that solid and straight streams have been so effective at fire suppression over the years because they move relatively little air into the fire area. This limited amount of air can be exhausted through vented windows in the house. Consider that there are 1,536 ft^3 of air in a 12 ft × 16 ft × 8 ft room. The smoothbore is only moving about a third of that volume every minute into the room. More importantly, because this line is flowing a decisive amount of water (180 gpm), the nozzle person will only have to operate it for a few seconds to darken down the room. This limited water application time and limited air flow does not overpressurize the room with air from outside the fire area, and the vent (most desirably ahead of the hoseline) can relieve this amount of entrained air and fire gases and smoke (out the vent opening), allowing the nozzle team to move through this room to continue the fire attack.

The lack of air movement from a smoothbore nozzle also assists in keeping the thermal balance of the room intact. The superheated 2,000-degree gases hanging near the ceiling are not swirled, turning the room into firefighter's soup. As you may have guessed, the fire attack team is the main ingredient in this painful appetizer.

Many firefighters still advocate the use of fog streams for interior fire attack at house fires. This is because their experience tells them that fog tips work really well for one-room fires. But you can do almost anything with a one-room job and get away with it. On a good day with a bit of luck, you could use a garden hose and be successful. As firefighters, we often must move through the one room of fire to additional rooms. If we make firefighter's soup in the first room, it will be difficult or impossible to make the next room.

In summary, the smoothbore sends a solid stream of water against the ceiling, breaking it into large drops of water that penetrate the heat on the way down to the burning material. These large drops (we prefer the word "globs") of water extinguish the material in a decisive manner. Again, because of the high flow, often this line is operated only 20–30 seconds on a typical room-and-contents fire to achieve complete knockdown of the fire. Remember, we don't want this to be a fair fight; we want to win and win as quickly as possible. The lives of other firefighters depend on this rapid extinguishment.

Straight streams off combination nozzles

Many firefighters argue that straight streams off combination nozzles are not as good as smoothbore tips. In terms of airflow into the fire area, our testing showed there is little difference. Some argue that the water drops are smaller, making the nozzle less effective. We think people are splitting hairs at this point and that other factors are much more important than this.

In a 1995 *Fire Engineering* article, District Chief Andrew O'Donnell had this to say, based on his valuable experiences from his early days on the busiest engine companies in Chicago: "I have been involved in many comparison tests of nozzles without the help of salespeople or other biased interference and conclude that, first, all nozzles vary from manufacturer to manufacturer (depending on quality of design and workmanship in eliminating water turbulence within the device) and, second, in the real world of interior structure firefighting, the difference between smooth bore and straight stream, in terms of direct attack effectiveness, is negligible" (O'Donnell 1995).

Fog nozzles for house fires

The use of a fog nozzle for an interior attack at a house fire is questionable at best. A nozzle set to a fog pattern is unsuitable for house fires because of the large amount of air that it will move into the fire area (fig. 7–24). The fire area and cubic volume in most homes is usually quite small.

Consider what happens when we use a fog line in a well-involved room of fair size (12 ft × 16 ft × 8 ft) in a home. This room contains 1,536 ft³ of air. In the testing mentioned above (Knapp, Pillsworth, and Flatley 2003), the authors measured with certainty that 2,000 ft³ of air per minute moved into the fire when a 1¾ in. fog nozzle flowing 150 gpm at 100 psi was used. They further estimated conservatively that

the figure was actually closer to 6,000–10,000 ft³ of air per minute.

Using the conservative estimates (6,000 ft³ of air per minute), the nozzle person introduces 2,000 additional ft³ of air every 20 seconds they operate the fog nozzle. Thus, in 20 seconds you have more than doubled the air that can be contained in that space. These estimates were in the same range as UL's measurements for fog stream air entrainment as part of the study *Impact of Fire Attack Using Interior and Exterior Streams on Firefighter Safety and Occupant Survival.*

Fig. 7–24. A fog nozzle is being directed into a room with a fully open door. At the arrow you can clearly see how the stream is driven back toward the nozzle team because the room, even with the large vent, is overpressurized by the massive amount of air drawn in from behind the nozzle person and directed forward.

The massive inrush of air from the fog nozzle creates the following dangerous conditions for firefighters:

- It can move fire and superheated products of combustion onto other firefighters working inside. Remember, the floor plan of most homes includes an open stairwell. The most current instrumented live burns done by UL and NIST clearly show (via both video and thermal image footage) that fog nozzles are very capable of pushing fire into uninvolved areas.

- The negative pressure at the outermost parts of the fog stream will draw heat and smoke back onto the nozzle team.

- Fog nozzle ricochet will occur because the overpressure in the room is driving more superheated air and steam back onto the nozzle team.

- This inrush of air may force fire or additional air to support combustion into void spaces, spreading the fire and causing the building to die a slow death.

- The inrush of air that continues as long as the fog stream is operated thoroughly mixes the thermally stratified air in the fire room. Thus the 2,000°F temperatures near the ceiling are now mixed with the cooler 100°F air at floor level, creating a heat-containing moist atmosphere in the room that is untenable for firefighters, who at the least must stop their aggressive advance to the seat of the fire and more likely are forced from the building, allowing the fire in rooms beyond the original to continue to burn.

The fog stream itself has other disadvantages for use at house fires:

- The small drops of water, so efficient at absorbing heat, do just that and create steam that is driven back onto the nozzle team (by the aforementioned overpressurization) and usually the backup firefighter, stopping the fire attack.

- These little drops of water will be vaporized long before they can reach the burning fuel, creating steam and, more importantly, not extinguishing the burning fuel in a final manner. Partial fire extinguishment causes incomplete combustion and large volumes of smoke, further slowing the nozzle team's progress and possibly making the search much more difficult.

- The fog stream has little reach. In the case of a long hallway with fire roaring in rooms on either side, the stream does not penetrate the heat, allowing the fire to continue. Straight or solid streams deliver large globs of water that can penetrate the heat and be bounced off walls and ceilings to control the fire in adjacent rooms or down the long hallway.

You may have noticed that this discussion does not mention the use of fog tips for ventilation. Nozzles are selected and used based on how effective they are at delivering decisive amounts of water to your target flow, not on their value as a ventilation tool. Do you ask how many gallons per minute the vent saw delivers before you purchase it?

Actually, the fog tip is an excellent tool for hydraulic ventilation after the fire is knocked down. As previously stated, it will move huge volumes of air, actually in the same range as the small electric smoke ejectors we are all familiar with. However, you must select your nozzles for their fire attack capabilities, not their ventilation uses.

The recommendation for an aggressive interior fire attack at house fires is to attack with zero-degree fog (straight stream) then use the fog pattern to ventilate *after* the fire is knocked down.

Fog nozzle misinterpreted. The confusion over the use of fog nozzles for interior fire attack is due to the fire service's misinterpretation of two books written by Chief Lloyd Layman in the 1950s. Chief Layman was a shipboard firefighting expert during the Second World War. He had great success using fog streams on shipboard fires. In essence, he found that if you injected a water fog stream into the closed hold of a ship, then sealed the compartment, the steam would extinguish the fire. He then attempted to transfer the success he'd had with fog to land-based structural firefighting, and had similar success with building fires where there was no ventilation by the fire or the fire department.

The misinterpretation was primarily related to the following two points:

1. Modern aggressive interior attack for house fires includes aggressive ventilation of windows, doors, and roofs. This obviously releases any steam produced by the fog lines, rendering it nearly useless for controlling the fire. Remember, in 1955 SCBA was not a constant companion for firefighters. Chief Layman's tactics kept the building as closed up as possible. Leaving the building sealed up (unvented) makes the death of any occupants a certainty and ensures that no SAR operation can be performed.

2. Chief Layman never said to take a fog line into a structure. Many pictures in his book show firefighters directing fog streams into windows from *outside* the structure. In fact, the following quotations from his books clearly state that it is dangerous to use fog inside structures: "A doorway is an undesirable type of opening due to the size, and usually the nozzleman will have to discontinue the attack and retreat to avoid envelopment by heated smoke and steam." He also said, "Where the area of major involvement is on upper floors, it may be possible and practical to attack from an interior stairway below the involved floor. The nozzleman may have to discontinue the attack temporarily to avoid the downward movement of heated smoke and steam." Recall also that turnout gear of the 1950s and '60s was a rubber coat with a flannel liner. (Layman 1955, 46)

When considering using fog streams inside a structure, remember the following:

- Don't base your fire attack plan for house fires on your successes at one-room fires.

- Don't base your fire attack on how the nozzle you select performs in the parking lot.

- Evaluate your fog stream and the massive amount of air it moves in a training building to fully understand the disastrous effects of a fog pattern in an aggressive interior fire attack.

Scenario Discussion

1. What type of fire attack strategy is appropriate and why?

This fire situation dictates an aggressive interior fire attack: an offensive strategy. The house is salvageable, and there are no apparent structural weaknesses such as trusses or laminated floor beams. With a reasonable risk to your firefighters you stand a good chance of minimizing the loss of this family's home and possessions. Also, the sooner you put out the fire, the less risk there is to your members.

The fire definitely has control of two rooms that are well vented. The critical factor in this fire is whether the fire has entered the attic in a significant way. Assuming for the minute it has not, this fire should be knocked down with one or two handlines inside. The lines will go through the front door, up the stairs, down the hallway to the seat of the fire (one or two bedrooms). If luck is with you today, the doors to these rooms are closed and still at least partially intact, holding back the fire and keeping conditions in the hallway tenable with good visibility. An option at this fire is to go offensive for a short time and see if you can get a good knockdown on the fire, then determine if it is in the attic in a big way. Then two options exist: keep up your interior attack by pulling ceilings and applying water on the involved attic, or withdraw your interior forces, pull the gable vents (which are usually plastic or at best aluminum), and apply water from ground ladders or the aerial device to achieve knockdown.

If you see that the fire has heavily extended into the attic via the soffit and maybe through failure of the interior ceilings, the offensive attack may not be your first choice. There may be tons of fire over your head once you get inside. This could easily extend downward in a flashover in the attic.

2. Describe your fire attack system.

The fire attack system is made up of the following, like links of a chain:

- Water supply
- Supply hose
- Engine/pump
- Pump operator
- Attack line
- Nozzle
- Officer
- Nozzle person
- Backup firefighter
- Door firefighters

If any of these links are less than efficient, the fire attack system will be less effective. Should any of these links fail, it is very likely that the fire attack will fail. This will endanger trapped occupants and the firefighters we have intentionally put in harm's way, assuming the task of fire control is being accomplished.

My department's fire attack system is our most basic and important play. We practice it often and realistically under simulated stressful conditions. Our goal is to apply decisive amounts of water at a house fire within 60 seconds of arrival on the fire scene. We are often diverted from training for this important mission by lots of well-intentioned people and regulations. For example, we have mandatory annual training on blood-borne pathogens, hazardous materials, safety, bailout ropes, and other important but not as vital operations.

From leaders in the fire service down to the most junior probie, we must know what the fire attack system is, know how often we should train to perfect it, and, most importantly, understand and be able to execute the responsibilities of each position.

3. What is your department's target flow?

Our target flow for private dwellings is 180 gpm via a 1¾ in. line with a ¹⁵⁄₁₆ in. smoothbore tip. We have flow tested all our discharges and preconnected lines to verify the correct pump pressure to account for the various and wide ranges of friction loss in both the pump and the hose. For our preconnects, we have marked our gauges so the pump operator does not have to figure it out at zero dark thirty. It is too important a safety issue not to do this.

4. How did you establish your target flow?

Target flow was established based on several things. First we wanted to flow as much water as possible but still have the line manageable with an engine crew made up of an officer, nozzle person, backup firefighter, and door firefighter. We tested lower flows but settled on 180 gpm as the best mix of high flow and low nozzle reaction.

The 180 gpm was also based on a literature search of best practices used by other fire departments and the fact that, as we all know, the fire load has increased due to more synthetics in the home environment. Additionally, we have 500 gal tanks on all our engines and this gives us a full 2 minutes of flow, which does not sound long until you have the nozzle in your hands. The 180 gpm ensures that we are not going to have a fair fight, that we will quickly overwhelm the fire and reduce the hazard.

Realistically, we know that on a real scene our flow may be less than 180 gpm due to kinks, elevation differences, and other variables. But if we shoot for 180 gpm and get 150 we are still in the safe range. Had we targeted 150 and got 120, we would likely fall back into the fair-fight arena where we don't want to be.

Unless you flow test, you will be using theoretical numbers. All the departments whose fire attack systems I have tested have not flowed what they thought they were flowing. Often this is because they use classroom-calculated and theoretically derived pump pressures. Test your fire attack system.

5. How do you train your attack team?

Getting your first line into position and delivering decisive amounts of water is a perishable skill but is also our most important skill. We routinely (four to six times per year) do live hose stretches up to and including flowing water and pushing lines into a training building. We have a written SOP with photographs explaining the responsibilities of each member in the stretch, advance, and attack. Sometimes we use live fire at our training center but at least half of our drills are done without fire or smoke so we can observe, evaluate, and coach our engine company members in a positive way.

6. What type of nozzles does your department use for this type of fire attack and why?

We use both combination nozzles and smoothbore nozzles. The smoothbore nozzles give us great reach and penetration, delivering those large globs of water to the burning material at a low nozzle pressure and low nozzle reaction. Our combination nozzles are the new low-pressure variety (150 gpm at 50 psi), which we use on straight streams for interior fire attack. We tell our people, "Turn right to fight" for a straight stream for an interior attack or "Turn left for lobster" and use a fog stream, if you want to steam-burn yourself and your fellow firefighters during the interior attack. These nozzles operate at the same low pressure and reduced nozzle reaction as our smoothbores, which makes our pump operators more effective in delivering the correct pressure and volume to the attack team that has their face in the fire.

References

Comella, Jay. 2003. "Planning a Hose and Nozzle System for Effective Operations." *Fire Engineering* 156 (4). http://www.fireengineering.com/articles/print/volume-156/issue-4/features/planning-a-hose-and-nozzle-system-for-effective-operations.html.

Fang, J., and J. Breese. 1980. *Fire Development in Residential Basement Rooms*. NBSIR 80-2120. Washington, DC: Department of Commerce. http://nvlpubs.nist.gov/nistpubs/Legacy/IR/nbsir80-2120.pdf.

Kerber, Steve. 2010. *Impact of Ventilation on Fire Behavior in Legacy and Contemporary Residential Construction*. Northbrook, IL: Underwriters Laboratories. http://www.ul.com/global/documents/offerings/industries/buildingmaterials/fireservice/ventilation/DHS%202008%20Grant%20Report%20Final.pdf.

Knapp, Jerry, Tim Pillsworth, and Christopher Flatley. 2003. "Nozzle Tests Prove Fireground Realities, Part 2." *Fire Engineering* 156 (9): 71–76. https://nozzleforwarddotcom1.files.wordpress.com/2013/03/nozzle-tests-prove-fireground-realities-part-2-2003.pdf.

Layman, Lloyd. 1955a. *Attacking and Extinguishing Interior Fires*. 2nd ed. Boston: National Fire Protection Association.

———. 1955b. *Fire Fighting Tactics*. Reprint. Boston: National Fire Protection Association.

National Fire Protection Association. 2005. *NFPA 1410: Standard on Training for Initial Emergency Scene Operations*. Quincy, MA: National Fire Protection Association.

———. 2016. *NFPA 1710: Standard for the Organization and Deployment of Fire Suppression Operations, Emergency Medical Operations, and Special Operations to the Public by Career Fire Departments*. Quincy, MA: National Fire Protection Association.

O'Donnell, Andrew. 1995. "Choosing a Nozzle." *Fire Engineering* 148 (9). http://www.fireengineering.com/articles/print/volume-148/issue-9/features/choosing-a-nozzle.html.

Pillsworth, Tim, Jerry Knapp, Christopher Flatley, and Doug Leihbacher. 2007. "How Kinks Affect Your Fire Attack System." *Fire Engineering* 160 (10). http://www.fireengineering.com/articles/print/volume-160/issue-10/features/how-kinks-affect-your-fire-attack-system.html.

Stakes, Keith. 2017. *Impact of Fire Attack Using Interior and Exterior Streams on Firefighter Safety and Occupant Survival*. Northbrook, IL: Underwriters Laboratories.

8

Ventilation

Summary

Ventilation is a key component of successful rescue and fire attack operations at modern house fires. The where, when, and why factors for horizontal and vertical ventilation specific to house fires will be explained, as well as how to control the ventilation so as not to negatively impact operations. We will highlight the latest changes and recommendations for ventilation practices based on scientific research and live burns of fully furnished homes specifically built for the full-scale tests. We will finish with a review of traditional ventilation practices to provide perspective on how ventilation has evolved and why strategies have changed.

Introduction

As we are writing this book, new and exciting ideas and philosophies are being developed in the area of ventilation by agencies like the Underwriters Laboratories (UL) and National Institute of Standards and Technology (NIST). Many we presented in previous chapters, and there are more in this chapter. These revelations are derived from research, the real scientific and experimental kind, the kind of research that collects data; embraces our intuition, experience, and knowledge from going to fires and figuring out what went wrong; and applies science to confirm the results. No two fires are exactly alike so the results may be similar but not the same. The skilled analysis of fire conditions done under laboratory conditions, like those conducted by UL and NIST, control as many variables as possible to derive actionable data. These revelations have had far-reaching consequences: Many fire departments across our nation have modified their procedures and enjoyed new tactical fireground success utilizing new procedures. FDNY has updated *Firefighting Procedures* to reflect the concepts of flow path. Two of the most important changes are below:

1. Amending the concept of "vent for fire" to "ventilation for extinguishment": "the controlled and coordinated ventilation tactic which facilitates the Engine Company's extinguishment of the fire" (FDNY 2013, 12).

2. Defining "vent for life" as "ventilation for search": "a horizontal ventilation tactic performed to facilitate the movement of a member into an area in order to conduct a search for a life hazard" (13).

We will begin this chapter with a scenario. As with previous chapters, answer the questions based on your current knowledge. Since this chapter focuses on ventilation, we will only mention other critical tactical actions (search, attack, etc.) in relation to ventilation. Following the scenario we will highlight the new information derived from the work of UL and NIST and finish with a discussion of what we'll call the "traditional" ventilation techniques. After you have read the chapter, revisit your answers and compare what you may (or may not) do differently based on the information in the chapter.

Scenario

Fig. 8–1. Scenario photo

Situation: It is Sunday morning, 1023 hours, and you are tapped out to a fire at 2 Veterans Street. A 911 caller reported smelling smoke in the area.

Radio Report: Dispatcher reports multiple 911 calls.

Resource Report: You are on the scene and your first-due piece of apparatus is approximately 2 minutes from the scene. There is a good hydrant 200 ft before the fire building.

Based on the photo and information above, answer the following questions and use this scenario as a basis for the topics examined in this chapter.

1. Is the fire ventilation limited?
2. What type of ventilation is already in place and what is possible?
3. How will your ventilation plan affect the search-and-rescue (SAR) operation?
4. What negative effects could improper ventilation have on this fire?
5. What is your ventilation strategy and timing?

Ventilation of Modern Home Fires

Although the basic principles of ventilation in support of fire attack and SAR at nonresidential buildings have not radically changed, there is a great deal of excellent, groundbreaking research done by Steve Kerber and the UL regarding ventilation practices specifically at house fires. The executive summary of their study *Impact of Ventilation on Fire Behavior in Legacy and Contemporary Residential Construction* describes the work:

> Under the United States Department of Homeland Security (DHS) Assistance to Firefighter Grant Program, Underwriters Laboratories examined fire service ventilation practices as well as the impact of changes in modern house geometries. There has been a steady change in the residential fire environment over the past several decades. These changes include larger homes, more open floor plans and volumes and increased synthetic fuel loads. This series of experiments examine this change in fire behavior and the impact on firefighter ventilation tactics. This fire research project developed the empirical data that is needed to quantify the fire behavior associated with these scenarios and result in immediately developing the necessary firefighting ventilation practices to reduce firefighter death and injury. (Kerber 2010, 3)

This study, which is specific to residential fires, should have a profound impact your preparation, training, tactics, and strategies for rescue and fire attack operations at house fires. The results should alter the way firefighters tactically approach house fires and the way officers and incident commanders strategically approach house fires, including size-up and developing and executing plans for rescue, ventilation, and fire suppression. The following points are critical to our effectiveness and safety at modern house fires:

- Structural framing and house construction (size and shape) have changed over the past 30 years.

- House fires burn differently today than they did 20–30 years ago, because legacy and contemporary furnishings react very differently under fire conditions. We must reevaluate traditional thoughts such as venting as we go and other time-honored tactics.

- It is critical to our safety to get decisive amounts of water on the fire quickly.

- SAR operations may not be possible without a hoseline.

- Forcing the front door is ventilation and will drastically affect the fire behavior, starting the flashover countdown clock.

- Traditional "venting for life" may result in the search team's death as a result of untenable conditions.

In essence, the study adds science to our conjecture and confirms some of what we "know" from our fireground experience while updating us on the modern fire growth curve and providing a future direction for strategy and tactics to reduce firefighter fatalities and injuries at house fires. More importantly, it provides, through realistic full-scale live burn experiments, significant hard data for us to use to become safer and more effective at our next house fire.

One note of caution: as firefighters, we tend to ignore the high-end science and research as unrealistic and not applicable. Often the reports are written at the PhD level and beyond our ability to apply to everyday fires. This study is not like that. It is based on full-scale, live burn tests, in houses built, furnished, and burned realistically. The live burns were done under controlled conditions with state-of-the-art photography, data collection, and instrumentation. The report is written for firefighters by a group of subject-matter experts led by a world-class researcher. UL has extensive experience with both fire modeling and full-scale burn experiments and has produced very applicable and valuable information for our use in the street. This chapter is one of those products. It is our intent to interpret the findings with an emphasis on the tactical and strategic changes they suggest for house fires.

Houses have changed

According to data from the US Department of Housing and Urban Development's *2016 Characteristics of New Housing* survey, the average size of single-family houses built in 2016 was 2,422 square feet (sq ft), up from 1,525 sq ft in 1973

(Office of Policy Development and Research 2016, 345). Houses keep getting bigger and now McMansions, large homes (3,000+ sq ft) on small tracts, are often the rule rather than the exception. These larger houses tend to have more open floor plans and higher ceilings. The open floor plan allows the fire to spread unrestricted by compartmentation. Higher ceilings allow fire to grow more rapidly due to larger amounts of oxygen while masking the severity of the fire due to the heat rising and accumulating high above our heads.

Homes that are more energy efficient and better insulated retain more of the fire's heat while concealing outward signs of a deadly working fire, especially during the exterior size-up phase. As we have heard so often, synthetic materials burn hotter and greatly reduce times to room flashover.

This study contains a short video of side-by-side room burns, one with legacy furnishings and the other with modern furnishings. The video is available on the UL website and is an excellent training tool to dramatically visualize what hotter and faster flashover means for us on the fireground. In summary, the living room furnishings are ignited simultaneously, and the modern furnished room flashes over at 3 minutes and 40 seconds, while the legacy room lasts 29 minutes and 25 seconds (Kerber 2010, 309). Obviously other factors will influence fire development but this is an excellent base from which to develop our understanding of fire development in modern house fires.

Lay these timelines over your own department's response times and estimate where you or your members will be when the flashover occurs (figs. 8–2 and 8–3).

Fig. 8–2. Live burn in the legacy room, 5 minutes after ignition

Fig. 8–3. Live burn in the modern room, postflashover, 3 minutes, 30 seconds

So what does this mean for the modern house fire? We should expect a flashover at house fires especially if or when the fire gets enough oxygen. We should not think of flashover as a rare occurrence; it is part of the natural life cycle of the modern fire. We should not be surprised on the fireground. Strategically, we must anticipate flashover and have hoselines positioned to protect members on search teams and deliver decisive amounts of water to prevent or kill the flashover. The study states the following:

> There is a continued tragic loss of firefighters' and civilian lives, as shown by fire statistics. It is believed that one significant contributing factor is the lack of understanding of fire behavior in residential structures resulting from natural ventilation and use of ventilation as a firefighter practice on the fire ground. The changing dynamics of residential fires as a result of the changes in construction materials, building contents and building size and geometry over the past 50 years add complexity to the influence of ventilation on fire behavior. (Kerber 2010, 8)

UL Study Details

The study examined the fire environment of the modern house fire and the effect of traditional ventilation tactics (taking windows and doors) on fire development. The experiments included 15 full-scale burns in the large UL fire facility. Two homes were constructed: 1,200 sq ft, 1 story, 3 bedroom, 1 bath with 8 rooms total and a 3,200 sq ft, 2 story, 4 bedroom, 2½ bath with 12 rooms total. The homes were furnished with appropriate types and amounts of furniture typical of these homes. Fires were ignited in the furniture of the living room or family room. The parameters of the live burns can be seen in figure 8–4.

To examine ventilation practices as well as the impact of changes in modern house geometries, two houses were constructed in the large fire facility of Underwriters Laboratories in Northbrook, IL. Fifteen experiments were conducted varying the ventilation locations and the number of ventilation openings [see below]. Ventilation scenarios included ventilating the front door only, opening the front door and a window near and remote from the seat of the fire, opening a window only and ventilating a higher opening in the two-story house. One scenario, each for the one-story and two-story structures were conducted in triplicate to examine repeatability.

Experiment #	Structure	Location of Ignition	Ventilation Parameters
1	1-Story	Living Room	Front Door
2	2-Story	Family Room	Front Door
3	1-Story	Living Room	Front Door + Living Room Window (Window near seat of the fire)
4	2-Story	Family Room	Front Door + Family Room Window (Window near seat of the fire)
5	1-Story	Living Room	Living Room Window Only
6	2-Story	Family Room	Family Room Window Only
7	1-Story	Living Room	Front Door + Bedroom 2 Window (Window remote from fire)
8	2-Story	Family Room	Front Door + Bedroom 3 Window (Window remote from fire)
9	1-Story	Living Room	Front Door + Living Room Window (Repeat Exp. 3)
10	2-Story	Family Room	Front Door + Family Room WIndow (Repeat Exp. 4)
11	2-Story	Family Room	Front Door + Family Room WIndow (Repeat Exp. 4)
12	1-Story	Living Room	Front Door + Living Room Window (Repeat Exp. 3)
13	2-Story	Family Room	Front Door + Upper Family Room Window
14	1-Story	Living Room	Front Door + 4 Windows (LR, BR1, BR2, BR3)
15	2-Story	Family Room	Front Door + 4 Windows (LR, Den, FR1, FR2)

Fig. 8–4. UL study experimental series. *Source*: Kerber 2010, 82.

Fires were allowed to grow and ventilation took place at 8 minutes for the one-story and 10 minutes for the two-story test homes. This time frame allowed for a typical fire department response and represented when the fire became ventilation limited in each size home. The various ventilation actions (doors and windows) were taken at the times planned for each test burn. Fires were allowed to grow to flashover and extinguished when they entered the decay phase.

Tactical considerations and strategic changes

Although it is impossible to review all the important details of this 400+ page report, there are several findings that should have an impact on your tactics and overall strategy at your next house fire.

The new fire curve. Understand the new fire growth curve as shown on the right-hand graph below. Based on these tests, because the fire is ventilation limited, there are two periods of rapid fire development (fig. 8–5a and b).

A key element to fire behavior training is in the fire growth curve that is used to demostrate where the different stages occur in relation to each other. The basic fire growth curve shows fire ignition followed by the growth stage, flashover, fully developed stage and finally the decay stage [1]. This curve can be misleading. If an item like a sofa was placed in a large room or outside, where plenty of oxygen is available, then the growth curve will look like [1]. A more realistic explanation of the modern fire environment where there is usualy not enough oxygen available would be ignition followed by a growth stage, decay stage, ventilation (either by fire service or by an opening created by the fire like window failure), a second growth stage, flashover, fully developed stage and finally the decay stage. [2] depicts what this would look like. Taking an average of all the fire room ceiling temperatures for both houses shows the shape of the time-temperature curve for all of the ventilation scenarios [3].

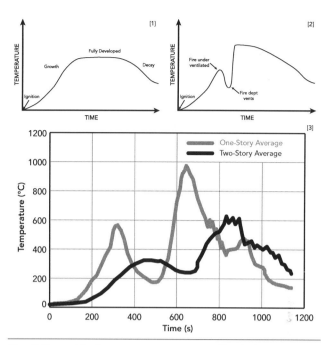

Fig. 8–5a and b. Average temperatures of both the one-story and the two-story home. It is clear that there is a decay phase in each, then a rapid growth phase (flashover) stimulated by opening the front door and other ventilation. This graph is vastly different from the basic fire growth curve we are used to and may have been taught in probie school.

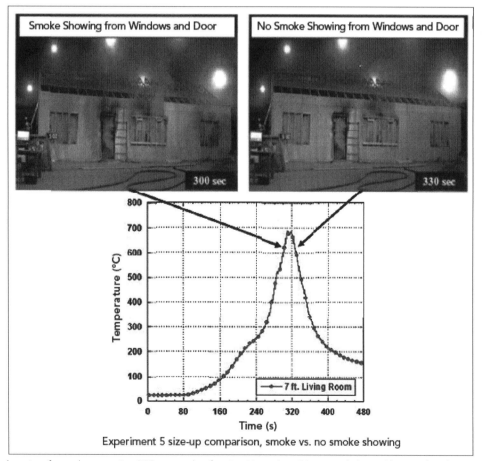

Fig. 8–6. Smoke showing from the exterior 300 seconds after ignition. Just 30 seconds later (during the decay phase) there is "nothing showing" from the exterior

For clarification and discussion purposes, during this chapter we will examine potential changes in tactics and strategies that the report results suggest are applicable for house fires.

For discussion purposes, "tactical consideration" means a review of tasks being carried out by a firefighter or group of firefighters (e.g., forcing the front door) and "strategic command consideration" means the considerations for change of the overall plan (strategy) designed by the officer in charge or the incident commander at the time based on size-up and other factors.

Tactical consideration. Expect a flashover or rapid fire development at house fires. We tend to think of house fires as routine, but that doesn't mean they aren't very dangerous. Consider what phase the fire is in and what effect your actions will have on the fire growth.

Strategic command considerations. During the decay phase (assuming the fire is ventilation limited, no doors or windows are open or have failed due to fire), your exterior size-up may not show any signs of fire, which may cause you to send out the infamous "nothing showing" radio call. In actuality, the fire has loaded the gun and is just waiting for a load of oxygen to light up and kill our members Temperatures drop drastically during decay, resulting in lower pressures that do not push smoke from the house (fig. 8–6).

Opening the front door. Sometimes we can't see the forest for the trees. This is certainly the case with forcing or opening the front door of the fire building. Most firefighters do not recognize opening the front door as ventilation. We view it as forcible entry and the first active step in SAR or advancing the line. This action (tactic) has a profound effect on the fire growth (fig. 8–7).

Tactical consideration. If you open the front door, adding air to the fire, how long do you have until the room becomes untenable or goes to flashover? Do you have time for a search operation? How far in should you go?

Strategic command consideration. In order to protect search teams from rapid fire development, hoselines must be operational immediately following or simultaneously with the search team's deployment.

Examining fire room temperatures at 5 ft above the floor in both houses during the experiments where only the front door was opened shows the increase in temperature following ventilation. In the one story experiment (Experiment 1) the temperature was 180°C (360°F) at ventilation (480 s), exceeded the firefighter tenability threshold of 260°C (500°F) at 550 s and reached 600°C (1110°F) at 650 s. In the two story experiment (Experiment 2) the temperature was 220°C (430°F) at ventilation (600 s), exceeded the firefighter tenability threshold of 260°C (500°F) at 680 s and reached 600°C (1110°F) at 780 s. Both of these experiments show that opening the front door needs to be thought of as ventilation as well as an access point. This necessary tactic also needs to be coordinated with the rest of the operations on the fireground. A simple action of pulling the front door closed after forcing entry until access is ready to be made as part of the coordinated attack will limit the air to the fire and slow the potential rapid fire progression.

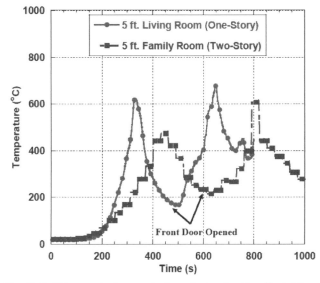

Fig. 8–7. This graph demonstrates what a drastic effect this simple unrecognized action may have on our survival.

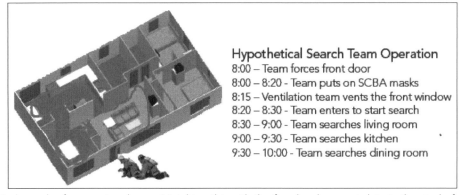

Hypothetical Search Team Operation
8:00 – Team forces front door
8:00 – 8:20 - Team puts on SCBA masks
8:15 – Ventilation team vents the front window
8:20 – 8:30 - Team enters to start search
8:30 – 9:00 - Team searches living room
9:00 – 9:30 - Team searches kitchen
9:30 – 10:00 - Team searches dining room

Fig. 8–8. Hypothetical search of a one-story home. Matching this with the fire development data in the study, firefighters would be forced to bail out from windows approximately 2 minutes into their search. This scenario assumes: modern furnishings in a one-story home, fire in the living room, entry, the only ventilation from the front door, and no water on the fire. (Kerber 2010, 290)

Decisive water application

Application of decisive amounts of water on the fire is more important than ever before, especially at house fires. The preceding explanations and facts show just how deadly the fire will grow, trapping occupants and killing firefighters.

For the first time we have quantitative estimates of how long the fire will take to devolve to untenable conditions in common (one- and two-story) homes. We must understand these facts and use these data in developing SOPs for operations at house fires. From the executive summary of the report:

> If you add air to the fire and don't apply water in the appropriate time frame the fire gets larger and safety decreases. Examining the times to untenability gives the best case scenario of how coordinated the attack needs to be. Taking the average time for every experiment from the time of ventilation to the time of the onset of firefighter untenability conditions yields 100 seconds for the one-story house and 200 seconds for the two-story house. In many of the experiments from the onset of firefighter untenability until flashover was less than 10 seconds. These times should be treated as being very conservative. If a vent location already exists because the homeowner left a window or door open then the fire is going to respond faster to additional ventilation opening because the temperatures in the house are going to be higher. Coordination of fire attack crew is essential for a positive outcome in today's fire environment. (Kerber 2010, 4)

Search safety

It is clear that we have to alter our SAR strategies. If we don't, we will continue to create fire conditions that injure and kill firefighters at house fires. The UL study does an excellent job of describing a hypothetical search of a one-story ranch home (fig. 8–8). This is a realistic representation of how a fireground primary search may actually be executed. It is very liberal in that it has a two-firefighter search team doing the primary search in three rooms in just 2 minutes. This is certainly possible but the reliability of the search may be questionable as they will spend only 40 seconds in each room. Regardless, if the search takes longer the only difference is where firefighters are in the house when the conditions become untenable and they are forced to bail out (fig. 8–9). Additionally, our entry point is no longer a viable escape route. Remember that bailout in these conditions is a last-ditch survival effort that is not as reliable a tactic as we would like to believe. Studies of case histories of firefighters caught in flashovers reveal that often it is luck and not the skill of a firefighter that determines if they survive or are

burned to death or forced to jump from an upper floor to their death. (See the case histories in ch. 3 for a review of this important point.)

Tactical consideration. If you enter the house before the hoselines are in place, monitor your radio for progress in getting water on the fire and reports of fire conditions.

Strategic command consideration. It is critical to the safety of your members to protect their means of egress with a hoseline or to quickly knock down the fire. In limited-staff situations (not enough first-alarm members to stretch and conduct a search), it may be necessary to knock down the fire before primary and secondary searches can safely be done.

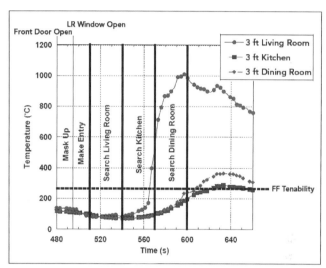

Fig. 8–9. The timeline on this graph links the hypothetical search shown above, the temperatures from the live burn testing, and when the search team encounters untenable conditions. (Kerber 2010, 290)

Vent for Life

We have, over the past few years, been bombarded with statements like "Modern fires burn differently, hotter, faster." As previously mentioned, the video comparing legacy and modern furnishings clearly demonstrates this, and the data support it. It is important to link these facts to one more ventilation tactic: venting for life, or venting as you go.

This traditional action came out of desperation during primary search to improve conditions so you can make a rapid and thorough primary search and make a hasty exit with your victim. With legacy construction and contents, this was a viable tactic. However, with modern construction and furnishings, this tactic may actually backfire, making conditions worse and decreasing the possibility of a safe exit. Remember, closing the door to the room you are in will hold back fire for a significant time when you are venting-for-life.

The UL study shows the following results when venting for life tactics are used:

- Faster fire growth: Flashover occurred 40 seconds faster in the one-story and 80 seconds faster in the two-story.

- More ventilation does not make cooler temperatures.

- Higher smoke- and heat-release rates of synthetics result in less lifting of smoke than traditionally experienced.

A portion of the executive summary of the report contains critical survival information:

Can You Vent Enough?

In the experiments where multiple ventilation locations were made, it was not possible to create fuel-limited fires. The fire responded to all the additional air provided. That means that even with a ventilation location open the fire is still ventilation limited and will respond just as fast or faster to any additional air. It is more likely that the fire will respond faster because the already open ventilation location is allowing the fire to maintain a higher temperature than if everything was closed. In these cases, rapid fire progression if highly probable and coordination of fire attack with ventilation is paramount. (Kerber 2010, 4)

Vent, enter, isolate, search (VEIS): a safer tactic

The traditional search plan—entering through the front door and searching the most likely places to find victims (often second-floor bedrooms)—is a good, solid plan. However, depending on the fire situation, it may be too dangerous, especially if staffing limitations prevent the first-alarm assignment from searching and stretching the hoseline simultaneously.

VEIS, or searching from the outside in, may be a better option, especially with limited first-alarm personnel. Although it is certainly more time consuming than traditional through-the-door search methods, VEIS provides a viable alternative. The main advantage is that members are in an area of refuge when they start the search. This vastly increases their safety while still allowing the first-alarm members to conduct an aggressive search. The UL study has this to say

in regard to this topic, specifically the impact of a shut door on safety and search operations:

> Conditions in every experiment for the closed bedroom remained tenable for temperature and oxygen concentration thresholds. This means that the act of closing a door between the occupant and the fire or a firefighter and the fire can increase the chance of survivability. During firefighter operations if a firefighter is searching ahead of a hoseline or becomes separated from his crew and conditions deteriorate then a good choice of actions would be to get in a room with a closed door until the fire is knocked down or escape out of the room's window with more time provided by the closed door. (Kerber 2010, 4)

It is also important to look at the big picture. Even with "reliable" reports of victims trapped, there may not be anyone really in the house. The case histories of firefighter fatalities while searching for occupants who were not inside are too numerous to mention. However, once the fire department arrives and begins SAR and fire attack operations, there is 100% certainty that people are in the building: these people are fire department members, and they are without a doubt in significant danger presented by rapid fire development and other factors. Firefighters are occupants (although temporary) and deserve the same amount of life-safety consideration as the full-time occupants.

We are not saying not to search and be aggressive; firefighters *should* take sensible, manageable risks. What we are saying is that you must evaluate the survivability of victims, use every tool you have available (VEIS in this case), consider the overall risk versus gain in every operation, and understand the risks of venting at a house fire.

One risk to the VEIS concept is that searchers are directed to where people may have been before the fire began; it does not take into account the possibility that they may have attempted to escape and may be located near the exit. Remember back to basic search skills and always check behind the door when you enter! This requires that the likely pathways from those areas to the exit be searched as quickly as possible.

Energy Efficient Windows (EEWs)

We know from our own fireground experiences that when we ventilate the fire building, adding air to the fire, it will intensify. Experimental data and fire modeling by the UL have proven and put important numbers to this fireground reality (Kerber 2010). By these same methods we know that when the fire self-vents a window in a home, the result will be the same: rapid fire development. This rapid fire development may manifest itself as a flashover, trapping and killing our members.

There are four important points regarding self-ventilation of EEWs that are critical safety factors that we must be aware of:

1. There is a common misconception that all modern EEWs will provide greater fire resistance than single-pane windows.

2. A false vent can be created by failure of the inside pane of an EEW.

3. It appears that EEW failure is primarily dependent upon the frame the glass is contained in.

4. If the windows are made of tempered glass (glazing) there could be a total failure of the window (like the shattering of a side or rear car window) caused by heat from the fire at some point during fire.

An EEW is a window designed to reduce heat loss from the structure in winter and minimize heat gained by the home in summer. EEWs are two or three panes of glass ($\frac{3}{16}$ or ⅛ in. thick) spaced about ½ in. apart and sealed together. The space is filled by air or an inert gas. Modern windows generally have some type of spring balancer in them to counter the weight of the sashes, making it easy for occupants to move the sash to the desired position.

Legacy or single-pane windows are generally one pane of glass contained in a wooden frame. These frames may contain one large piece or as many as nine pieces of smaller glass (called lights) with wood mullions in between holding the glass in each top and bottom sash. These windows may have a counter balance system with a rope or light chain, pulley, and sash weight (hidden in the frame) to balance the window sashes.

Fire resistance

There has been a misconception about EEWs circulating around the fire service for several years. This misconception is that all EEWs hold back heat, smoke, and fire and result in faster flashovers because they have greater fire resistance than single-pane windows. And to be fair, this has been the experience in several major northeastern cities. In an article in *Fire Engineering*, Chief Tom Kennedy of the FDNY recounts several fires in multiple-dwelling structures where EEWs contained the heat, smoke, and fire. Unquestionably, in these case histories, EEWs caused a delay in transmitting the alarms,

made locating the fire difficult, caused high heat and smoke buildup in apartments, drove hot gases into the cockloft quickly, reduced the time available for a primary search, resulted in high levels of CO, and, most importantly, caused pressures to rise inside the apartments, resulting in forceful explosions and flash fires when windows were vented (Kennedy 1991). It is important to note that these case histories all had EEWs with frames made of aluminum and were about fires in multiple-family dwellings, not single-family homes.

Without hard experimental and repeatable data, firefighters across the country jumped to the conclusion and assumed that all EEWs reacted to fire loading as related by these experiences. In our defense, this assumption seems logical. And it is certainly true that because the windows are new and retrofitted to conserve energy, air leakage is minimized by insulation, weather stripping, and sealers applied during the installation process. This prevents leakage of smoke, which forestalls passersby from sounding the alarm and prevents firefighters from even suspecting an active fire during exterior size-up.

However, overall, the experience with EEWs under a fire load in house fires has been completely different from the experience in other types of structures. In 1997, the Rockland County Fire Training Center conducted realistic live burns directly comparing the fire resistance of modern EEWs and wood-frame legacy windows (Knapp and Delisio 1997). The goals of this testing were to compare the visible performance of either type of window, determine at approximately what temperature each failed, and determine how long it would take each window to fail and add significant amounts of air to the fire room. Window failure was defined as when significant portions of the glazing or frame fell out, melted, or otherwise allowed enough air into the fire compartment to cause noticeable increase in fire intensity. The information that follows comes from Knapp and Delisio 1997.

All live burns were conducted using the Swede Survival Flashover Simulator as the fire room. This room is 8 ft long, 7 ft high, and 8 ft wide. Both end doors were opened and a residential-style wall was framed out with typical 2 × 4 in. wall studs and sheetrock. Thermocouples recorded temperatures during the test burns. Fuel loading was kept as consistent as possible for all burns and consisted of typical residential living room furniture.

Tests 1 and 2. These burns compared a single-pane window to a modern EEW. The single pane was a 1930s vintage wood-frame, double-hung sash with six over six lights. The EEW was the most common type of residential replacement window: vinyl extruded frame and sash. It was double glazed with ½ in. air space between panes (fig. 8–10).

Fig. 8–10. Live burn testing showed how important frame material is to actual performance in real fire situations.

At 5 minutes and 58 seconds after ignition of the furnishings and at 700°F, approximately 75% of the interior pane of the EEW fell inward. A firefighter could easily mistake this sound for the outside vent person taking a window. The exterior pane remained intact until 7 minutes and 30 seconds after ignition, when it failed with about 85% of the glass falling outward. As you may expect, this was followed by an inrush of air that caused rapid fire growth and ceiling temperatures peaking near 1,800°F.

If firefighters were inside the home and heard the glass break, they would assume the outside vent firefighter had taken a window. In reality, the second pane was still trapping explosive products of combustion, further loading the flashover and rapid fire development cannon with the unsuspecting firefighter below. With conditions remaining relatively constant, unsuspecting members may decide to continue the search.

When the exterior pane failed, the sudden inrush of air lit the room up dramatically. It is important to note that when the EEWs failed, they generally failed with large pieces of glass falling out of the frame at once. This obviously allows an significant inrush of fresh air leading to rapid fire development and possibly the injury or death of firefighters inside.

The single-pane wood-frame window began to make tinkling sounds around 600°F. At 6 minutes and 48 seconds, one light of the window cracked and broke out about 20% of its area. This was the only failure of the single-pane window whereas the EEW failed completely during the remaining burn time. The burn was terminated by a hose stream at 10 minutes and 45 seconds.

Both of these tests had similar results. This led to the conclusion that apparently the type of framing the energy efficient pane is contained in determines its performance under a fire load. Regardless, the false vent (the interior pane's breaking followed by the large failure of the exterior pane) is an important safety consideration for firefighters operating inside house fires.

Test 3. This test used aluminum-clad wood-framed EEWs. These performed significantly differently from the vinyl extruded sashes and frames of tests 1 and 2. As in the first test, the interior pane fell inward at about 5 minutes, but it was only a small percentage of the glazing area. Failure of the exterior panes of both sashes was considerably slower and less dramatic than before. The personnel door of the test unit was opened at 6 minutes and 48 seconds, the room flashed over at 9 minutes and 48 seconds (1,448°F at the ceiling), and at 10 minutes and 50 seconds, the EEW top pane completely failed. These test data are somewhat more consistent with the fireground experiences mentioned earlier. The single-pane legacy window failed (50% area) at 10 minutes and 6 seconds.

Test 4. This test utilized two high-performance EEWs with tempered glass. High performance denotes that the amount of heat transmission through the window is less than ordinary EEWs. These had wood frames clad in aluminum. These windows performed more in line with the reputation of EEWs. They concealed the fire very well from the outside and withstood very high temperatures (up to 1,400°F) before failure at 8 minutes and 46 seconds, but failed rapidly and totally, leading to almost instant flashover.

These tests are generally supported by the fact that the UL study utilized earlier in this chapter draws roughly the same conclusion: Legacy windows will withstand a residential fire longer than a modern EEW. Two charts from the study summarize the large amount of data collected on window failure caused by fire development (fig. 8–11a and b).

Desgn.	Description	Legacy (L) or Modern (M)	Size Width (in) x Height (in) / Glass thickness (in)
A	Wooden Frame, Two Pane, Single Glazed Storm	(L)	30 x 46 ½ / .093
B	Vinyl Clad, Wood Frame, Two Pane, Double Glazed	(M)	30 ½ x 56 ⅞ / .087
C	Wood/Metal Frame / Nine Pane over One Pane, Single Glazed	(L)	25 ⅝ x 59 ½ / .115
D	Premium Plastic Frame, Two Pane, Double Glazed	(M)	28 x 54 / .087
E	Plastic Frame, Two Pane, Double Glazed	(M)	28 ½ x 54 ½ / .088
F	Premium Wooden Frame, Two Pane, Double Glazed	(M)	29 x 57 / .089
1	Door – Flush Oak, Hollow Core	N/A	30 x 80
2	Door – Six Panel Molded, Hollow Core	N/A	30 x 80
3	Door – Six Panel Pine, Solid Core	N/A	30 x 80

In addition to time, the temperatures at which failure occurred were analyzed. [Fig. 8-11b] shows the average gas temperatures just inside and just outside the upper pane of each window just prior to failure. Similar to time, the temperatures for the legacy windows were higher than those of the modern windows. The legacy windows failed when the furnace side temperature was between 650 °C (1200 °F) and 790 °C (1450 °F), while the modern windows failed between 540 °C (1000 °F) and 650 °C (1200 °F). The corresponding outside of the glass temperature was between 370 °C (700 °F) and 380 °C (720 °F) for the legacy window and between 80 °C (180 °F) and 205 °C (400 °F) for the modern windows.

Experiment #	Window [mm:ss (sec)]		
	A (L)	B (M)	C (L)
1	6:34 (394)	4:24 (264)	11:49 (709)
2	10:06 (606)	4:38 (278)	14:30 (870)
3	12:11 (731)	3:56 (236)	16:00 (960)
Average	9:37 (577)	4:19 (259)	14:06 (846)
	D (M)	E (M)	F (M)
4	3:58 (238)	5:16 (316)	3:39 (219)
5	3:39 (219)	4:26 (266)	5:49 (349)
6	5:05 (305)	5:55 (355)	4:02 (242)
Average	4:14 (254)	5:12 (312)	4:30 (270)

Fig. 8–11a and b. Study results

Conclusions

1. EEWs will not all react the same to fire conditions.

2. Type of glass, sash, and frame construction appear to have an effect on performance.

3. Legacy windows will show the effects of fire sooner than modern EEWs but failure and its effect of letting air into the fire compartment are more severe for modern EEWs.

Traditional Venting

Ventilation is one of those tactics that in the sequence of fireground activities has no hard start or end time: it is part of the success of the overall operation. Ventilation must be coordinated with the search and rescue and fire attack and support the overall operation. It can be conducted early in an operation to allow search crews to move to an area to search for life, or it can be done to provide a path for heat and smoke to exit the building ahead of an advancing hoseline. In all cases, ventilation will be ongoing tactic until units leave the scene (fig. 8–12).

Fig. 8–12. Ventilation allows heat and smoke to exit the home and allows the hoseline access to the seat of the fire for rapid extinguishment. Courtesy of Brian Duddy.

Critical knowledge

Ventilation operations (both horizontal and vertical) are critical components to the success or failure of the entire fireground operation and to firefighter safety. Battalion Chief John Mittendorf, Los Angeles City FD (retired), is one of the nation's premier experts on truck company operations and ventilation. Mittendorf describes how understanding modern house construction methods is a cornerstone to safe and effective ventilation operations.

Unfortunately, new construction methods are not usually designed to assist fire suppression operations. Considering the cost of labor, equipment, and building materials, it is not economically feasible to construct a structure the same as the early 1900s. Today, heavy timbers have been replaced by two-by-fours and TGI trusses. Petrochemical-based compounds have replaced conventional interior materials, regardless of building type or size. As modern architects reduce the mass of a common structural member and change the chemical composition of building contents, we are losing one of our most valuable factors—fireground time!

Unfortunately, we are fighting structure fires the same way we did 40 to 50 years ago. Hose lines are taken inside an involved structure, and ventilation operations are initiated to improve the fireground environment. But are modern buildings the same as the buildings constructed during the 1920s and 1930s? They are not even close. Therefore, firefighters who can recognize and evaluate the strengths and hazards of buildings will increase their efficiency and safety. A working knowledge of building construction provides not only the necessary expertise to conduct a quick and accurate size-up of a structure, but also the foundation for effective, timely, and safe fireground operations. (Mittendorf 2011, 90)

Overhaul Ventilation

Ventilation during overhaul is a very important operation often overlooked. The postfire control phase of the operation, often called overhaul, is when we are opening up wood trim, walls, ceilings, and sometimes floors to find fire that may have extended into void spaces out of view of the naked eye and even the thermal imaging camera (TIC). During overhaul, the products of combustion from the smoldering debris produce some of the worst carcinogens on the fireground. Firefighters must continue to use SCBA or provide enough fresh air to dissipate the smoke and toxins. Simply monitoring for carbon monoxide (CO) or providing limited ventilation does not protect you from the numerous cancer causing agents.

Lieutenant Thomas Healy of the FDNY, who runs a cancer foundation in his father's name, says, "Can you smell that? What you are smelling is cancer," when he wants to impress upon his firefighters the dangers of smoke inhalation during the overhaul stage of the fire. You obviously cannot literally smell cancer, but his point is that the smell comes from the off-gassing of the burned materials that are in the air and dangerous to your health.

Types of Ventilation

Ventilation during the SAR or fire attack operations falls into two broad categories: horizontal and vertical. Horizontal ventilation is done by using windows and doors on the same level as your operations and vertical ventilation is where the smoke and heat are coaxed up and out of the building, usually through a roof opening. We will also discuss positive pressure ventilation.

The purpose of ventilation is to make conditions better inside for either the search teams or the engine company. Ventilation will lift the smoke and heat for search teams so they can conduct a safe, thorough, and effective search operation. For the engine company pushing the hoseline, it improves conditions (reduces heat and steam), allowing them to be aggressive and either control the fire or provide safety for the search team. Either way, it is critical for the engine to get in and get water on the fire. The old adage "If the engine can't get in, the fire can't go out" remains true for house fires. As much as we all love our truck company members, you can't put the fire out with a hook or a Halligan tool. Ventilation helps the engine company make it safer for everyone on the fireground by quickly controlling the fire.

Horizontal ventilation

We will begin with horizontal ventilation since it is often the venting that is done in the early stages of the fire. Horizontal venting is usually first because members can reach first- and second-floor windows from ground level or with a 12 or 24 ft ground ladder (fig. 8–13).

Fig. 8–13. Horizontal ventilation. Courtesy of Brian Duddy.

Horizontal venting is often accomplished with a ladder or tool to break windows, providing an escape route for smoke, heat, and toxic and flammable products of combustion. (See ch. 2 for dangers of products of combustion.) The timing of horizontal venting is critical since it will allow heat and smoke out, but will also let fresh air in and rapidly increase the fire's intensity. We are taking a calculated risk and balancing the negative effect of increasing the growth rate of the fire with the positive effect of increasing visibility and decreasing heat conditions by venting for life (taking windows as we search). These are the most basic risk-reward decisions an initial incident commander and search team must make. All firefighters, especially those inside, must understand this risk and the effect on the growth rate of the fire. As previously stated, if the timing is correct, as conditions lift and become better, the engine company has already applied or immediately applies decisive amounts of water to the fire. This obviously increases the safety of all members on the scene and any occupants who may be trapped or attempting to escape. Whenever this technique is employed, careful consideration must be given to planning an alternate means of escape or not cutting off your primary entry. Improper use of this method has driven firefighters to jump out the windows for their own survival. Bailout kits are last resort fireground survival techniques and should not replace a solid awareness of fire conditions.

Correct timing for horizontal ventilation is easy to describe but, in the chaos of the initial fire attack, can sometimes be difficult to accomplish, especially if a search operation is required or staffing is limited. It is IC's responsibility to do the best job possible to coordinate the search, ventilation, and hoseline placement and actions. It is also the responsibility of the search team to report interior conditions, their locations, and difficulties that may affect them or prevent them from completing their assigned tasks. These conditions, actions, needs (CAN) reports reenforce strategic objectives and help develop tactical plans in an evolving situation. Captain Bill Gustin from Miami Dade Fire Rescue puts it more simply: when reporting to the next level of command, "Tell 'em what ya got, what ya doin', and what ya need."

It is during this phase of the operation that the timing of the ventilation is most critical. Done improperly, commanders can at the least hinder or stop the search operations (the most important operation at this time) and at worst burn firefighters to death in flashover or rapid fire development conditions inside caused by the additional oxygen the fire will receive from the improperly timed vent or delayed hoseline stretch and operation.

Often times an officer is not available to provide timing guidance for ventilation and hoseline operations. Firefighters may have to decide on their own when the proper time to

vent or take a window is. A good practical fireground rule is that horizontal ventilation should be executed when the hoselines are in place and ready or flowing water. A good rule for timing ventilation is that when the hoseline is charged, it is time to start taking the windows. The time it takes to clean out the window should be equal to the time it takes the engine company to push into the fire area. But which windows do you take? The best place to vent is ahead of the hoseline if possible. These are windows near where the stream is hitting, will hit, or is being aimed toward. This tactic allows the air that is being driven into the fire area by the stream to be at least directed toward the vent.

As noted in chapter 7, do not be confused by those who say a fog stream directed toward an open (vented) window will drive all heat, smoke, and steam out that vent hole. This is simply not the case. The fog stream brings with it a huge volume of air that in fact overpressurizes the room, driving heat, steam, and products of combustion back onto the nozzle team and those following behind them. If you don't believe this or if you want to test this and see for yourself, direct a fog stream toward an opening about 15 ft or more from the nozzle without any fire or smoke in an acquired structure or burn building. You will draw huge amounts of air in the open door behind you and the fog stream will overpressurize the room, driving air and water droplets back over the nozzle person's head. You have just experienced fog nozzle ricochet (fig. 8–14). Imagine how interesting this would be at a house fire when those water droplets move through the 2,000°F temperatures at the ceiling of the fire room!

Fig. 8–14. Overpressurizing this room forces superheated steam back on the nozzle team.

A solid or straight stream from a combination nozzle will draw a relatively small amount of air into the fire room or area that can be vented out the open window. A fog stream will overpressurize the room, driving superheated steam back on the members and causing the hoseline advance to immediately stop. For a one-room fire, fog streams may be successful, but if the line has to advance through the firefighter's soup that fog stream just made, regardless of the ventilation accomplished, the line will likely not make the additional rooms of fire.

All firefighters must report any adverse effects of horizontal ventilation, such as the fire increasing, exposing other rooms, autoexposing the floor above, or blocking the exit of the search team.

The advantages of horizontal ventilation for life safety cannot be overstated. The fresh air let in as a result of horizontal ventilation will displace toxic air at the floor level where the unconscious fire victim may be. This fresh air may be all the victims need to take a few more breaths to stay alive until you reach them and remove them to safety. As we will see, this fresh air can also be all the fire needs to turn into a vicious killer.

Searching during and after horizontal venting. What you will see inside when horizontal ventilation is accomplished is the smoke layer lift or rise a few inches to a few feet. This is a huge improvement and one we must always remember to take advantage of: when the smoke lifts, get under the smoke layer and shine your hand light across the floor. When you do this, especially in very bad fire, smoke, or heat conditions, you may be able to conduct a very effective mini-search of the room without entering beyond the limits of safety. Remember, unconscious fire victims will most often be on the floor. They are not floating around the room or stuck to the ceiling or walls. Since furniture is usually placed around the walls of a room in a home you will be able to visually search for unconscious fire victims efficiently and safely. The addition of a thermal imager enhances this technique. If conditions permit, remember to look on couches or beds for victims. Also recall that children tend to hide in closets, in bathrooms, under furniture, and sometimes in kitchen cabinets.

Methods for horizontal ventilation. When windows are used for ventilation, first take the glass with your tool or a ladder tip. You want to start to relieve the bad conditions inside as soon as possible for the engine company. The curtains, blinds, shades, and sashes must be completely cleared. This not only provides the maximum vent opening, it also provides a second exit for interior crews. It additionally provides a viable opening for outside crews to provide assistance to the inside crews. The old adage about making a door out of a window pretty well sums up this technique and its advantages (fig. 8–15).

Fig. 8–15. Make a door out of a window. Courtesy of Jeff Arnold.

Double-hung windows are the most common in residential construction. When the two sashes are closed, the top of the bottom sash and the bottom of the top sash meet. This is where the locks are installed; this part of the window is called the "meeting rail." Some departments have an SOP that the presence of a meeting rail in a vented window is a sign to the interior crews that no ladder is below the window. We feel this is too much to expect people to remember and also provides no safety. The benefits of a large escape opening far outweigh the need to indicate the lack of a ladder. If a firefighter needs to exit the home due to untenable conditions, they can use their bailout rope or a hang technique after calling for a ladder via radio. Since a ladder was necessary to completely vent the window initially, a better SOP is to not remove the ladder once it's placed. This avoids overventing ahead of the search crews, drawing fire into areas that have not been searched.

Another reason to remove the entire sash is that the multiple-layer (two to three) thermal pane glazing in the modern house windows does not fall out like single-pane glazing, and the multiple shards that remain are dangerous and limit ventilation. Removing the entire sash may be the only way to clear the glass. Often modern windows are made of extruded or folded vinyl and are relatively easy to remove with a few sharp pulls with a hook or Halligan.

It goes without saying that if your department makes a door out of a window during ventilation, ladders should be placed to windows to provide an alternate means of escape in case a firefighter needs to make a hasty exit.

What would cause the need for an emergency escape? Loss of water on the attacking hoseline is one of the most common causes. Hidden fire suddenly showing from closets, knee walls, or balloon-frame construction and lighting up the room is another instance. In the modern home, fire attacking lightweight structural members that fail more quickly than their dimensional lumber counterparts can create a situation where the sagging or spongy floor is a sign of imminent collapse. A firefighter may need to move quickly to an outside wall near a window to avoid falling through the floor. Another common cause for emergency egress from upper floors of homes is the attic lighting up, causing the second-floor ceiling to fail and filling the occupied room with fire from a well-involved attic.

Vertical ventilation

Preventing backdraft. If the possibility of backdraft exists, ventilation must be performed at the highest levels of the house. It is critical that the heat be released before air is introduced at lower levels, as the inrush of air can cause an explosive event.

Roof operations. Cutting or opening up a roof on a peaked-roof private dwelling is not an easy task. The pitch of the roof, roof covering, and roof construction, not to mention the fire underneath you, make it very dangerous.

Cutting the roof for a fire that has not extended to the attic is ineffective. If the fire is in the attic and you can cut a hole near it safely, okay. If the fire is on the top floor and is contained to the room or rooms then cutting the roof will provide little relief of smoke and heat. If you cut the hole you must push down the top-floor ceiling so the smoke and heat can rise through the attic to vent. All too often we see plywood floors in attics for mechanical equipment and storage, which will prevent any venting (fig. 8–16).

Fig. 8–16. Roof coverings like this ceramic tile roof may impact vertical ventilation.

The gable end vents may be easier and safer to remove and will provide some relief (fig. 8–17). Don't overlook the possibility of opening the gable end wall for additional ventilation. The gable end of a home may have a vent already in it, in which case simply removing the louvered vent will increase

the ventilation of the space and enlarging the hole will increase the effectiveness.

Fig. 8–17. The gable vent is plastic or light aluminum and easy to pull out for immediate ventilation.

Before proceeding with roof operations, you need to size-up the roof. What can you tell from the smoke or fire conditions? Is the roof a truss? Is there a ridge vent? Smoke from a ridge vent can provide information on the heat in the attic. Absence of smoke may not be an indicator. High heat may have melted the plastic ridge vent material and sealed the shingle caps back to the roof, concealing any fire or smoke. If the vent is intact, pulling the cap shingles and the vent material will provide an easy vent. It will also indicate if the roof is a truss. If a ridge beam is present, it indicates that the roof is "stick framed"; the absence indicates a truss system.

If roof operations are required, proceed safely. Working from a tower ladder bucket is safest, with the aerial ladder being a good second choice, a portable roof (hook) ladder third, and walking directly on the roof surface being the least preferable. You can see why this is the last option since there is nothing to distribute your weight across what may be a weakened roof deck and your footing is not secure (fig. 8–18).

Fig. 8–18. If the roof sheathing is weakened by fire or simply rotted, you could fall through into the fire below. Courtesy of Brian Duddy.

The roof team must be prepared. Have a clear understanding of what it is you are going to do, get it done, and then get off. Know where you are going to operate, communicate the plan to the entire roof crew, and only bring the crew necessary to complete the task. This is not the place for spectators. You need to be prepared with an escape plan should something go wrong. Have two ways off the roof. Don't position your vent opening to cut off your escape route.

We don't care what kind of hole you make. Square, triangle, it doesn't matter, it's not our job to make the repair easier for the roofer. We just need to make it safe, so don't cut a truss (fig. 8–19)!

the location of the hole, be sure it is venting the proper attic space.

When the house was constructed and the roof was sheathed, the plywood or oriented strand board (OSB) was put on from the bottom up (fig. 8–20). Depending on the length of the rafter a small strip of wood may be needed to reach the ridge. Pulling the shingle caps will expose the edge of the wood and you may be able to get a pry tool, Halligan, or Halligan hook and pry the boards or thin strip of plywood, not a full sheet, off the rafters. Notice I said to pry off, not pull. If you are pulling on a sloped roof you can lose your balance and fall.

Fig. 8–19. Firefighters cut the roof for vertical ventilation. Courtesy of Brian Duddy.

Fig. 8–20. Pulling the roof sheathing or planks allows fire to vent upward. Courtesy of Brian Duddy.

Oh, and by the way, if you made the hole in the right spot, smoke and fire should be coming out of it, so you need to be breathing out of your SCBA. Put your facepiece on; you don't need it dangling in your way and getting tangled in your tools, especially a power saw.

If a cut is your plan, a rotary saw or chain saw should be your personal preference. The key to using power tools effectively is that you must be experienced with them and able to use them under these extreme conditions.

One large hole is better than several smaller ones, because fire will take the path of least resistance: a single large hole keeps fire from popping up all over. You should be able to expand it if necessary. You should also cut the hole high toward the ridge, the higher the better, as this will release the most heat. Many modern houses have several smaller attic spaces that are not interconnected. When determining

Positive pressure ventilation (PPV)

PPV techniques have been used in areas of the country for many years; other areas provide some of the staunchest opposition and harshest critics. PPV is something you either believe in or don't, but as the culture of the fire service changes even as we write this book, we are seeing traditions, beliefs, and ventilation procedures change in numerous large and small fire departments across our nation.

This west coast practice has gained some acceptance even in New York City, where tests were conducted on wind-driven fires and PPV fans were used by the FDNY for smoke control in high-rise buildings. But the move from field testing to using PPV in actual firefighting operations may not be applicable for every department. PPV is a concept that you should not disregard just because you "don't do it that way." It may not work everywhere, but in departments that embrace it, the tactic has had resounding success.

This discussion continues the concept of this text: encouraging you to look outside the box (what your procedures call for and what you are comfortable with), explore the research, and apply the policies, procedures, and science—if they work for you. This is not to say you have been doing it all wrong for so long, just understand that you should be informed and make a conscious decision to have more of an open mind. By educating yourself, you will be able to have meaningful debate and make informed decisions. For many departments it will be a monumental shift in their firefighting procedures that may cause some of the reluctance. For others, it may be a way to embrace the future, to address training, staffing, and safety issues.

PPV may be seen as one of those fads in the fire service, like green fire trucks. But just as the research and statistical data on accident rates of green fire apparatus have led to more visible warning light packages and increased use of reflective striping, so might PPV be an evolution in keeping firefighters out of the most deadly environments.

The concept of PPV is simple: you set up a large fan to blow air into an area from the door you enter. The pressure created will drive heat and smoke away ahead of the advancing entry team. The air injected into the space lowers temperature and allows the entry team to move into an area by lessening the adverse effects of the heat and smoke. By controlling and limiting the exhaust opening firefighters can control and "channel" the smoke exiting the building.

The concept may be simple but the application requires training and education. (For more information and several reports developed by the National Institute of Standards and Technology [NIST] go to http://www.nist.gov/fire/ppv.cfm.)

Years ago, pushing heat and smoke ahead of the entry team was done by operating a hoseline on a 30-degree fog pattern. Horizontal ventilation was required to allow the heat and smoke to escape. The size of the vent opening was a topic of research by the authors to determine how much was enough. However, the hoseline option can disrupt the "thermal balance," which is when the heat and smoke layer stratify with the hottest air at ceiling level and cooler air near the floor, and that level is recognizable. When the water from the operating hoseline turns to steam, it will obscure visibility and create a hot, humid atmosphere that can burn firefighters or civilians.

Additionally, with the hoseline option there is a risk of "overcooling" the environment. This is when the water reduces the ambient temperature to the point where the heat does not make the smoke and steam lift, further obscuring visibility. Overcooling can have the additional effect of cooling all the contents to a similar temperature, reducing the effectiveness of TICs. This was not a consideration when this type of fire attack was popular since TICs for the fire service had not yet been developed. For departments still employing this type of attack today, the adverse effects of the misapplication of water must be understood.

The authors are proponents of the use of a straight stream or solid stream for interior structural fire attack operations to reduce the negative effects described above.

Entire books have been written and millions of dollars have been spent on studying the effects of positive pressure ventilation on various type fires. Our attempt in this section of the chapter has been to provide a solid overview and background of this critical tactical tool for use at house fires.

How Much to Vent: When Is Enough, Enough?

Breaking windows and cutting big holes in the roof are often seen as destructive measures, but most of the time ventilation allows the fire to be extinguished faster and lessens the overall damage to the structure. The damage caused by ventilation appears worse than it is. Windows can be the cheapest thing to replace in the home when compared to replacing structural elements, girders, joists, and rafters, or a priceless possession of the homeowner. Remember, this is a house with regular windows, not a cathedral with Tiffany stained glass. We know it doesn't answer the question completely, but when venting windows, if you need to, take them.

As we have said, ventilation must be coordinated with two other critical fireground activities: the SAR and fire attack operations. Interior conditions will dictate when to vent. If you enter through the front door and the smoke is down to the floor, you need to vent. You will not be able to find victims, find the fire, or find your way out if conditions do not improve, even slightly.

The vent has to support the other fireground activities. Advancing the hoseline to extinguish the fire is one of those activities. How much air the attack team lets in will affect how much the ventilation team needs to open.

Ladders

Do you have enough ladders to place one at every window cleared for ventilation? In the case of the initial alarm assignment, the answer may be no. If this is the case, then the location of portable ladders must be communicated to inside crews. Before the fire, plan with your mutual aid companies for additional ladders. Consider automatic mutual aid and get the help coming based on the initial dispatch. The assistance is free and if you are stuck in that upper floor room and need to escape quickly, that ladder will be very valuable right then, not after a 10-minute response time. We recommend a minimum of one ladder for every room firefighters are operating in above ground level.

Chapter Conclusion

Ventilation is a critical component of your overall strategy at modern house fires. It is a tactical operation with several options that must be carefully and properly selected. Ventilation must be executed and timed properly as an integral part of your fireground success.

Scenario Discussion

Fig. 8–21. Scenario

Situation: It is Sunday morning, 1023 hours, and you are tapped out to a fire at 2 Veterans Street. A 911 caller reported smelling smoke in the area.

Radio Report: Dispatcher reports multiple 911 calls.

Resource Report: You are on the scene and your first-due piece of apparatus is approximately 2 minutes from the scene. There is a good hydrant 200 ft before the fire building.

Questions

Answer the following questions based on the photo and information above, and use this scenario as a basis for the topics examined in this chapter.

1. **Strictly from a ventilation point of view, what is your size-up of this fire?** This fire is very complex. It is free burning in the bedroom on the A-D corner and has been for some time. It is unclear from this view whether the bedroom door is closed or the fire has entered the hallway and is extending toward the front door. Even if the door is closed, the amount and duration of the fire suggests that the door has been or soon will be burned through. The window frames are completely burned away and fire has consumed the vinyl siding around the window openings.

It appears that the fire has entered the attic via auto-exposure through the soffit over the windows on the front. From the smoke showing along the A side toward the B side, it appears the fire is ventilation limited and just needs some air to light up the attic. The black smoke may be from the roof shingles burning.

2. **What type of ventilation is already in place and what is possible?** Obviously the fire has horizontally self-vented what appears to be the room of origin. If possible, other horizontal ventilation should be attempted via windows and doors. On this type of home, there usually is a deck on the rear with sliding glass patio doors on one or sometimes both floors.

There appears to be a substantial body of fire in the attic so vertical ventilation may be desired, but may not be an option, especially if the roof is supported by truss construction. Even if this home is stick-built, putting members on the roof would be questionable at best.

3. **How will your ventilation plan affect the SAR operation?** Venting the front door, whether for search, fire attack, or both, may create a flow path. If you cannot get water on the fire quickly the entire place may light up and do so very quickly. Putting members inside ahead of the line will severely endanger them. If other members have started a VEIS operation on the rear, the fire lighting up rapidly due to the air you allowed in through the front door may have the same effect. Since the fire appears to be vented already, getting the line in through the front door and to the seat of the fire should not need any additional ventilation support.

4. **What negative effects could improper ventilation have on this fire?** Simply, the more air we give this fire, the faster it will grow, and it looks like this would mean a short time to full flashover. Taking any windows or doors will create a flow path resulting in rapid fire growth. It is unclear how charged the rest of the home actually is, but you will find out when you open the front door. You have to assume, with this much fire in the bedroom, that the remainder of the home is charged and just waiting for some fresh air.

The patio doors on the rear are an inviting opportunity especially for truck company members. They can vent them easily and the large doors provide a large, easily utilized vent hole. That same vent hole is a large source of air for the fire to feed on. Taking these doors before the line has water on the fire would be disastrous. It is likely you could end up with another fire like the Cherry Road fire in Washington (DC) in 1999 that killed two firefighters in a lightning-speed flashover (Madrzykowski and Vettori 2000).

5. **What is your ventilation strategy and timing?**
   Timing is critical for ventilation, especially at this fire. Until the SAR is completed, all ventilation decisions and efforts must support the rescue team(s). Appropriate strategy here is to not vent until there is good control of the fire in the bedroom and wherever it may be extending. Getting the line down the hallway quickly to get water on the fire is critical to the success and safety of the search teams and keeping this fire from running through the entire house. This line protects the means of egress for occupants and firefighters. Consideration of the fire in the attic is next and action should be taken immediately after the bedroom is knocked down. Venting the gable vents will provide some limited ventilation for the attic fire.

Getting water on the fire quickly is a key tactical action to accomplish your overall strategy of first saving lives and then saving property while minimizing threats to members. However, how long will it take to get the line down the hallway and to the fire room after the front door is opened? Will the fire light up the entire house in this time? Will venting the front door, though well intentioned, cause an undesirable outcome? This could manifest itself in injured members and lost occupants and property.

Since getting water on this fire (the sooner the better) is a critical task, why not apply water from the outside through the vented windows? As the UL has suggested, if you keep the nozzle close to the outside of the house (in this case, in the driveway) and direct the stream up toward the ceiling of the fire room *without* moving the nozzle, you will create a 180 gpm sprinkler-head water pattern. This will provide excellent water distribution and knock down the fire. Since the main body of fire is suppressed, this may result in the fire in other areas not flashing over and endangering your members, and may improve conditions for victims in the house.

One of the key factors to success with this tactic is to not move the nozzle. The UL has shown that moving the nozzle (as we have traditionally been taught) results in entraining about 7,000 ft^3 of air per minute into the fire area. This may be why we feel like a hoseline is pushing the fire at us: the experience of some of this heated and moist air being pushed toward us because we're using this nozzle technique (or a fog line) could easily be mistaken for the exterior team pushing fire on us.

It is also critical to recall that use of this exterior stream is an option, another tool in our tactical tool box. No one—we repeat, *no one*—is saying that this is the way you have to operate every time.

References

Fire Department of the City of New York. 2013. *Ventilation.* Vol. 1, book 10 of *FDNY Firefighting Procedures.* http://www.firecompanies.com/modernfirebehavior/governors%20island%20online%20course/story_content/external_files/ffp_b10.pdf.

Kennedy, Tom. 1991. "Energy Efficient Windows in Multiple Dwellings." In "Training Notebook," special issue, *Fire Engineering*, January.

Kerber, Steve. 2010. *Impact of Ventilation on Fire Behavior in Legacy and Contemporary Residential Construction.* Northbrook, IL: Underwriters Laboratories. http://www.ul.com/global/documents/offerings/industries/buildingmaterials/fireservice/ventilation/DHS%202008%20Grant%20Report%20Final.pdf.

Knapp, Jerry, and Christian Delisio. 1997. "Energy Efficient Windows: Firefighter's Friend of Foe?" Firehouse.com (blog), July 1. http://www.firehouse.com/news/10544872/energy-efficient-windows-firefighters-friend-or-foe.

Madrzykowski, Daniel, and R. Vettori. 2000. *Simulation of the Dynamics of the Fire at 3146 Cherry Road NE, Washington D.C., May 30, 1999.* NISTIR 6510. Gaithersburg, MD: National Institute of Standards and Technology.

Mittendorf, John. 2011. *Truck Company Operations.* 2nd ed. Tulsa: PennWell Corporation.

Office of Policy Development and Research. 2016. *2016 Characteristics of New Housing.* Department of Housing and Urban Development. https://www.census.gov/construction/chars/pdf/c25ann2016.pdf.

9

Basement and Garage Fires

Summary

This chapter's purpose is to examine and discuss strategic and tactical considerations, options, and outcomes (positive and negative) for operations at fires in basements and garages. As in the previous chapters, this chapter discusses the practical application of the appropriate skills and development of strategic plans by chief officers, as well as plans and considerations and their execution by company officers and firefighters. For firefighters reading this book and building your experience base, this chapter provides an understanding of specific tactical operations you will execute as part of the tactical side of successful operations at house fires, specifically basement and garage fires. This chapter provides fire officers with in-depth discussion of strategic options and tactics supporting these strategies. (The techniques for tactical operations were covered in previous chapters and will not be included here unless specifically applicable.)

Introduction

Basement and garage fires present unique challenges to firefighters across the nation. If you decide on an interior search or fire suppression operation, members are operating above a fire that may have weakened structural components of the house. Fire attack crews may be working in the basement in close proximity, with limited ingress and egress capability and below weakened, collapse-prone structural elements. Basements and garages are areas that often contain multiple utility dangers (gas, oil, electric), stored hazardous materials, unfinished or unprotected structural components and shoddy, do-it-yourself construction, and even unexpected life hazards by virtue of illegal occupancy. In addition, both areas are often dark and crowded with stored materials. These below-grade locations have limited access and egress with small windows that limit ventilation. They also have a sad history of killing firefighters.

Basements are often accessed by an unprotected interior stair located under the main stair in the home. This stair may be narrow and not intended to support the weight of an advancing fire crew. It may be compromised by the fire, creating a potential for collapse. If there is an exterior entrance through a covered stair enclosure, it may be more advantageous to use it for ventilation, which would be consistent with an attack strategy of attacking the fire from the unburned side.

Important Notes

Strategic and tactical considerations included in this chapter must be applied to the conditions, staffing, types of apparatus, types of homes, response time, and other variables specific to your department and region of the country. What works well under specific conditions in the northeastern United States may not work well in the Pacific Northwest. The specific conditions in which you work may dictate departures from or modifications to the strategies and tactics presented in this chapter. The goals here are to familiarize you with the hazards of basement and garage fires and to give you a systematic method with practical exercises to examine, discuss, and apply strategic and tactical alternatives. (When discussing scenarios in this chapter, we assume that a reliable and continuous water supply has been or will be established via tanker operations, drafting, or municipal hydrants.)

Although we have made extensive efforts to include important details in scenarios to provide a common operating picture of the scenario, please remember that we will all see the same scenario differently, based on our experiences. There are many judgment calls to be made when using scenarios. Firefighting is not like baking a cake, where if you do everything in the right order and in the right amount, the cake is perfect every time. Don't get hung up on the details; it is the process and the examination of strategy, tactics, alternatives, and effects that are important. The fires examined in this chapter never happened, so we don't have to argue over how to extinguish them.

Basement Fires

For the years 2010–12, there were approximately 6,500 basement fires in one- and two-family residential buildings per year in the United States. These fires result in approximately "65 deaths, 400 injuries and $278 million in property loss" (National Fire Data Center [NFDC] 2015, 1). Basement fires account for 3% of fires in one- and two-family homes (1). Utilities for the home are often headquartered in the basement, so it is not surprising that the leading causes of basement fires are "'electrical malfunction' (19 percent); 'heating' (14 percent); 'appliances' (12 percent); and 'other unintentional, careless' actions (12 percent)" (1). January, December and February are the peak months for basement fires, with 12%, 10%, and 10%, respectively (1).

A basement is the lowest floor of a home that is partially or wholly underground (figs. 9–1 and 9–2). Usually in homes that have basements, the landscaping defines how much is underground. Cellars are usually defined as having 50% or more underground, according to several building codes.

Fig. 9–1. At first glance, this home appears to be built on a slab without a basement.

Fig. 9–2. The rear view shows a full basement, with three windows and a full-height access door on the right side of the picture. Note the difference in the grade: one floor in the front, two in the rear.

Construction is always a concern in basement fires. Usually the structural components (first-floor joists, I-joists or parallel cord trusses) are directly exposed to the fire from below (fig. 9–3). For firefighters operating in the basement, a collapse means heavy structural components trapping them in the fire area. For firefighters above, a collapse often throws them into a now well-involved basement fire with fresh air supplying the heavy fire load of stored materials. According to an article in *New Science Fire Safety*, even in "basement fire experiments, where the variables were systematically controlled, there were no reliable and repeatable warning signs of flooring collapse" (Underwriters Laboratories [UL] 2014, 19). One of the recommendations in the article, then, is that "when possible, the floor should be inspected from below prior to operating on top of it," because "signs of collapse vary by floor system:

- Dimensional lumber should be inspected for joist rupture or complete burn through.

- Engineered I-joists should be inspected for web burn through and separation from subflooring.

- Parallel Chord Trusses should be inspected for connection failure.

- Metal C-joists should be inspected for deformation and subfloor connection failure" (20–21).

For additional information, please refer to the LODD report from 2007 that was detailed in chapter 2 (National Institute for Occupational Safety and Health [NIOSH] 2007).

Even finished basements present problems during fire situations. Often the unsightly undersides of floors, with pipes and wiring runs, are covered with drop ceilings. During a fire, these can come down on crews in the basement, resulting in a very deadly trap.

Fig. 9–3. Structural members are often exposed to direct flame impingement during basement fires.

Access and egress is always another major concern at basement fires. In homes, the interior stairs to the basement are often narrow and winding and may have just the bare essentials, without hand rails or other safety features (fig. 9–4).

Fig. 9–4. Basement stairs are often unfinished, old, narrow, and unreliable, especially under the load of several firefighters in full turnout gear.

Opening the interior door from the first floor to the basement stairs provides a vertical ventilation opportunity for the fire to extend to the first floor. If a basement window is broken by the fire or firefighters and the first-floor interior basement door is opened, a flow path will likely be created, leading to very rapid fire development often with tragic consequences for members operating both above and below the fire. Exterior stairs aren't much better, however: when they are available, they come in the form of fold-up (clamshell) doors that are usually locked from inside and force firefighters to climb down through the rising heat and smoke just like interior basement stairs.

The least urgent of our numerous problems at basement fires is visibility. Remember and anticipate that the lighting will always be exceptionally poor to begin with, then add to that the fact that smoke and heat have limited opportunities to exit. Always remember that electric circuit breaker boxes and other utilities may be exposed, and be wary of unintentionally contacting live wires, unsecured gas lines, and other hazards with your hand tool.

The insidious and inherent hazards to firefighters mentioned above often combine with deadly consequences during basement fire operations. The NIOSH fatality investigation summary below provides details about one such instance that resulted in the line-of-duty death of a volunteer firefighter.

On March 5, 2008, a 35-year-old male volunteer Fire Lieutenant (the victim) died while fighting a basement fire. About 30 minutes after the fire call had been dispatched and the crews had been evacuated from the structure and accounted for, a decision was made to re-enter the structure to try and extinguish the fire. The victim, an Assistant Chief (AC), and a Captain had made their way down an interior stairway to the basement area where the victim opened a 1 ¾-inch hoseline. Shortly thereafter, the Captain told the AC that he had to exit the basement stairs. A few seconds later, the AC told the victim to shut down the line and evacuate the basement because the fire was intensifying. The AC was second up the stairs and told a fire fighter at the top of the stairway landing that the victim was coming up behind him. The AC exited the structure while the fire fighter stayed at the top of the stairway and yelled several times to the victim, but received no response. The fire fighter exited the structure and informed the AC that the victim had not come up from the basement. The AC then notified the Incident Commander who activated a rapid intervention (RIT) team. The RIT made entry into the structure but was repelled by the intensity of the fire. After several more rescue attempts, the victim was removed from the building and later pronounced dead at the hospital. Four other fire fighters were treated for minor injuries and were released from the hospital. The following factors were identified as contributing to the incident: an absence of relevant standard operating guidelines; lack of fire fighter team continuity; suboptimal incident command and risk management; and lack of a backup hose line. (NIOSH 2009, 1)

Basement fires in modern homes, especially the larger (3,000–7,000 sq ft) estate homes (McMansions) being built in more affluent suburbs and often rural areas, present us with a unique hazard (different grade levels) and a potential bonus (English basement):

> Some estate homes will have what is termed an English or walk-out basement, present when land at the rear of the house slopes or is graded downward. The home will likely be three stories tall at the rear and only two stories tall at the front.... The English basement will also have an entry door similar to the front door, and full-size windows or bigger at the rear grade level. Although these may have more heavily fortified locking mechanisms, both access and the ventilation should be more easily accomplished when an English basement is present when compared to a common residential basement. Don't rule out an attack from the rear when the fire is in the basement. As at all residential structures, make sure, even if attacking from the rear, that the initial line protects the first floor and basement stairway before seeking alternative hoseline entry points. Reconnaissance of the rear is essential at all estate homes. (Avillo 2015, 254)

Let's begin our in-depth look at basement fires by examining a typical basement fire with a life hazard (fig. 9–5).

Fig. 9–5. Scenario

Situation: It is 23 June, a Tuesday night, at 2330 hours. A police officer on routine patrol finds smoke coming from the basement windows of this home at 20 Guadalcanal Drive. Smoke is issuing from the first- and second-floor windows. The house is in a good residential neighborhood known for young families with children. Your fire department has been dispatched and is responding with your normal apparatus and personnel. You are the first-arriving fire chief. Using RECEO VS as a guide for our discussion, let's examine the strategic considerations at this fire (fig. 9–6).

Fig. 9–6. Basement fire

Rescue

Your overall impression is that there is a high life hazard here. The home appears to be occupied—there are cars in the driveway, and the home is well kept and in a good neighborhood. Their survivability profile is urgent (refer to ch. 6); since it appears that fire has not reached the upper floors, victims are likely there and salvageable. The immediate concern is that there is significant fire in the basement and that any occupants of the home are above it and subjected to toxic products of combustion and extending fire. The overall initial size-up is that this is a basement fire in an occupied, 20 × 40 ft Cape Cod house with possible victims trapped above. The 360-degree size-up shows a clamshell back door on the first floor leading to the basement in the rear. Smoke is showing from the clamshell door and a small basement window on the D side shows smoke and fire.

Situation update: The police officer has questioned the neighbor, who states that a family consisting of a mother, father, and three children lives there and that all members are home. Her son was playing with the family's son earlier in the day.

Option #1. The fastest way to get into the house and access the victims is through the front door. Supporting this search team with a hoseline to the top of the basement stairs to contain the fire is a good option. It is important to note that if this line is directed down the interior stairs from above, little water will make it onto the burning fuels and the basement ceiling (joists and sub floor), which are likely involved. Additionally, pushing this line down the stairs has a high risk factor with four main advantages:

1. It is the quickest way into the house and allows a continuous path to safety for victims.

2. It allows a rapid size-up of interior conditions and structural stability by the search crew, and will tell the IC if the floor has burned through or collapsed and if the door between first floor and basement has burned through.

3. Forcing the front door paves the way for the hoseline to protect the interior exposures, the first floor, and the second floor.

4. Placing the hoseline between the victims and the fire is always a good tactical decision. This line can later advance down the stairs if necessary.

There are also three significant disadvantages to entering through the front door:

1. Interior firefighters will be exposed to significant risk operating on the floor above the fire. Unprotected floor joists, trusses, or engineered lumber may fail, plunging them into the fire below. Obviously a Mayday or injured firefighters freezes the effort to rescue occupants if personnel is limited.

2. If the front door is heavily fortified, there may be a delay in forcing the door (although this is usually not the case in single-family homes).

3. Opening the front door and leaving it open creates a flow path for the fire, which may endanger both victims and firefighters if water is not applied quickly.

Option #2. Commit all your firefighters to SAR. This may be a strategic option, but it depends on how low your staff level is. If victims are at the windows and in imminent danger, these rescues will be made first and consume your personnel and by default limit your strategic options. You may choose to assign all your limited personnel to SAR via interior routes or VEIS. No matter how you do it, be aware that your entry into the building will be much more hazardous if water is not being applied to the fire. There is no easy or cookbook solution for inadequate staffing. As a wise old fire chief once told us, "You know what you get with trying to do more with less? Less." Sometimes you can't save everyone (or anyone) because of inadequate staffing.

Option #3. VEIS the upstairs bedroom windows. This method has the following three advantages:

1. It provides immediate access to a place where victims are probably located: the bedrooms.

2. A single firefighter can search a bedroom using this technique with relative safety.

3. If the firefighter closes the bedroom door, they do not create a flow path.

There are also two disadvantages to this option:

1. Victims who tried to escape in hallways or are on the first floor may be missed or their rescue delayed.

2. The removal of unconscious victims from second-floor bedrooms out small windows and onto a steep roof will be very difficult. It appears that the

long driveway may make ladder truck access difficult or impossible.

Exposures

The most critical exposures are the floors above the fire where the victims are likely located. Is the fire extending up the interior stairs from the basement? Has the door held or is it or about to burn through? What is the condition of the basement steps? How do you protect the exposures?

Option #1. A good general rule for both fire suppression and rescue is to place the first hoseline between the victims and the escape route or the victims and the fire. In this case, the hoseline to the door at the top of the interior basement stairs will keep the fire from extending upward and endangering both occupants and firefighters. This method has the following advantages:

1. It utilizes the front door (forced open by search team), providing immediate access to a strategically important position: between the fire and the victims (and firefighters) and their escape route. The strategy is to conduct an aggressive SAR operation and this tactical position supports that strategy. Placing the first hoseline between the fire and the victims is a time-tested tactic.

2. This hoseline could be directed to proceed down the stairs to the seat of the fire, conducting a search operation and controlling or extinguishing the fire, minimizing all other problems. Note that it is important to be alert for changes in the conditions that may affect the other activities occurring while the attack is conducted.

3. This line may protect the stairs to the second floor, especially if they are directly over the stairs to the basement from the first floor and fire is extending upward.

There are also some disadvantages:

1. The time it takes to get the line stretched inside may allow the fire to grow unchecked.

2. Getting this line into operation inside will take more than one firefighter.

3. If the crew inside gets injured and cannot perform this task, the fire is uncontrolled and a flow path may have been created, leading to rapid fire extension inside the building that endangers both occupants and firefighters.

Confine

You should expect this fire to extend rapidly upward to the first floor and on to the second. The hoseline at the top of the stairs (first floor) is in the best position to confine the fire and protect the means of egress for occupants. It is ahead of or in the path of the expected fire travel. It supports our first priority—rescue—and our overall priority, the safety of firefighters on the search mission. It is also the most dangerous position for firefighters on the hoseline but this manageable risk is justified if your size-up shows there is a reasonable potential for salvageable human life. However, the fire will also spread up through vertical voids and penetrations for electric, heating and cooling, and supply and waste pipes. The basement fire is a worst-case scenario: almost everything you want to protect (people and the remainder of the house) is above and directly in the path of the fire. Confining the fire to the area of origin is critical to your overall strategy and success.

It is important to note that the research done by the UL of the data collected during the joint FDNY-UL-NIST (National Institute of Standards and Technology) live burns at Governors Island (2012) proves that the water applied from a hoseline from the first floor into the basement fire does little to control it (fig. 9–7; Kerber 2014, slide 57). The obvious question then becomes, "Is it better to apply water from a different tactical position?"

It is important to recall that at this point, early in the fire attack, the life safety of both our firefighters and the occupants is the top priority. So although this line may not be the best for fire extinguishment, it is the best choice for life safety.

Obviously, if other firefighters can get water directly on the seat of the fire, things get better and dangers are reduced for everyone on the scene.

Option #1. Stretch additional hoselines (we are assuming some type of attack has already been directed on the fire in the basement) to the first and second floors to check for extension. If it is a balloon-frame home, pull the base molding along exterior walls to check for vertical fire extension. Get to the attic quickly; the fire may already be there. There are two advantages to this method:

1. It is obviously a good option if staffing levels and apparatus permit.
2. It confines the fire to the basement if successful, saving property and protecting the search crews.

There are also two disadvantages:

1. It is personnel and time intensive. You will need a truck crew to open up places were fire is suspected to be advancing in walls and ceilings and an engine crew to extinguish any fire you find.
2. This house is not very big and adding all these crews may well impede your SAR strategic objective, especially early in the fire.

Option #2. Put water on the fire directly through the small basement windows. This alternative, like most confining-the-fire tactics, leads directly into extinguishing phase. Especially if staffing is short, forcible entry is delayed

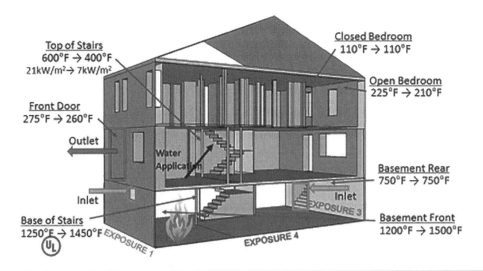

Fig. 9–7. Data collected from live burns show that water directed down the interior stairs onto a basement fire does not improve fire conditions. *Source:* Kerber 2014, slide 57.

by high security, or you cannot quickly get the line inside and to the top of the interior basement stairs, putting water directly on the fire through the small basement windows may confine the fire to the basement. This method has four advantages:

1. Puts water directly on the fire, limiting its growth and slowing it down.

2. Can be done by one firefighter in a safe position from the outside through multiple windows.

3. May control the fire, preventing weakening of unprotected first-floor structural members.

4. Upward movement of steam through voids will help control the fire spread.

There is one disadvantage to this method:

1. Water may not reach the seat of the fire if blocked by walls or stored material.

Extinguish

Option #1. If you have a hoseline in place at the top of the interior first-floor door to the basement stairs, you can push this line down the stairs into the basement. The advantages to this method are as follows:

1. The line is already inside the building, flowing water and staffed.

2. It can make a direct attack on the seat of the fire.

However, this method also has four disadvantages:

1. This crew is essentially pushing down the chimney of the fire. They are in the flow path of the fire.

2. The interior stairs may be weakened by fire or of shabby construction and condition.

3. Dangerous position: Once this attack team makes it to the bottom of the stairs they have only one way out: the way they came in. If the fire flashes over and the line cannot contain the fire, their vertical exit path may be filled with fire. Attempting to exit through an involved basement has caused numerous firefighter LODDs.

4. The team may already be fatigued and low on air since they were operating to support the SAR operation.

Option #2. Force the clamshell doors (if available) at the side or rear and advance a hoseline to the seat of the fire. This method has the following advantages:

1. Firefighters have complete access to the seat of the fire.

2. Firefighters can continue advance to the seat of the fire to complete extinguishment.

3. The advance down the stairs will be shorter and easier than trying to push down the interior stairs.

4. Exit will not be blocked by fire.

5. They are fighting the fire on the same level as the fire.

There are two primary disadvantages:

1. Will take time and personnel to force the steel doors, and there may be a secondary door at the bottom of the stairs that needs to be forced as well.

2. Interior walls or piles of debris may prevent water from reaching the entire basement.

If there is heavy fire in the basement, firefighters are essentially crawling down the chimney (vent point) to gain access to the seat of the fire.

Overhaul

Overhaul is required to open areas to examine for extension; this will prevent rekindle and further property loss. Overhaul may occur after the fire is knocked down and before final extinguishment. It does not mean we can't take steps to protect property (salvage) in the meantime.

Ventilation

Forcing and opening the front door to let in search crews is ventilation. Early in the fire this can be good or bad. Since this fire is in the basement and it appears that it has not extended into the living space, smoke that has made its way to the first and second floors should lift for you, improving visibility. Ventilation options for basement fires are limited by small windows and a generally closed-off fire compartment (basement) separated from the rest of the house. It is important to recall that providing air to the fire before water is applied will increase the fire intensity. Timing (coordination of truck and engine company actions) is everything and a command responsibility.

Option #1. Take the basement window(s). This option *does not help your strategy* of aggressive search operations supported by protecting the exposed parts of the home and containing the fire with a hoseline unless the line is going into the basement and already has water on the fire. It is likely that the fire is ventilation limited (has more fuel to burn and just needs oxygen to increase the size, intensity, and heat release rate of the fire). Adding air from this ventilation option will make the fire worse, increasing smoke production and increasing the threat to members operating inside. However, taking the windows may give you a clear indication of the exact location and extent of the fire in the basement since the added air will cause the fire to increase in size and speed.

Option #2. Vent first- and second-floor windows from the exterior. Venting second floor windows to increase survivability of victims trapped there may improve search conditions. This action may lift smoke and allow the interior crews to operate in better visibility. However, if the fire has penetrated up through the first-floor basement door or vertically in walls or floors, a flow path may be created, vastly and rapidly increasing the danger to members and victims.

Option #3. The third option is to open the roof. Early in the operation, it is unlikely you will have enough personnel to attempt this. Again, a flow path may be created, worsening the overall situation. In this style home, there is no attic, so opening the roof (and pushing down ceiling material) will directly open up the bedrooms. Since the seat of the fire is in the basement, opening the roof is of little value.

Salvage

As soon the fire is out, the cleanup can begin—this is what salvage was once thought to be. In the modern house fire, there are two concepts of salvage, one as property conservation and another as evidence preservation. Much can be done during the fire attack to minimize the damage and maximize what can be salvaged. Our missions are life safety and property conservation. If we do more damage putting the fire out than it would have or has on its own, we have not successfully completed one of our most important tasks.

Property conservation. We have sworn to save lives and protect property. Saving lives is always the first priority, which may be why we give less consideration to our second duty. The fire service has developed a reputation for cutting big holes in the roof, breaking every window, and getting everything wet. However, applying the new understandings of flow path and oxygen-limited fires may change some of these techniques. Extinguishing the fire, supported by forcible entry and ventilation, is the first step in property conservation. Stop the destroying action of the fire.

To those concerned about water damage, and in defense of the fire service, renowned fire service instructor and author Rick Fritz gave some advice on the matter: "I have seen a lot of stuff dry out; not a lot of stuff unburn." We need hoselines to deliver decisive amounts of water for quick fire knockdown, reach, and for our own safety.

However, water damage is a major concern for house fires. Consider how you would feel if someone came into the room you are in right now with a hose stream of 180 gpm. Sure, if that room was on fire, you would really appreciate it. If, however, the fire room was on the second floor of your home, you would probably just be thinking about what the runoff water would destroy below or in adjacent rooms, should the nozzle give an errant, couple-second burst. If staffing allows, we

must return to the basic process of throwing salvage covers for items at risk for water damage, assuming the other priorities of life safety and fire suppression operations are being executed.

Additionally, only use the amount of water necessary for a given situation. If you only need 90 gpm to knock down the fire, only operate the 180 gpm line for 30 seconds. The result is a decisive amount of water flowed to kill the fire without unnecessarily soaking the room. Shutting down the line at the appropriate time and not overcooling the room will allow convected air currents to vent the smoke in the desired ventilation path.

The simple concept is, damage not done does not have to be repaired. Salvage operations in a residential home are different from those at retail establishments where everything is up for sale at the right price. In commercial occupancies, the occupants likely have no sentimental attachment to the items, because the items do not belong to them.

In house fires, personal belongings are at risk. These may be antiques, heirlooms, or mementos. They may be irreplaceable items of the family's history, such as photo albums. They are valuables. Consider also the amount of electronics in the modern home and how expensive and easily damaged by water they are—things like flat screen TVs, microwaves, computers, cameras, and other common household electronics.

Saving this property for the homeowner can be accomplished by something as simple as placing it on the bed or couch, or even in the center of an uninvolved room and throwing a salvage cover over it for protection. Think of how grateful you would be if someone did this for you. Imagine your pictures of deceased friends and relatives, your wedding album, and other treasures saved by some simple acts of kindness and professionalism. At house fires we have a tremendous responsibility to help the homeowner or occupant in any way we can. Even a one-room fire is a traumatic event. Consider the loss of your favorite sneakers, your most comfortable jeans, and treasured photos, diplomas, or awards. A home is a collection and reflection of the occupant's life, their family, their very soul. We need to do everything we can to soften the blow of a house fire. Salvage is not fun or exciting, but can be very, very rewarding.

If items need to be removed from the house, don't just toss them out the window into the front yard. The fire itself is traumatic for the owners. No one wants to see their personal items piled in the front yard for everyone to see. Be respectful of their property.

Evidence preservation. Every fire will be litigated in some way. The facts of the fire may not be used as evidence in a criminal case but it depends on the circumstances. Initially, these may be part of a cause-and-origin determination. Often, sadly too often, a house fire is the scene of a suicide. Even more tragic is a murder-suicide. Unfortunately, we all can recall these horrific events in our towns. Sometimes the fire was set to cover a burglary, in which case you will be dealing with a crime scene. Before your members open up window and door frames and strip off the sheetrock, check to be sure there is nothing suspicious.

If a piece of equipment is involved as the source of the fire, it could lead to product recalls. The "equipment involved" box on the NFIRS report is used to track overall product safety. The make, model, year, serial number are valuable for overall product safety. These details can be useful for homeowners, assisting them in cost recovery. This requires a little attention to detail.

Arson. Firefighters need to be observant for suspicious indicators during the fire attack. They will be the first to see indicators of arson. An alert firefighter may be able to recall conditions on arrival. Were there multiple fires in the home separate from each other? Were doors or windows open or locked? Was there heavy fire in unusual places? Were deceased victims laying face up or face down? These and other questions will be important to arson investigators.

Cause-and-origin determination is the duty of all incident commanders and a legal requirement in some states. If the cause cannot be determined, the fire must be considered suspicious. Suspicious fire can be referred to fire investigators for further investigation. That investigation will depend on how well the scene has been preserved. If the evidence is in a pile in the front yard along with everything including wood trim, wall coverings, and the contents of the room, it will be much more difficult to trace burn patterns, depth of char, and other fire dynamics indicators.

Arson for any reason is, by definition, a crime. Once all accidental causes are ruled out, the fire is considered "incendiary" and a criminal investigation opened. This is when how well you protected the evidence becomes extremely important. If the material that can be used to determine whether the fire is accidental or incendiary is destroyed by sloppy overhauling, this can cause a delay in payments to the homeowner or a delay in prosecution.

The reputation of the fire department is often based on the homeowner's interaction with the department during an emergency, what the owners are left with, and how easy it was for them to recover. Neighbors will be watching, and many will have video or cellphone cameras. Your actions will be recorded and documented and easily shared via various types of media. We would like to offer three recommendations for dealing with homeowners during and after a fire in their house:

1. Be respectful of their property. Everything in their home means something to them. It has been saved for a reason.

2. Be professional. Save the conversations between you and your fellow firefighters until you get back to the station. You don't know who is listening; it could be family members or the media, and you don't want to have to deal with a comment taken out of context.

3. Be compassionate for their loss. For you this may have been a good job, but for them it is a terrible day. Being able to direct them to municipal services or other nongovernmental organizations for assistance takes professionalism to the next level.

SLICERS

The SLICERS acronym has become popular throughout the International Society of Fire Service Instructors (ISFSI) as a result of UL/NIST research and the American fire service's recognition that modern house fires behave differently than house fires from the 1950s and 1960s. SLICERS is a memory jogger for strategic and tactical considerations at house fires: **S**ize up, **L**ocate the fire, **I**dentify and control the flow path, **C**ool from the safest location, **E**xtinguish the fire, **R**escue any occupants, and **S**alvage. (Note that rescue and salvage are actions of opportunity and should be done at any time.)

Modern house fires are fuel rich and ventilation limited. Most often, all they need is more oxygen, which we so willingly introduce when we force the door (chock it open) and vent the windows. In chapter 3, we looked at how different fire development in modern homes is now, with faster times to flashover, higher heat-release rates of contents, and rapid exterior fire envelopment due to flammable siding, insulation, and sheathing. The recent UL and NIST studies on modern house fires have also helped drive a more firefighter-safety-focused approach to house fire tactics.

Keep in mind that any acronym is a very condensed version of the full understanding of our entire operation: fire development, the fire building, and of course fire department actions. If there were a miracle tool we would all have one in a holster hanging off our hip. Our greatest safety, strategic, and tactical tool is our brain. With that in mind, let's use the SLICERS acronym to consider tactical options and positive and negative effects for this basement fire.

Size-up. If you are the chief officer or the first officer on the scene, you need to develop a strategic plan; it all begins with a good size-up. Your on-scene report should sound something like this: "I'm on the scene at 20 Guadalcanal Drive, a 20 ft × 40 ft Cape Cod home with smoke showing from basement windows and open first-floor windows, probable basement fire, probably occupied, offensive operation initiated, send a second alarm." The 360 will provide additional information, such as availability of a rear door, access to bedrooms via second-floor windows, and in this case clamshell doors providing access to the basement.

Also part of your on-scene transmission will be establishing or assuming command. As discussed in the command and control chapter, this formally assigns responsibility to you for the operation. It is worth repeating that as a leader you have the responsibility to thoroughly prepare yourself before the alarm, follow your department's procedures during the alarm, and protect lives, both civilian and firefighter, then property from beginning to end. You are more than assuming command, you are assuming responsibility for the lives of people on the scene. It is a legal, moral, and ethical obligation of the most serious nature.

Strategically, the plan will be for an aggressive interior search operation focused on rescue and supported by hoselines to suppress the fire. Let's look at some tactical considerations to support the strategy.

Locate the fire. A key part of any size-up is figuring out where the fire is and where it wants to go. In this scenario, it appears to be in the basement and obviously can and might be extending upward into living and sleeping areas. The strategic plan is to focus on SAR and simultaneously protect the search crews with a hoseline by containing or suppressing the fire either from inside or outside.

Identify and control the flow path. The most likely flow path is up the interior stairs from the basement to the first and second floors. Controlling the flow path may already be done if the door is not burned through. Making sure by placing a charged hoseline at this point is a tactically correct option. Following the SLICERS concept, not controlling the flow path typically will lead to the fire moving quickly toward unburned areas (upstairs) where members are searching and victims are likely located.

During the RECEO VS years, we simply used the term "exposure" for identifying the flow path. It was assumed that we all understood that we would identify (during size-up, as best we could) where the fire was and where it wanted to go by identifying both the interior and exterior exposures. It was obviously necessary to know where the fire was and what areas were exposed (i.e., how the fire would spread to interior or exterior exposures). We did not know the science of flow paths or know them by that name yet. Thus our tactics and strategy were developed based on our observations during actual fires. It is important to remember that these observations based on experience are irreplaceable and invaluable. They are, however, made during a physically and mentally busy period and not supported by actual measurements of important factors such as heat flux, heat-release rates, air movement, or temperature.

These observations, especially those on fire dynamics and flow paths, are solely based on what you are able to see and feel in the immediate area. Conditions are not uniform throughout the home, so an action that makes conditions better in your immediate area (such as improving how far you can see in the smoke) may not improve conditions in other areas. Our mental perceptions and measurements of time, temperature, and other factors are often skewed during these intense search and fire suppression operations. If your action improved conditions for a few seconds, did it make them worse a few seconds later? How long did the improvement last and did it have a positive effect on the operation? Did the long term degradation in conditions outweigh the short-lived improvement?

Cool the space from the safest location. Strategically, how to cool the fire is a tactical decision made by the IC. (C stands for confine in RECEO.) Here is the leap: in the RECEO VS days the accepted tactic to put water on the fire was always from the inside. The facts from modern research show beyond any doubt that in the modern house fire environment, putting water on the fire from the outside does not make conditions inside any worse for firefighters or civilians, and actually improves temperatures throughout the structure (Kerber 2014, slide 59). This assumes you are using a straight stream from a combination nozzle or a smoothbore, and are flowing adequate amounts of water. (A fog pattern from a combination nozzle will drive huge amounts of air into the fire area and beyond, possibly pushing and extending the fire into uninvolved areas.)

The advantages of applying water to a basement fire via a basement window are numerous (figs. 9–8 and 9–9):

1. In low-staffing situations, one firefighter can put water on the fire, protecting the search crews inside.

2. Water is applied quickly to the fire. (This assumes there is direct access to the fire through the window. Usually, multiple windows are available around the foundation.)

3. This outside operation is much safer for members, with less chance of injuries that result in diversion of on-scene personnel to firefighter rescue.

4. Members operating this line are not in danger of falling into the basement and being injured or killed.

5. Members are not attempting to push down dangerous interior basement stairs.

6. If the fire vents out one or more basement windows and exposes flammable siding, insulation, and sheathing, this exterior line will prevent exterior fire envelopment. (See case history in ch. 3.)

Fig. 9–8. New research has shown that knocking down the basement fire via a nozzle in a window then following up with an interior attack may be the best tactic. Many large fire departments have updated their procedures.

Disadvantages to this option include:

1. The fire may be behind partition walls or protected by stored materials in the basement and the stream may not reach the seat of the fire.

2. There is a danger in not knowing what is on fire. If the fire is gas fed or involves energized electrical equipment, applying water may not be appropriate. Consider controlling the utilities prior to implementing this tactic.

3. If the fire is burning through the interior door from the basement to the first floor, the flow path is not controlled, endangering members searching on the first and second floors. If personnel is available, a line should be simultaneously stretched to the interior door.

4. Without members in the basement, the SAR effort will be delayed, reducing the chances of a successful rescue.

An important force multiplier here is this: interior search crews should be trained to provide the IC with critical information such as interior conditions. If first-in crews report fire extending through a partially burned-out basement door, the obvious place for the hoseline is inside.

Another option to get water on the fire from the safest location is to force the exterior clamshell doors, descend the steps, and apply water from there. Forcing these can be difficult and staff and time consuming. Like opening the front door (for a first-floor fire), this may also provide fresh air and a flow path for the now-developing fire, especially if your stream cannot reach the seat of the fire.

Extinguish the fire. If the SLICERS process is going smoothly, our basement fire has been knocked down by an exterior stream. Certainly it can and will rebound quickly so you must be prepared to aggressively go inside and conduct final extinguishment. If you have adequate staff, the second line (the hoseline at the top of the interior stairs) can now proceed down and get up close and personal with the fire with the intention of killing it. The clamshell doors can be forced to provide ventilation and windows can be taken to provide

Water through the Basement Window

Flow Path from Open Basement Window and Bilco Door, exits through Open Front Door
Basement 3 (644C) — 60 seconds of water
Flowing Water on the Fire Improves Conditions Everywhere in the Structure

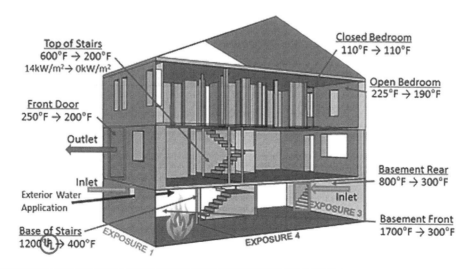

Fig. 9–9. Data from Kerber 2014 show the cooling effect of hose streams directed into basement windows.

further lifting of smoke and heat for members in the hot basement.

Rescue. The late, great fire attack expert Andy Fredericks was fond of the expression, "More lives are saved by a well-placed hoseline than any other method." This is supported by our fireground experiences, but is not substantiated by hard statistical facts. Often, the fire attack does not go smoothly or quickly for any number of good and not-so-good reasons. Good reasons (we prefer "reasonable reasons") include highly fortified doors delaying entry, water supply failures like frozen or vandalized hydrants, and illegal construction. Unreasonable reasons include inadequate staffing, poor training, inadequate target flow, and poorly maintained automatic nozzles.

According to ISFSI's SLICERS process (explained here by Division Chief Eddie Buchanan),

> the IC should consider the potential for rescues at all times. Firefighters should always be prepared to remove trapped or endangered occupants. Reinforce that often the best action the fire department can take is to suppress the fire. The IC and fireground officers must make a rapid and informed choice on the priority and sequence of suppression activities vs. occupant removal. As life safety is the highest tactical priority, rescue shall always take precedence. The IC must determine the best course of action to ensure the best outcome for occupants based on the conditions at the time.

> In other words, we still go get them! The concept of vent-enter-search has been updated to include "isolate," referring to the importance of door control and compartmentalization. Now deemed "VEIS," truck companies play an important role in quickly placing ladders and searching rooms of probable rescue. Additionally, once the thermal threat has been managed, normal interior search operations should occur. This is one of the more controversial positions in the "new method." But it makes sense when played out on the fireground. VEIS missions can be carried out for immediate rescues, and truck company crews can be ready to open the door and create a flow path once the thermal threat has been controlled. (Buchanan 2013)

Ventilation. Regarding the conspicuous lack of ventilation in IFSI's process, Buchanan explains the following:

> You've probably noticed by now that *Ventilation* is missing. Given the research, ventilation has been reclassified as a specialty action. It requires direct orders from the IC and generally occurs after the main body of fire has been subdued. No longer can anyone break anything at anytime for no particular reason. Every ventilation opening can influence the flow path, and that requires the approval of the IC. Yes, there will be times when windows must be taken, but take care to match that opening with a closed door whenever possible. (Buchanan 2013)

Windows cannot be unbroken, but if the ventilation techniques employed create negative consequences, the fix can be as easy as closing a door and altering the flow path.

Modern research

Throughout this book, we have presented current strategic and tactical concepts. We would be remiss not to include the following part of the executive summary for the great work that has been done by the UL on basement fires. This is presented to provide you an up-to-date study of modern basement fires, their hazards, and potential new strategies and tactics you may want to consider for use in your department. We strongly encourage you to dive deep into this work so you can gain an understanding of the complexities of residential basement fires. It is especially important to use the data from this report to improve firefighter safety and survival at basement fires.

It is also critical to again ask this question: what are *your* department's strategies and tactics for basement fires based on? Are they based on what has been passed down from previous generations? Houses and basements have changed over the years as we have described early in this chapter. Is it time for a review of your tactics and strategy? The most important question is this: how can you use the data provided by fire protection engineers gathered at realistic live burns to improve your fireground efficiency and simultaneously improve firefighter safety?

Excerpt from the Executive Summary of *Improving Fire Safety by Understanding the Fire Performance of Engineered Floor Systems and Providing the Fire Service with Information for Tactical Decision Making* (Kerber et al. 2012, 3–4)

Experiments were conducted to examine several types of floor joists including, dimensional lumber, engineered I-joists, metal plate connected wood trusses, steel C-joists, castellated I-joists and hybrid trusses. Experiments were performed at multiple scales to examine single floor system joists in a laboratory up through a full floor system in an acquired structure. Applied load, ventilation, fuel load, span and protection methods were altered to provide important information about the impact of these variables to structural stability and firefighter safety.

There are several tactical considerations that result from this research that firefighters can use immediately to improve their understanding, safety and decision making when sizing up a fire in a one or two family home.

Collapse times of all unprotected wood floor systems are within the operational time frame of the fire service regardless of response time.

Size-up should include the location of the basement fire as well as the amount of ventilation. Collapse always originated above the fire and the more ventilation available the faster the time to floor collapse.

When possible the floor should be inspected from below prior to operating on top of it. Signs of collapse vary by floor system; Dimensional lumber should be inspected for joist rupture or complete burn through, Engineered I-joists should be inspected for web burn through and separation from subflooring, Parallel Chord Trusses should be inspected for connection failure, and Metal C-joists should be inspected for deformation and subfloor connection failure.

Sounding the floor for stability is not reliable and therefore should be combined with other tactics to increase safety.

Thermal imagers may help indicate there is a basement fire but can't be used to assess structural integrity from above.

Attacking a basement fire from a stairway places firefighters in a high risk location due to being in the flow path of hot gases flowing up the stairs and working over the fire on a flooring system which has the potential to collapse due to fire exposure.

It has been thought that if a firefighter quickly descended the stairs cooler temperatures would be found at the bottom of the basement stairs. The experiments in this study showed that temperatures at the bottom of the basement stairs where often worse than the temperatures at the top of the stairs.

Coordinating ventilation is extremely important. Ventilating the basement created a flow path up the stairs and out through the front door of the structure, almost doubling the speed of the hot gases and increasing temperatures of the gases to levels that could cause injury or death to a fully protected firefighter.

Floor sag is a poor indicator of floor collapse, as it may be very difficult to determine the amount of deflection while moving through a structure.

Gas temperatures in the room above the fire can be a poor indicator of both the fire conditions below and the structural integrity of the flooring system.

Charged hoselines should be available when opening up void spaces to expose wood floor systems.

During all of these controlled experiments where the variables were systematically controlled there were no reliable and repeatable warning signs of collapse. In the real world, the fire service will never respond to two fires that are exactly the same. On the fire ground there are many variables to consider and most of the parameters being considered are often unknown which makes decision making that much more difficult. Information such as how long the fire has been burning, what type of floor system, was it built to code or altered at any point, is it protected with gypsum board, what is the loading on the floor and how long is the span are all unknown to the responding firefighters. There are also no collapse indicators that guarantee the floor system is safe to operate on top of. Sounding the floor, floor sag, gas temperatures on the floor above and thermal imager readings even when taken all together do not provide enough information to guarantee that the floor will not collapse below you. Flooring system components and floor covering materials are composed of materials that work to limit the flow of thermal energy through them. As a result flooring materials could be on fire on the bottom side (basement side) while only exhibiting modest temperature increases on the top side of the floor.

In addition, rapid changes in fire dynamics can result from flow paths created by ventilating the basement and first floor of a structure. These flow paths combined with the fast spreading fire that results from the ignition of an unprotected wood floor system can place firefighters on the floor above the fire in a vulnerable position with little time to react. It is acknowledged that there are times where firefighters may choose to operate on top of a basement fire to carry out their life safety mission however this decision must be made understanding the potential for catastrophic consequences. There are also alternative tactics to consider in order to control the fire without first committing crews above the fire such as suppression initiated from a basement window or doorway. Coordination to control the basement fire prior to opening the first floor and committing crews on the first floor is essential. This report summarizes the results from each of the experimental series and provides discussion and conclusions of the results. Each series of experiments was also documented and analyzed independently and these documents are attached as appendices of this report. There is also an online training program that was developed for the fire service based on all of the material included in this research project. It can be accessed for free at www.ul.com/fireservice (Click on "Basement Fires")

Basement construction

Basement fires expose all the structural members and contents of the floors above to the possibility of rapid fire spread. Each type of residential construction has vulnerabilities that must be recognized by fireground commanders. This information is critical to planning an effective fire suppression strategy.

Platform construction. Despite the fact that this type of construction has built-in fire resistance from the floors all the way to the outside walls, penetrations for utilities provide multiple opportunities for rapid vertical fire spread (figs. 9–10 and 9–11).

Fig. 9–10. In this platform-frame home, gas, waste, and dryer vent lines penetrate from the basement to the first floor.

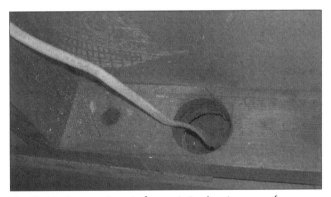

Fig. 9–11. Penetrations in fire-resistive barriers are often left unprotected in residential construction, allowing rapid vertical fire spread especially in void spaces.

Balloon frame. Chapter 2 provided details of balloon-frame construction and a case history of a classic balloon-frame fire. In the event of a basement fire in a balloon-frame home, cover all parts of the building quickly. What appears to be an attic fire may well have started in the basement. Expect basement fires in balloon-frame homes to extend rapidly to the attic via the exterior wall stud channels, which act like vertical chimneys. Depending on the conditions of the walls on both first and second floors, fire may break out there as well on its march to the attic.

Pulling baseboards on exterior walls of first and second floors will provide you an excellent opportunity to see if fire is extending into and up the walls. Injecting water into the wall cavity is a very good tactic. Always keep in mind that the fire may have already extended above where you are operating. Basement fires in balloon-frame homes require excessive amounts of personnel to open up on all floors and the attic simultaneously with an attack on the main body of fire in the basement. Where truck company members are opening up, a line must be in position to kill any fire. If you are looking for fire, you had better have a line in position to deal with it when you find it (fig. 9–12).

Fig. 9–12. Basement fires in balloon-frame homes require multiple truck and engine companies to search for fire extending upward into all floors and especially the attic.

I-joist. As we have seen, I-joists and other lightweight construction materials will fail rapidly and without warning when subjected to a basement fire (fig. 9–13). Consider the dangers to interior crews if the selected tactic is to push a hoseline down the interior stairs. Will they fall through the fire-weakened floor shortly after entering the front door?

Fig. 9–13. Lightweight construction like the home above does not resist a fire load for very long. Consider alternative fire suppression tactics for a basement fire in a home where you have identified this vulnerable type of structural member.

Western duplex. A typical western duplex actually contains two separate but attached housing units (fig. 9–14). In this case, a basement fire threatens two homes under the same roof. This of course doubles your forcible entry problems, creating a need for at least two hoselines above the fire and a truck company to open each side that is threatened by the fire below.

Fig. 9–14. Duplex homes double the exposure problem of a classic single-family home. Courtesy of John Mittendorf.

It is impossible to cover all the types of homes that are built and occupied across our nation. The specific types shown above provide a solid generalized understanding of the complexities of basement fires at various homes.

Garage Fires

Generally speaking, the storage in a household garage is similar to that in a basement, so we have included the discussion in this chapter. We acknowledge the hazards are not exactly the same; however, this will allow us to focus on the differences, not the similarities. What is similar is that the garage may be under the home and that a fire there will threaten the entire structure.

The main tactical difference between garage and basement fires is that in a well-involved garage fire, the wide opening of the door provides your engine companies with direct access to the fire through the opened or burned-through door. The reach of your stream will almost always get your water on the entire fire while you operate outside the structure in relative safety. Obviously the basement fire is a close quarters combat operation if an interior attack is chosen.

A garage may be filled with a variety of hazardous materials. We are not speaking of the toxic industrial chemicals, though it is possible. In areas where real estate prices are high it is not uncommon to find very dangerous businesses run out of garages in homes. In Rockland County (NY) several years ago, a local department responded to a garage fire in a high ranch home. The home and business owner ran a

seamless vinyl flooring company out of his garage. Highly toxic and flammable industrial-grade solvents were burning furiously upon arrival.

Residential garages contain things like propane cylinders, pesticides, fertilizer, pool chemicals, paints, oils, gasoline, and other assorted hazardous materials (fig. 9–15). They may actually be present in greater amounts due to the easier access the garage affords, as compared to carrying them to the basement. (This same line of thinking makes lawn sheds dangerous.)

Fig. 9–15. This home was destroyed and an occupant killed after a gasoline spill ignited in the garage. Courtesy of Tom Bierds.

Garage spaces were intended for vehicle storage. In fact, in some communities, vehicles *must* be parked in the garage because homeowners association (HOA) rules or local ordinances do not permit parking in the driveway or the street. In these cases, a car in the garage may not be an indication that the occupants are home, as in a multicar family.

Vehicles parked in the garage may have caused the fire, and also present a large fuel load, not only from the synthetic components of the car itself but also from the fuel to run it. More and more automotive components are plastic. A plastic fuel tank exposed to high heat from a garage fire may leak. Leaking fuel that creates a flammable-liquid fire or contaminates the runoff from the fire attack are all complications that need to be considered.

Door safety

While other variations of sliding or swinging doors exist, overhead doors are the most common, and can be a major safety concern. Entry through the overhead door could be challenging depending on how it is lifted—manually or electrically. The primary safety concern is when it is in the raised or overhead position, because the weight of the door hanging overhead is a tremendous risk. It can come off one or both tracks and collapse on firefighters, or roll down, trapping them on the interior of the garage and disrupting the water supply to the attack line, all possibly creating a Mayday situation (fig. 9–16).

Fig. 9–16. Garage doors and their heavy, tension-loaded springs can be a threat to safety. Courtesy of Tom Bierds.

Our recommendations are to not to be under the door when it is overhead and to secure a second means of egress before entering the garage through the overhead door. Additionally, if entry can be made through another access door, leaving the overhead down until the fire is controlled and it is needed for ventilation, that would be the best option.

Securing the door after you open it and push it up is critical to the safety of members (fig. 9–17). You must prevent this door from coming down after it has been opened. It will let in significant air, causing rapid fire development, and will trap firefighters inside. If it comes down, there is the possibility of rapid fire development possibly injuring and trapping members. If not secured by a firefighter's action, the door may suddenly drop for the following reasons: the springs, connections, and cables may be damaged or destroyed by fire, an electric door closer/opener may operate if wires are cross-connected by fire-damaged insulation, or it may just slide down the tracks because of gravity.

Fig. 9–17. Place a pike pole or clamp a pair of locking pliers on the track of the door to prevent it from coming down and trapping members operating inside. Courtesy of Tom Bierds.

Consider the reach of your hose stream. You don't need to be in the garage for your stream to penetrate all areas of the involved garage. Positioning your nozzle crew off to one side of the garage provides numerous safety advantages. First, should a propane cylinder or other closed container overheat and launch out of garage, you have maximized firefighter safety by simply staying out of the line of fire of these fast-moving cylinders and shrapnel from exploding cylinders and valves. Second, should a flashover or rapid fire development or a violent reaction to your hose stream occur in the garage, again, you are likely out of the danger zone.

Location

Where the garage is located with respect to the life hazard is a key point in the discussion. Attached garages are a more significant concern than standalone detached structures (fig. 9–18).

Fig. 9–18. This garage fire extended upward, threatening the bedrooms and occupants above. Courtesy of Tom Bierds.

An attached garage located below the living space creates a more significant challenge for fire spread, as fire travels vertically and the most dangerous area is above the fire (fig. 9–19). The adjoining area is the second most significant area. When the fire spread is contained and not allowed to extend vertically, it will mushroom and spread horizontally to adjoining areas.

Fig. 9–19. A fire in this garage directly exposes the victims on the floor above. Darkening down the fire quickly provides the greatest chance for either self-rescue or rescue by search crews. Courtesy of John Mittendorf.

The fire break between the garage and the residential portions of the house are a concern. In spite of the increased fire hazard in the garage, the fire protection envelope in a garage may not be any different from the main house, depending on local building codes. Generally ½–⅜ in. sheetrock is used in the main house with variations for particular rooms, water resistant (green or blue) for bathrooms, mold resistant (purple) for basements and damp locations, and abuse-resistant sheetrock, which has more durable paper and a reinforced core for high-traffic areas and playrooms. Fire retardant or resistant board can resist flame for 45–60 minutes (doityourself.com 2016). Another type is type X, which is fire rated and will provide 30–45 minutes of protection to the structure. Many local building codes only require type X for the ceilings and walls on the garage side of ceiling-floor assemblies and common partition walls.

The door between the attached garage and the lower level of the home is most likely the only fire-rated door in the home. Keep this door closed and utilize its built-in fire-resistive strength. Opening the door from the inside thinking you are going to "protect" the opening with a hose stream is not a good tactic. Put water on the fire quickly from the outside, use the reach of your stream, kill the fire quickly, and all your other problems will be minimized.

Sure, you want to get inside with a hoseline to be sure this door is holding and the fire is not pushing through and exposing the basement or house side of the door. Opening this door creates a flow path for fire to extend from the garage through the house. This is especially possible if you have opened the door to the fire area and given it access to what it needs: air (the smoke is fuel rich), a place to travel from high pressure (fire area) to low pressure (yet unburned), and a flow path (burned area through garage and following your hoseline back through the open front door). The only thing missing to create a failure chain with drastic consequences is a problem with the hoseline (such as a burst length), injured firefighter, failure of the pump or pump operator, or loss of water source.

What is above the garage—whether this bonus room is a master bedroom suite, home office, or "man cave"—changes the search considerations. The lack of this information can complicate the search and make it difficult to identify the areas of highest life hazard. These rooms may be on the opposite end of the house from the other sleeping areas. Where to assign search teams, particularly in a low-staffing situation, must be carefully considered.

Detached garages

These structures can be completely detached or partially attached, connected by a breezeway or car port. The breezeway can create a void space for fire extension and must be examined. Detached garages may not have a code requirement for sheetrocking the interior. This could expose the structural members to direct flame contact, leading to early collapse. Detached garages often have attic spaces used for storage. What is hidden in the rafters creates an overhead hazard in addition to the roll-up door. The distance the garage is detached from the main house can present or prevent an exposure problem. Radiant heat from a fully involved garage may require additional handlines to protect the main house.

There are many variables in developing an attack plan for garage fires. Carefully consider your options before committing firefighters to what may only turn out to be a noble but futile effort. We will take manageable risks for significant gain (lives and property saved) but firefighter safety is our primary concern.

The following scenarios will help you reinforce some of the critical considerations that will help you develop or execute your strategy and tactics at your next basement or garage fire.

Basement Fire Scenario

Fig. 9–20. Scenario

Dispatch

Saturday, 1423 hours, 7 December, reported basement fire at 41 Hickham Drive.

Size-up

The homeowner meets you upon arrival and confirms that everyone is out of the house. Smoke issuing from a window behind the shrubs and light smoke is coming from the front door. Heavy smoke is seen from the D side and rear.

Your 360 shows heavy fire from both windows on the D side (fig. 9–21). The vinyl siding is becoming involved and threatening to enter first-floor windows. The C side shows good access through a personnel door but heavy fire out the rear window (fig. 9–22). It appears there is heavy fire in the basement on the D side of the house.

Fig. 9–21. D side of the structure

Life hazard

There is none at this time. At least until the fire department arrives on the scene.

Fig. 9–22. C side of the structure

Hoseline positioning

You are faced with two immediate problems in controlling this fire. First, if you don't control the exterior fire, it will quickly weaken the vinyl window frames, cause the glass to fail or completely fall out, and enter the first floor and the attic, again through a vinyl gable vent. Second, this appears to be a well-involved basement fire with heavy fire threatening to burn through the first-floor door to the upstairs and spread vertically via electric, pipe, and other penetrations in the floor and walls. There are three options for hoselines.

1. Stretch through the front door, find the basement stairs, and push the line down into the basement. The advantages to this include cutting off the fire from a likely vertical extension path, and the fact that the front door is the quickest access into the house and thus the shortest stretch.

    Disadvantages to this option include the following:

    – Firefighter safety: Heavy fire in the basement may have weakened the floor and a "through the door, through the floor" scenario may develop, with Maydays. As we have all experienced, when a Mayday happens, all our effort is concentrated on saving firefighters' lives, so the goal of saving property is usually lost. Not a good outcome for anyone.

    – Fire attack: Opening the interior door to the basement will create a flow path that the fire desperately needs and wants. This fire wants to come up the stairs to the fresh air and fresh fuel. Pushing this hoseline down the stairs is a dangerous job, essentially crawling down the chimney of a well-involved fire. If the engine crew makes it down the stairs, they have one way out: back the way they came, through the flow path that could easily be filled with fire. Additionally, at the bottom of the stairs, if the basement is subdivided by walls or storage, this line may not have direct access to the seat of the fire. Sure, the team can move in and get to it, but that puts them in ever-increasing danger. An aggressive interior attack is a wonderful thing, but there was no life hazard here until we arrived, and now the entire engine crew is in a fair degree of danger trying to make the coveted "good stop" that we are all so (justifiably) proud of.

Consider for one minute, that one of the engine crew is killed in this fire attack option. Hindsight, which we have the unrealistic but hugely beneficial option of utilizing here (during training), would summarize the situation something like this: One of our number was killed making an aggressive fire attack on a residential fire. During the FAST team operation, we were able to recover them in about 15 minutes, which was a really good rescue operation. During this time, fire did extend to the upper portions of the home. However, we did not consider the financial future of the home when making this attack: the insurance company determined it was cheaper to declare a total loss and rebuild new. In the end, our department is deeply divided over the tactics chosen that day and the newspaper's editorial asked if and why we traded a firefighters life for such an insignificant gain.

2. Stretch to the rear of the home and conduct an interior attack through the personnel door. This is a good option. It combines an interior attack with good access to the fire. Again, we do not know if the interior of the basement is divided or how far the engine crew will have to push in to extinguish the fire. Disadvantages to this option include the fact that unless we extinguish the exterior fire first, this fire will continue to be a threat and extension source because of the flammable fuels on the outside. Assuming we get water on the fire from the inside, some steam will exit the window and hopefully (never a good plan!) control some exterior fire, unless the wind is blowing or the fire has extended beyond the steam's reach. Placing our engine crew in the basement with only one way out puts them in danger of being caught by a collapse and of course the hazards of gas, oil, electric utilities, and dangerous stored materials that may be found in the basement. Additionally, there is very little ventilation at this fire and opening this door will create a flow path allowing an inrush of oxygen to increase the fire intensity.

3. Stretch the line around the D side to control the exterior fire and apply water through the two windows on this side. Flow water long enough to darken down the fire at each window. This is a great option because it both controls the threat of extension that will ultimately cause the house to die a slow death and applies water directly to the seat of the fire. In this case utilizing both windows for water application improves your chances of getting water on the fire even if the basement is divided.

After a good dash of water in both windows, the line can make its way around to the rear and do the same at the C-side window if the previous dash of water has not killed the fire there. Obviously, this is a judgment you can make at the time. This same line can then make its way to the rear door and push in and extinguish any remaining fire.

This option does not needlessly or excessively risk firefighters' lives. In the time it took to apply water from the outside, you had time to assess the collapse hazard and to allow any bad things to happen in the basement without firefighters in there. Broken gas lines and exploding fuel oil tanks and portable propane tanks are all likely hazards for the interior attack under heavy fire conditions. Sure, these can still occur after the fire is knocked down but they are not as much of a threat to firefighters as in combination with heavy structural fire and very limited visibility.

Obviously, if you have enough personnel for multiple lines, another option is to operate these lines simultaneously in the windows (solid stream or straight from a combination nozzle), then shut these down and push a line in through the rear door.

Ventilation

It is difficult to vent a basement fire. The good news is that the small windows are generally located near the ceiling of the basement so they are in a good position to vent out the hottest and worst products of combustion. We must remember this fire is ventilation limited so when attack crews open doors to put water on the fire, they also add huge amounts of air that will increase the intensity of the fire and create a flow path.

Another consideration

You may have noticed the garden hose on the D side that was stretched to the C side. The homeowner reported everyone was out of the house but was not aware that a police officer tried to extinguish the fire with the garden hose through the rear door. Are they still inside? The police will often be on the scene before us and, without a drop of disrespect, police officers do not understand fire development, flashover, or other fire dynamics. Always consider what other first responders or well-intentioned neighbors are doing or have done to "help" us.

Detached Garage Scenario

Fig. 9–23. Detached garage fire. Courtesy of Tom Bierds.

Dispatch

322 New Zealand Avenue, Sunday at 1323 hours, 17 February. Neighbor reports a garage fire across the street.

Size-up

From this view it is difficult to tell if the garage fire has extended into the house because the fence between the house and garage and smoke and steam obscure your view. A report from the rear will be extremely helpful.

Life hazard

If your information gathering leads you to believe a search of the house is required, you may choose to place the first hoseline in support of your search team to protect the occupants' and firefighters' means of egress. Once you are sure that all occupants are out of the house, the hoseline can continue to knock down any fire that may be extending from the garage. This line could make its way through the house and operate on the garage fire through the open side door of the house, but this would expose the interior of the home to a large body of fire.

Hoseline placement

An interior size-up of the house and whether or not you need to go interior to conduct a search is critical to determine where to position your first hoseline. If it is not needed inside, the burned-through garage door provides an excellent opportunity to quickly and safely knock down the main body of fire. Depending on how much fire has extended to the home this may be the best initial option. Extinguishing the main body of fire generally makes everything else better, safer, and faster. The need for a search operation and firefighter safety considerations will help influence your decisions.

Ventilation

Certainly the garage is well ventilated by the fire and your hose stream will demolish the remaining wood frame on the garage door. The black smoke looming behind the steam may indicate fire has gotten into the attic of the home. Vertical ventilation may be required. At this point it appears the home's windows and doors facing the detached garage fire are holding, so you do not want to open these avenues of fire extension to the home.

Garage Fire Extending to Home Scenario

Fig. 9–24. Garage fire is extending. Courtesy of Tom Bierds.

Dispatch

232 Ned Hobbs Drive, July 15, 0023 hours. Police department reported the fire while on patrol.

Size-up

Through the smoke and flames you can see a picture window on the right side of the house. This usually means the living room is behind the window and also means the bedrooms are on the left side and are a priority for search (based on time of day). Fire has complete control of the two-car garage on the A side and D side and a report from the rear says the rear outside deck is burning.

Life hazard

Based on the time of day, searching the bedrooms and the rest of the house is a priority. Due to the heavy fire, you may consider VEIS via as many outside windows as you have personnel to execute. This will be a safer alternative to entering the front door and passing by heavy fire in the living room. The front door looks viable and the stairs should be close behind the front door, but the heavy garage fire is extending into the living room through the picture window. Also this heavy fire in the garage is extending via the soffit eaves and getting into the attic. All this extending fire presents serious hazards to interior crews. If a search team enters the front door and tries to make the bedrooms on the left side of the house (B side) they will need a hose team to protect their means of egress. Alternatively they could make the bedrooms via the interior, shut bedroom doors behind them to buy some time, and evacuate occupants via ladders to the windows. Sounds easy, but as we all know, this can be very hard to do, especially if occupants are panicked, elderly, infirm, or obese.

Hoseline placement

Ideally, if you have enough personnel, one line goes inside to protect the search teams and the means of egress and one or more lines operate on the fire. The line operating on the fire from the outside may be a 2½ in. or a portable master stream from the driveway or front lawn into the garage. This line can knock down the main body of fire in the garage with the goal of slowing the fire extension into the main parts of the house, at least until we can get the search done and all firefighters out of the building. After the main body of fire is knocked down in the garage a decision can be made to remain on the offensive inside, or remain defensive and not put members back inside due to collapse or other concerns. Note that in the photo below, before the hoseline went inside it was used to darken down the garage fire, buying some time for the search operation.

As a reminder, if staffing level permits, deploy a 2½ in. line when you have this much fire upon arrival. We don't want this to be a fair fight. We want to knock this fire down as quickly as possible (fig. 9–25).

Fig. 9–25. Stretching the attack line. Courtesy of Tom Bierds.

Ventilation

From the smoke seeping from the eaves and the amount of fire impinging on the soffit and the overhang it is a good assumption that fire has entered the attic and may be under the first floor in the joist bay. Opening the roof may be necessary to get the attic fire under control.

Rapid Fire Development Scenario

Fig. 9–26. Scenario. Courtesy of Tom Bierds.

Size-up

You should always be extra vigilant at a garage fire. Look in the photo above. There are two 1 gal jugs of some product just outside the door on the driveway. On the wood block is another gallon of something. There is also some work-related debris near the shrubs. All of this should stimulate you to ask questions: Is the homeowner running a business out of the garage? Does the business (or hobby) use flammable or toxic chemicals? Is there propane, acetylene, or other compressed gases? If the homeowner is around, ask them what they think started the fire. This may tip you off to large quantities of chemicals or dangerous processes in the garage.

Hoseline placement

Access to this fire is limited until you get the garage door up. You could force the front door of the house and snake a line down through the interior of the house, open the fire-rated door between the garage and basement, and attack the fire from the interior. This takes time and personnel.

Be aware that when you open the garage door and let in huge amounts of air, the fire will likely increase in intensity rapidly, as in the photo below. After the door was opened, the fire increased and again threatened to race up the exterior. Look at the eaves. The plastic covers have melted off, providing the fire direct access to the attic (fig. 9–27). When a garage fire gets into the attic, it is a difficult fight and will require more staff, which you may not have.

Fig. 9–27. Additional ventilation is not needed.

Modern homes have flammable exteriors (siding, insulation, and sheathing) and only light flammable coverings (thin plywood or plastic) keeping exterior fire out of overhangs, soffits, and eaves. We must recognize how important it is to extinguish the exterior fire in the overall battle to extinguish the entire fire.

Legacy homes sided with aluminum, wood or even asbestos shingles do not propagate fire as fast as modern homes.

Legacy homes, because of their design, do not have overhangs or soffits like modern homes.

Ventilation

The good news is that the large garage door(s) provide significant ventilation when opened, allowing steam and smoke to escape. If the fire is contained to the garage there is little need for additional ventilation.

Light Smoke Scenario

Fig. 9–28. An old garage. Courtesy of Tom Bierds.

Dispatch

Report of light smoke coming from the garage on 332 Iona Way, 23 July.

Size-up

It is a hot summer day. You approach the garage and the smoke does not look quite right, not like it is coming from a fire. Old garages should make you suspicious; you never know what can be left over in there. Common things include pesticides, pool chemicals, industrial chemicals that were not all used, and even war souvenirs like old hand grenades.

Life safety

If you can look in from a distance, do that. If you have to get up close to the window or glass in the door, make it a quick look and put your SCBA with your facepiece on! If the remainder of your size-up shows no life hazard and the smoke still does not look like it is coming from a fire, call your hazmat team. It may not be smoke; it may be toxic vapors of some kind. Obviously, monitor the situation but consider not acting. This garage is not in a populated area and the wind is blowing toward a large unoccupied area. You may want to ask the caller if they smelled or saw anything. What did it smell like? Did you hear any explosions or popping sounds? Did you see anyone suspicious enter or leave the garage? Consider nefarious activity, like a drug lab or terrorism. Have a detailed discussion with the police: Are there any known drug-making or terror groups in the area? All this sounds far-fetched, but we still don't know what is causing the vapors from the garage. The smoke has now stopped.

Life safety of firefighters is critical and this is a really good time not to risk any firefighters' lives for an old garage containing unknown materials. Does your bomb squad have a robot they could send in with either a video camera or air monitors that can send back readings? We like to try to resolve or fix every situation, but we need to know when we cannot and should not attempt to resolve certain situations.

Ventilation

Obviously this poses undue risk to firefighters, as we still don't know the hazards of the vapors pushing out of the garage. Venting the structure also lets out the vapors from inside and if they are toxic, this is not a good idea.

References

Avillo, Anthony. 2015. *Fireground Strategies*. 3rd ed. Tulsa: PennWell Corporation.

Buchanan, Eddie. 2013. "Rethinking RECEO VS: Breaking Up with an Old Friend." FireEngineering.com (blog), 5 November. http://www.fireengineering.com/ articles/2013/11/rethinking-receo-vs-breaking -up-with-an-old-friend.html.

doityourself.com. 2016. "7 Types of Sheetrock Explained." http://www.doityourself.com/ stry/7-types-of-sheetrock-explained.

Kerber, Steve. 2010. *Impact of Ventilation on Fire Behavior in Legacy and Contemporary Residential Construction.* Northbrook, IL: Underwriters Laboratories.

———. 2014. "Top 20 Tactical Considerations from Firefighter Research." Powerpoint presentation. Underwriters Laboratories Firefighter Safety Research Institute. https://www.dropbox.com/s/ g1fcomhxdkccbdz/FDIC%202014%20Classroom%20 -%20Top%2020.pptx.

Kerber, Stephen, Daniel Madrzykowski, James Dalton, and Bob Backstrom. 2012. *Improving Fire Safety by Understanding the Fire Performance of Engineered Floor Systems and Providing the Fire Service with Information for Tactical Decision Making.* Northbrook, IL: Underwriters Laboratories.

National Fire Data Center. 2015. "One- and Two-Family Residential Building Basement Fires (2010–2012)." *Topical Fire Report* 15 (10). https://www.usfa.fema.gov/ downloads/pdf/statistics/v15i10.pdf.

National Institute for Occupational Safety and Health. 2007. *Volunteer Fire Fighter Dies after Falling through Floor Supported by Engineered Wooden-I Beams at Residential Structure Fire—Tennessee.* NIOSH Report F2007-07. Atlanta: Department of Health and Human Services. https://www.cdc.gov/niosh/fire/reports/ face200707.html.

———. 2009. *Volunteer Fire Lieutenant Killed while Fighting a Basement Fire—Pennsylvania.* Fatality Assessment and Control Evaluation Investigation Report No. F2008-08. Atlanta: Department of Health and Human Services.

Underwriters Laboratories. 2014. "New Dynamics of Basement Fires." *New Science Fire Safety* 2:18–23. http:// newscience.ul.com/wp-content/uploads/2014/04/ NS_FS_Journal_Issue_21.pdf.

First- and Second-Floor Fires

10

Summary

This chapter's purpose is to examine and discuss strategic and tactical considerations, options, and outcomes (positive and negative) for operations at fires on the first and/or second floor of homes. This chapter utilizes the information from previous chapters (house layout, construction, hazards, search, command, ventilation, and fire attack options) in combination with the practical application and development of comprehensive and safe strategic plans and their tactical fireground executions. For company and chief officers, it provides an opportunity to think through development and consequences of strategic objectives and tactical alternatives. For firefighters this chapter provides an understanding of the strategies developed by chief officers so you know the rationale behind the tactical operations directed by company officers. Techniques for tactical operations such as search, ventilation, and hoseline advancement were covered in previous chapters and will not be included in this chapter except for some very special or incident-specific skills, important reminders, and critical tactical operations.

Introduction

Firefighting operations in occupied areas of homes (first and second floors) include size-up, search and rescue (SAR), ventilation, and fire suppression. At modern house fires, because of the rapid release of heat from modern synthetic contents and shorter time to flashover, tactical fire operations (search, ventilation, fire attack) must be very well coordinated, timed, and executed precisely to ensure the best and safest outcome for both occupants and firefighters.

Conducting a size-up and deploying firefighters balances the prerogative of saving civilian lives and property against an analysis of manageable risk for our members. For all levels of firefighters, from rookies to seasoned officers, this is where it all comes together, often with life-and-death consequences. It is the big game with very high stakes: the lives of civilians and firefighters. We must make decisions based on the best available information we can gain during our initial and ongoing size-up. It is, however, critical to remember that this is incomplete information and may be incorrect, as the truth may be obscured by darkness, smoke, or other conditions.

Success is not guaranteed, hence the need for evidence-based risk assessment. Firefighters have a limited amount of time think about what we did or did not do and make adjustments based on what our opponent is doing. Tactical adjustments are made on the fly, during the operation if necessary. Therefore it is incumbent on fire officers, company officers, and firefighters to continually learn, update, and perfect their skills. As we have seen in previous chapters, our house-fire battlefield has changed in numerous ways. It is time to review and adjust our strategies and tactics. This chapter will provide an opportunity for you to do just that. Remember, success on the fireground is the art of applying your training tempered by your experience and your understanding of the science of modern house fires.

Strategy, tactics and other critical considerations, plans, and techniques provided in this chapter must be adjusted and applied for the specific fire conditions, personnel levels, types of apparatus, types of homes, and response times common to your department. The specific conditions in which you work may dictate changes and modifications to the strategies and tactics presented in this chapter. If you have some very severe situations (your engines are operated by one driver or firefighter, you have very long response times, etc.), some of these suggestions may need further consideration.

The goal here is to give you an opportunity, through practical, real-life scenarios to size up, think, plan, and evaluate then work through strategies and tactics for both first- and second-floor fires. When discussing scenarios in this chapter, we assume you have or will establish a command structure, have a reliable and continuous water supply via tanker operations, drafting, or municipal hydrants, and apply your current staff and equipment as your on-scene resources.

Since we have covered SAR operations in detail in previous chapters, the emphasis in this chapter is on fire control, suppression, and extinguishment. Search will be mentioned and sometimes reviewed in detail in the scenarios but the emphasis is on fire control.

First-Floor Fires

First-floor fires are dangerous because they are in the living space of the home, often directly exposing people to hazards of fire, smoke, and heat. Most homes of multiple floors have an open stairwell that will provide a great chimney (flow path) for products of combustion and a perfect fire spread route. First-floor fires move upward quickly. This chimney effect combined with an open, broken, or fire-ventilated window will provide a lethal flow path for the fire. In any home, close proximity of the fire to the occupants exponentially decreases their chances of escape and survival.

We often forget a key hazard in first-floor fires: What is holding up the first floor? The fire is easily accessible, usually via the front door, so we don't perceive a hazard from a weakened floor structure. If the fire started in the truss space or has extended down into it, we can expect a weakened floor. Sounding the floor with a tool makes us feel better but does not actually ensure that the floor will hold our weight. We are not sure who coined the phrase but "through the door, through the floor" is always a consideration. Obviously this is more imperative at basement fires, but it is certainly a consideration here.

Front door, back door

We tend to always depend on the front door for access. It is a good choice because it is usually the closest to the rig parked out front. It also provides direct access to the fire area, so we can find and assist victims who are (or were, before they became incapacitated) on the normal egress path in the home.

Popular fire attack strategies long supported always attacking from the unburned side. The rationale was to get water between the fire and the undamaged area and kill the fire from that direction. Concurrent with this thought was we did not want to push fire into unburned areas.

Using this strategy, if the fire was in front of the home, the attack hoseline had to be stretched around to the back door of the home, the rear door forced, and the push inside made from the rear. Theoretically, this sounds good, but recent studies and experience have revealed several disadvantages, particularly at house fires.

First, stretching the line to the rear requires a much longer stretch. Often this is around cars in the driveway, through a fence gate around the rear, and finally into the home. This obviously takes time and personnel, which can be an issue, especially if you run with limited staffing on your engines.

Depending upon the location of the doors and the layout of the home, going through the rear may put you in almost the same place you'd be if you'd entered the home from the front door. You may not gain any significant advantage or better angle for the fire attack for all the time and work invested.

One of the most important reasons not to look first to the rear to stretch to and start your attack is the critical importance of protecting the means of egress. Unless you can very swiftly and decisively kill the fire, the mission of the first hoseline is to protect the means of egress. This is especially important at a two-story or greater home where the means of egress from the bedrooms during a nighttime fire is down the interior stairs. Considering the rapid fire development we are currently seeing in house fires, protecting the stairs is equally important for the safety of our search crews.

Let's examine some strategy and tactics with associated advantages and disadvantages at a realistic fire situation.

Scenario

0232 hours, Wednesday, 19 December, 23 Marine Drive, good neighborhood, house appears occupied, neighbors reported smoke coming from the house. Upon arrival, the front window fails and fire and smoke belch out (figs. 10–1 and 10–2). What are your strategic priorities at this fire?

Fig. 10–1. Scenario

Fig. 10–2. C side

Using the information presented in previous chapters tempered with your fireground experience and based on the information and pictures above, develop the following:

- **Size-up.** What is your size-up of this fire? Provide an on-scene radio report.

- **Rescue.** Is there a life hazard to occupants and what options are available to you to conduct an SAR operation?

- **Firefighter safety.** Are there obvious or extreme hazards present based on your size-up?

- **Fire suppression.** Where is the fire? Where does it want to go? What are the most important exposures? What are the likely flow paths? What is the best place and objective for the first line? Second line?

- **Ventilation.** What horizontal or vertical ventilation did you choose and why? What were your concerns for both of these?

- **Salvage and overhaul.** How important is it at this fire?

Stop, get out a pen and pad, and make some notes to answer the questions, then compare your thoughts to what we have come up with for this fire.

Size-up. The fire appears to be in the living room, typically indicated by the picture or bay window. It may also be in the kitchen behind the living room. We recognize that this is a straight ranch home similar to other floor plans in the area. The living room is behind and to the left of the front door, the kitchen is behind that in the C-D corner, and the bedrooms are on the B side. The rear of the house shows an addition that could be another bedroom or a recreation, TV, or playroom. With no smoke showing from the bedroom windows, it is likely that the door is shut. Probability of salvageable human life is high.

Your radio report should sound something like this: "I'm on the scene at 23 Marine Drive and have a 20 ft × 30 ft one-story straight ranch with fire venting from the A side. Life hazard is likely. Search and fire attack underway by first-due units."

The personnel door on the basement level is a clue that this house may be a mother-daughter or have a legal or illegal apartment in the basement. Obviously if you have enough personnel this needs to be searched, but the priority early on in this operation is searching for those occupants closest to the fire. In this case, these occupants are probably in the bedrooms. Search of the downstairs apartment is not the highest search priority. Clearly there is no guarantee of anything in the fire service, but statistics show that at this time of night, the odds are in favor of occupants being in the bedrooms, so searching the bedrooms is our best shot for a successful rescue.

Rescue. Life hazard appears high. SAR is a priority.

Option #1. Perform a vent, enter, isolate, and search (VEIS) operation through the windows on the B side, since it looks like they provide access to one or more bedrooms. The advantages of this method include the following:

- VEIS provides direct access to where occupants likely are at this time.

- There is a manageable risk to firefighters using VEIS and a high likelihood of success.

- A significant body of fire is between the occupants and the front door. Likely the hallway follows the ridgepole of the house and has occupants on one side, fire in the middle, and the means of egress on the opposite side. VEIS may be the best option for getting occupants out safely.

There are also disadvantages to this method:

- Firefighters will have to bring occupants out the window and down the ladder, a difficult task if they are unconscious or injured.

- The window air conditioner will slow members attempting to enter the window.

Option #2. Send the search team in the front door. This has the following advantages:

- Forcing the front door clears the way for the hoseline.

- The front door is often the fastest way into a home and provides an escape route for ambulatory trapped persons.

This method does have some disadvantages:

- The fire is between the egress point and the victims. This search team may have to wait for the fire to be suppressed before they can make it through the living room.

- Removal of the victims via this route will be challenging due to high heat, steam, and smoke conditions.

Firefighter safety. It appears the fire is the main threat. The fire just failed a major piece of the front window, allowing air into the fire. Opening the front door will create a flow path leading to rapid fire development. At this point there is no additional information on type of construction. If our department's building-information system had alerted us to lightweight construction or other hazards, these would be firefighter safety issues to consider.

Fire suppression. Fire is on the first-floor living room. It will go anywhere a flow path is created. If a firefighter forces and chocks open the rear door, the fire will move toward it and toward the open living room window. The open front door will also create a path for the fire to move toward.

The first hoseline must support the search operation. How you do that is up for discussion but the mission is not. Lives (civilians and firefighters searching) are in danger at this fire and everything supports life safety until that is resolved. Clearly, this does not mean we can't put the fire out; in fact, in this case that is a good option. The fire is easily accessible via the front door and a direct interior attack on the fire is an excellent option.

As mentioned previously, if the first line is taken to the rear with the goal of attacking from the unburned side, when you enter through the door into the rear of the kitchen you will be in almost the same place you would be had you entered through the front door. This move has gained little and cost a lot of time and personnel.

Ventilation. The correct choice for horizontal ventilation at this fire is to open the front door and, as soon as the engine company has water on the fire, finish taking the picture window on the A side, a window on the D side, and possibly the rear door. We need to be sure the fire is extinguished in the kitchen (C-D corner) before that door is opened. Until the fire gets into the attic—and there is no indication that it has yet—there is no need for vertical ventilation.

Salvage and overhaul. Until the life hazard is removed from this fire, all your efforts are directed toward saving salvageable human life. Then we can start thinking about property. The combination of the priority for search and rescue with the immediate application of water to either protect the means of egress or to quickly kill the fire are the guiding principles for all house fires with a life hazard. It really is not a difficult concept to understand but as we all know it is often complicated in the field by unknown or surprising on-scene circumstances.

This relatively simple example holds the basic strategy for most fires in occupied areas of the home: search quickly and aggressively with equally aggressive fire control or extinguishment.

Case history

Let's look at a case history for a first-floor fire in a two-story home (figs. 10–3, 10–4, and 10–5). This house is typical of many areas across our country. It is an older, classic 2½-story balloon-frame house built around 1900. These homes are prevalent east of the Mississippi River and as farmhouses in the Midwest. These homes are also common in older towns and sections of cities of the American west.

It is 2223 hours, 15 June, and you are dispatched to an electrical fire in the living room. The caller meets you outside the home and is 100% certain that he was the only occupant and that no one is in the home.

First impression of this fire is that it may not be so bad: small fire, easily accessible, no life hazard. However, do not be fooled. The amount of smoke and the several areas it is pushing from should tell you that this fire has entered the walls of the balloon-frame home. Once this happens, the firefighting operation is never simple and never easy. The fire will have easy access to multiple avenues of vertical fire spread. It will take a lot of staff to quickly open up the interior and exterior walls and get ahead of this fire.

Fig. 10–3. Note the smoke from several areas of the home. Courtesy of Tom Bierds.

Fig. 10–4. A closer look at the D side shows that fire has a strong foothold in the wall. Courtesy of Tom Bierds.

Fig. 10–5. Obvious fire is darkened down quickly from the outside. Courtesy of Tom Bierds.

Ventilation. The strategy for ventilation at this fire depends on two things. First, is there water on the fire? If there is and ventilation is required to improve conditions inside, it is time to vent. The priority at this fire is to find the hidden fire and cut off its vertical spread. If interior crews are working in low visibility, it is time for some horizontal ventilation. Depending on the how bad the smoke and heat are inside, ventilation options might include opening the windows normally as opposed to taking them completely. It is important to remem-

ber that if fire department actions do not cut off the fire spread in the walls, the building will continue to be damaged. Venting windows is worth the gain if we can save the remainder of the building. The disadvantages of not venting include firefighters working in adverse conditions that slow their progress, allowing the fire to move through void spaces and resulting in the building dying a slow death. Eventually the ladder pipes will come out and we'll have created another parking lot with a total loss to the homeowner.

The second consideration for ventilation is: Has the fire made it to the attic yet? If it has and you have water on the fire, it would be a good time to cut the roof. As we discussed in the ventilation chapter, cut over the fire after water has been applied. Roof ventilating at this fire before water has some degree of control of the fire will create a perfect vertical flow path, up the walls and into the now-vented attic with fresh oxygen. No one should be surprised: the house is like a wood stove and we just opened the flue (roof) and fire box door (walls; fig. 10–6).

Fig. 10–6. Fire shows after crews open up the exterior siding. Courtesy of Tom Bierds.

As members opened up the exterior siding, the fire found fresh air and lit up. Saving this house means getting members inside, opening walls, and applying water into the wall spaces from the upper floors. Crews will be needed at several places on the second floor, starting with the exterior wall. Remember, a good technique is to remove the base molding from an exterior wall. This easily and quickly opens several stud channels where fire could travel vertically from the seat of the fire to the attic. As we have stated several times, if you are looking for fire you had better be prepared to find it.

Remember that if this home has knee walls in the attic, this fire has direct access to it. As truck company members open the knee walls, very rapid fire development is possible. Consider application of water from below as best you can to control a very nasty fire hiding in the knee-wall voids. All it needs to erupt and trap members in the attic is some oxygen.

Fig. 10–7. Note the vertical route of fire spread. Courtesy of Tom Bierds.

Members at this fire did an excellent job of getting ahead of the fire. Several stud channels provided avenues for rapid vertical fire spread (fig. 10–7). Aggressive application of water and opening up both interior and exterior to gain access and kill the fire prevented this fire from consuming and destroying the entire house. Note also that the presence of wood siding and the absence of foam insulation resulted in little and slow exterior fire spread. The ribbon-ledger board that is notched into the exterior wall studs can be seen in figure 10–7. The second-floor joists are supported by this board and it and the second floor have lost structural sup-

port. The room shown here is no place for firefighters. If your department has a collapse/shoring team, this would be a good place to put them to work, and your local building department should be requested to the scene. The homeowner or occupants will want to reenter the structure, which could result in your next alarm being to a collapse rescue at this same address.

Modern House

As we saw and practiced in the first example in this chapter, attempting to save the lives of the occupants is always the first priority while minimizing risks to our members. Until the life hazard is resolved, all other operations support the search and rescue. In a legacy home, this can be accomplished with a variety of tactics: aggressive interior attack on the fire with the search team following immediately behind, or in some cases in before the hose team (which is immediately following them), or using VEIS, or even some combination of these tactics based on the conditions. The house in the first example was built in the 1950s and we are all comfortable with the strategy and tactics used there. The 20-minute rule makes sense in these homes; you have about 20 minutes before you have to consider collapse. The legacy house gives you time to operate and tactical options to choose from to accomplish rescue, fire attack, and ventilation. (Loading the legacy home with synthetic contents will, however, drive it to flashover much faster than the 20 minutes it will take to structurally fail.)

But it is 2018, 68 years and two generations of firefighters after 1950. Let's look at strategy and tactics for a house that has just been built. The modern house fire is complicated because the building (design, structure, contents, and floor plan) is complex and very different from a legacy home. It may have a nontraditional layout. Recent homeowner desires are for large open areas that must be made with trusses or engineered lumber. Large windows allow natural lighting for these spaces, often with vaulted ceilings. When these windows fail, they allow massive amounts of air to enter and accelerate the ventilation-limited fire.

Additionally, as real estate prices rise, homes are being made similar to the old duplexes and trips, and have limited access on the side where they are attached to their sister unit.

Scenario

Your department is dispatched for a reported house fire on the corner in a new development. The address is 44 Kaybar Way and time of dispatch is 0023 hours on 15 January (fig. 10–8).

Fig. 10–8. Modern house fire

Size up. Your size-up of this fire has been aided by the computerized preplans that showed up on your screen in the engine while en route. Among other critical information, it showed you the floor plan and warnings for parallel cord trusses in the floors and a truss system supporting the roof.

It appears to be a kitchen fire that has extended toward the A side of the building. The fire has involved the exterior of the building, the vinyl siding, and the foam insulation, and the chip board sheathing underneath will be involved very soon. The fire has entered the attic space (this is telegraphed by the dark smoke coming from the gable vents). The fire apparently has entered the soffits in two places, both above where fire is venting out the windows (fig. 10–9).

Fig. 10–9. Modern house

The 360 of the building you got on your way into this housing area showed one side door, which in this case is the front or main door, and rear patio sliding doors. We will designate this the B side; the A side is the one with the large garage door facing the street in front. Due to the retaining wall and fence, there is no easy access from the rear of the building for apparatus or firefighters. Firefighters can only access the rear via the narrow side yard. Since firefighters entering the

main door on the side of the house and those going around the back will all at one time be in or passing through this area, it could get a bit congested, especially if they have to bring nonambulatory victims this way, and ground ladders and hoselines are in play. Since this is an end unit of a triplex, there is no immediate access via the D side, since it is attached to the middle unit.

Search and rescue. The time of day—just after midnight—indicates that victims should be in their beds sleeping, but the main body of the fire appears to be in the kitchen. Could someone have been cooking and started the fire? The floor plan tells you that there is a master bedroom in the C-D corner and one upstairs over the garage in the A-D side (figs. 10–10 and 10–11). The question is, How will you gain access to search these rooms?

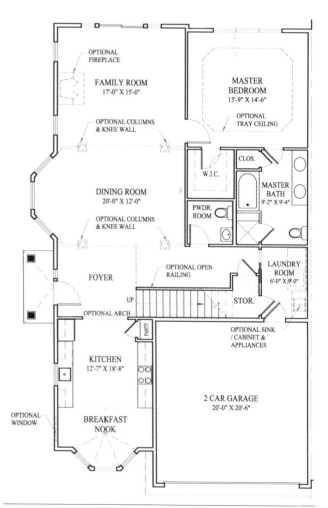

Fig. 10–10. End unit floor plan, first floor

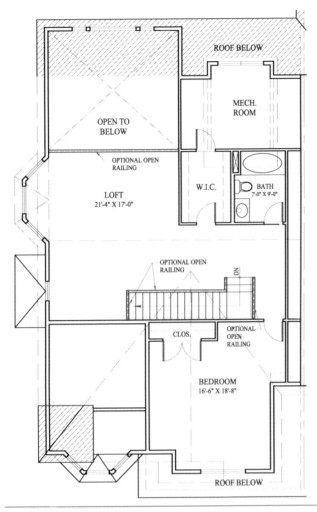

Fig. 10–11. End unit floor plan, second floor

One of the main factors helping you make this decision is the location of the fire. The main body of fire is in the kitchen, near the stairs. More importantly, it is near the high ceilings in the foyer, dining room, and family room. The fire will go to the air it needs. When you force the front door, the fire now has an excellent flow path: fresh air in the front door, into the fire, and out the fire-vented windows. The large areas of the foyer and dining and family rooms have been accumulating large quantities of flammable combustion products the entire time the fire has been burning. Is there fire rolling high near the vaulted ceilings above the smoke, where it cannot be seen? It's a good idea to take a look with a thermal imager to see just how far the fire has traveled from the kitchen.

Option #1. The first search team forces the main door (B side) and tries to make the first-floor bedroom. The second search team enters the same door and heads for the second-floor bedroom. A hoseline goes through the B side (main) door to protect the means of egress (and our members) and extinguish the fire if possible. This hoseline operation should be relatively easy and control the fire quickly, if everything goes well, and thus has several advantages:

- Gaining access through the main entrance and egress route (main door and foyer) provides a default search of this area where the line will enter and operate. Often, incapacitated victims are found in hallways that lead to the usual way they enter and leave the house. The hose team will also likely find the victim who may be in the kitchen because they were trying to extinguish the fire and became incapacitated.

- The hoseline provides solid protection for the search teams and may quickly be able to control the fire in the kitchen and small eating area in the front of the house. Once the fire is knocked down, an immediate and thorough secondary search should be initiated and completed.

However, this option does have a few disadvantages:

- If the hoseline is delayed or some event causes the line to lose water, such as a burst length, the search teams could be caught in flashover conditions and killed. The fire has a good flow path and lots of fuel and is well vented. Expect rapid fire development. It is uncertain if the fire has entered the dining room or family room area. Fire may be hidden above the smoke near the high ceilings of these rooms. If you have the staff on the first alarm, stretch a second line as a backup to the first. Even if you do not need this line to assist the first in controlling the main body of the fire, it is always good to send it to the floor above to stop the vertical fire extension.

- Another disadvantage of this search method is that to reach both bedrooms, members will have to navigate and pass through or close by the fire area. This exposes them to obvious hazards of the rapidly developing fire.

Option #2. Searching aggressively and searching the most likely places to find victims first are always good strategies. Since you know the location of the bedrooms, sending search teams directly to these rooms from the outside is an option. The master bedroom on the first floor has large windows on the C side that will provide easy access. A VEIS operation here has a good chance of success and easy victim removal. The "I" or isolate portion of this operation is critical to both civilian and firefighter safety and survival. Closing the bedroom door will isolate the flow path of fresh air into the fire and its race to the vent you made during your entry through the window. If the fire has built up dangerous conditions of heat, smoke, and flammable gases in rooms just outside the bedroom, it will not be long until conditions deteriorate rapidly in the bedroom you are searching. Rapid fire development will be a urgent concern if water is not on the fire.

The second search team could VEIS the front bedroom via a 24 ft roof ladder. Although there is no porch roof to work from, this is a relatively safe operation, even for one firefighter. Clearly, we are not advocating any single firefighter operations but we are all painfully aware that staffing levels are dropping for fire departments across our nation. We know you all are going to push the limits when lives are in danger, but you must always strive for safer, more effective operations under these drastic conditions. Just keep options open to accomplish your mission with manageable risks without sacrificing ourselves.

Controlling the fire is a high priority at modern house fires. But in this case, conducting a rapid search is also a high priority. There is no right or wrong answer here and there is no silver bullet that will make everything all right. Using VEIS is one option to get firefighters directly to the victims and effect immediate rescues. As we have noted in previous chapters we must also factor in the probability of salvageable human life in the other areas of the house. Some readers may decide that any victims in the kitchen area are not viable so searching that area is not a priority. Obviously we will not know that until we get inside, but if you don't have enough personnel on the first alarm to do all the required tasks, you have to prioritize them and make hard decisions. If you don't have enough personnel to both conduct the search and control the means of egress for your members (and/or knock down the fire), you must consider the safety of the members first and choose the safest search technique: VEIS. This obviously gives up, for the short term at least, search of the normal egress routes and the kitchen for victims.

There are, of course, several disadvantages to this method:

- Removing victims using VEIS can be difficult. Attempting to bring an incapacitated victim down a ladder is, as we all know, difficult at best. Even if they are ambulatory, getting a civilian out of a window, onto, and down a ladder is easier said than done. But getting inside via VEIS, closing the bedroom door, and getting the victim to the window may buy enough time for the victim to survive while second-alarm firefighters knock down the fire.

- If no water is on the fire it will continue to grow and ultimately be a huge threat to members searching. Again, there is no right answer and you must pick what you are not going to do, but remember, openings created by VEIS teams can lead to new flow paths and even more rapid fire development, so close the door!

No matter how you divide up your personnel, getting water on the fire is critical. In this case, it may be at the expense of getting primary searches done concurrently of all areas of the house that have a high probability of containing victims. There may be too many tasks and not enough firefighters to do everything you need to do for an ideal search and fire attack.

Fire attack. The ideal situation is to conduct an aggressive interior fire attack that will control the fire, protect the interior stairs, and result in default search of egress routes simultaneous with search operations on both floors. However, not many of us work in ideal situations, especially when it comes to staffing.

Since this house is a modern house with an open floor plan, if a hoseline is directed into the A-side window (by one firefighter), the stream would reach all the way to the rear of the structure. This line, operated from the outside, would, if not extinguish, at least control most of the fire on the first floor.

When discussing fire attack options it is critical to consider all the details. In this case the open floor plan is a key factor in determining the effectiveness of your fire attack. Again, considering all the factors (or at least those critical to this scenario), taking the line inside is probably the most desirable and effective technique. We recognize that the interior attack is the preferred method, especially if civilian lives are involved. However, other factors (often critical decision points and factors), such as staffing levels, delayed access into the building (high-security doors, bars on windows), and involvement of structural components (truss, I-joints) that may lead to early collapse, should always impact your decision to go inside. These other factors may lead you to reconsider putting firefighters inside. Consider also different situations at this same house:

1. A basement fire. Your building information from dispatch indicates a truss floor. You may want to knock down the fire from outside to keep members off the fire-weakened floor. If the occupants were out of the building, you already achieved your first objective: life safety. No one was in danger until your team entered.

2. Confirmation that everyone is out of the building. A knockdown from outside before entering makes our safety and survival more likely and puts the odds in our favor.

3. Floor plan. Your stream can reach most of the areas because there are no walls dividing the rooms.

Recent UL and NIST research on house fires has shown that directing a handline hose stream in through a window does not push fire through the other parts of the house. We will get into the data that support this later on but for now, let's assume you want to consider it as an option. If your personnel is limited, one firefighter could get a hoseline in operation at the A side of the building nearest the engine and direct that stream into the burning kitchen and eating area at the front of the house. This option uses only one member of your team and gets water directly on the fire, making conditions safer for everyone in the home by keeping the fire from flashing over and wildly expanding throughout the home. This hose stream could also be directed into the attic by using the stream to dislodge the thin vinyl eave coverings and get water into the fire that is extending into the attic. After putting water in to darken the fire, the nozzle operator can extinguish the outside fire and work their way with the stream into the attic. Sure, sounds unconventional but this is a modern house. Consider this, what is holding up the second floor?

If fire has penetrated the ceiling in the kitchen and exposed the parallel cord trusses, getting water on the fire sooner rather than later is a good tactic that will increase firefighter safety. Maintaining structural integrity is an important priority because you have members operating inside during the searches. You know these trusses don't stand the insult of a fire load very long and you don't know how long they have been under attack. If the stream is directed off the ceiling, this hoseline will act like a big sprinkler head: instead of flowing 10 gpm, it will flow 150–180 gpm. That is a very effective head. If the house had a sprinkler system, you would not think twice about operating inside. When you think of operating a hoseline from the outside, your thoughts probably go to pushing fire on members inside. The key to successfully operating a line from the outside is to not move the nozzle in the traditional way. Point the nozzle at the ceiling of the involved room from a position close to the building. You will create a very effective, large flow (180 gpm), sprinkler-head water pattern.

If additional personnel have arrived by now, these members on second-due apparatus can help advance this line or a second hoseline into the house for a continuation of your aggressive interior attack. What is more aggressive than getting water on the fire sooner rather than later?

Recall this important fact: This is a house, so your stream will reach in from one end to another, especially in this one with an open floor plan. You are not delayed by forcible entry issues and you can immediately get water on the main body of fire and its route up the outside of the building, preventing it from entering second-floor windows (vinyl-frame windows fail quickly), spreading fire rapidly and possibly trapping your members. (See the report about Kyle Wilson in chapter 3.)

What is imperative to recall here is that the nozzle operator outside the building cannot grow roots outside and remain there. The line has to, if at all possible (and strategically and tactically smart), get inside to finish off the fire as soon as possible. We don't want this to be a fair fight. Soften up the enemy from a distance and close rapidly for the kill. No mercy, just fast movement forward for the nozzle team.

Traditional House Fire Tactics

The concept of putting a flowing hoseline in a window was something we were all simply taught not to do. Most of you reading this book either taught it or had it taught to you by seasoned veterans of the fire service. Traditional tactics call for us to get inside and push the fire out of the house, not back in. But consider for a moment and ask yourself these critical questions:

1. What are traditional strategies and tactics based on? They are based on our collective experiences on the fireground and training manuals. These training manuals, programs, videos, seminars, and hands-on training sessions are done by very experienced firefighters. Certainly we should respect their experience and opinions.

2. Can you provide any data besides what you think you saw, felt, or experienced at a fire? This is the next question we should consider when we think of reviewing current strategies and tactics for house fires. In fact, before the UL and NIST studies and live burn experiments, there was very little hard data. What do we mean by hard data? Parameters important to us as firefighters, like temperature, heat release rate, heat flux, and so on. Experienced firefighters will often say things like, "I know you can push fire." But what does it really mean to "push fire"? Since our skin gets a very painful second-degree burn at around 130°F and the ceiling temperature of a living room fire can be approximately 2,000°F, is it possible that hot air or steam was "pushed" on our gear as it was getting saturated with heat? Is this what we thought pushing fire was? Smoke probably obscured our vision so it was very difficult to even see what air currents were moving and what we think may have been pushing fire. It is very doubtful that you can push actual fire (flames) with a hoseline at a house fire. Could you push fresh air into a room that just needed it to light up? Probably. Can this be considered pushing fire? Seems a fine line of semantics that we need to agree on for constructive discussion.

3. Have houses and contents changed since our tactics were developed? We remember the 20-minute rule: bosses could push us in for up to 20 minutes before they had to worry about structural collapse. Now that number can be as low as 4. The strategies and

tactics we're attempting to use today on modern houses were developed, perfected, and based on experiences in homes from 1950 or even earlier. Did traditional tactics come from fires in the cities that had houses of balloon-frame construction with thick plaster walls supported by dimensional lumber? Were these houses filled with legacy furnishings with slow heat-release rates?

4. What do we know now about house fires that we did not know before? Modern houses are furnished with synthetic materials that release heat faster, much faster than ever before. We know houses have gotten bigger and floor plans have changed to be more open. We know structural systems, exterior siding, and insulation materials have changed. We know from the research on fires that have killed numerous firefighters, burning them to death inside what should have been a routine call, that modern house fires are flashing over faster and not even giving our engine company members time to open the nozzle before they are burned to death.

5. What else has changed? Our turnout gear lets us get deep inside a dangerous house fire while it absorbs heat from the house fire, protecting us for a while. Then, when we need it most, it is saturated with heat and simply passes the radiant heat of a fast-moving fire through the gear onto our skin. The superheated air currents too are moving and have essentially the opposite of a wind chill effect, meaning our gear gets hotter faster. It is no match for a modern house fire.

These considerations alone should motivate us to review our departments' current strategies and tactics at house fires. We should at least have an open mind to look for alternatives to evaluate. Clearly, it is critically important not to throw the baby out with the bath water; new research is not saying that everything we have been doing is wrong. But we may be able to gain some valuable modifications to our strategies and tactics from the data from the UL and NIST research and live burns. After all, we have had very little hard data up to this point.

Most current research

Exterior water application. In 2012, a study by the joint FDNY-NIST-UL partnership resulted in the Governors Island Experiments (GIE). These experiments examined fire dynamics and water application questions that arose from previous experiments. It was the GIE that showed us that applying water to a house fire from the exterior had a positive effect on suppressing the fire, lowering temperatures throughout the home, and resulted in improvements in both firefighter safety and victim survivability.

In the words of the summary in the study:

> UL used these homes to test a variety of experimental scenarios, including a number of innovative exterior attack tactics. The exterior attack is an offensive approach—analogous to the military concept of "softening the target"—that requires an aggressive attack just prior to entry, search and tactical ventilation. UL benchmarked the exterior attack against a traditional offensive attack that is initiated by deploying hoselines inside the structure directly at the seat of the fire. The UL experiments showed that the traditional approach is not always the best. Several experiments were conducted in homes with different fire conditions. In one example, in a two-story house, fire was showing from a second floor window. (UL 2013, 4–5)

The summary continues by quoting an article by Steve Kerber:

> Traditional tactics call for the hoseline to be charged in the front of the house prior to entry but water is usually not flowed onto the fire prior to entry. Even if the interior path to the fire is known, flowing water directly onto the fire is faster from the outside than it is from the inside. In this experiment, temperatures were measured in the hallway just outside the room and in the other bedrooms on the second floor. Twenty-five gallons of water directed off of the ceiling of the fire room from the exterior decreased fire room temperatures from 1,792°F to 632 °F in 10 seconds; the hallway temperature decreased from 273°F to 104°F in 10 seconds. (Kerber 2013)

It is impossible and inappropriate here to provide a full summary of all the UL and NIST studies. It is, however, critical for the student of fire protection to spend as much time as necessary to understand these critical and groundbreaking studies. They are all available from the UL website. Additionally, there are a number of excellent self-paced training programs available, free of charge, that provide an overview of the research, live burns, and acquired-structure live burns as well as the data from the instrumented experimental full-scale testing.

Even more important than simply understanding the data contained therein is our task to integrate the information into both our daily operations at house fires and our fireground experience. Change comes hard to the fire service. Specifically related to house fires, the research, live burns, and training materials provided via the DHS grants deserve our attention.

The following examples of data presented at FDIC 2015 by Steve Kerber do provide an overview and explanation of how applying water aggressively from the *outside on our way into the structure* can provide outstanding fire-suppressing effects that improve firefighter safety and occupant survival.

In figure 10–12, the rear of the house is on the right and the front on the left. Fires were set in all three rooms. The window is open on the left side of the drawing and water was applied through a window on the right side or rear. These data show that even if there is an open window opposite the hose stream being directed in a window, the stream will not push fire through the structure. The data also show that applying water from the outside improves conditions inside.

Figure 10–13 shows data collected from another of the GIE live burns. In this case, water was applied through a front window (shown on the left side) into two rooms of fire. Temperature improvements can be seen on the graphic. It is also critical to note that fire was not pushed into the rear room, which actually experienced a reduction in temperature, as did the hallway.

As we have done in previous chapters, what follows is a brief summary of the study's conclusions. In the words of UL:

The key findings of our experiments show that the common belief about exterior fire attack pushing the fire is unfounded and that innovative fire attack tactics can improve the safety and effectiveness of firefighting efforts:

Water applied via exterior attack does not push the fire.

The anecdotal experience of firefighters can be explained by one of the following scenarios: (1) A flow path is changed with ventilation and not with water application. (2) A flow path is changed with water when the thermal layer is disrupted and steam moves ahead of the line, elevating the level of heat and creating the impression to those downstream that the fire is being pushed. (3) Turnout gear becomes saturated with energy, which begins to pass through to the firefighter. If this occurs in close proximity to when a hoseline is opened, it might appear that the hoseline caused the rapid buildup of heat. (4) One room is extinguished, allowing air to entrain into another room, which causes that room to ignite, burn more intensely or reach flashover.

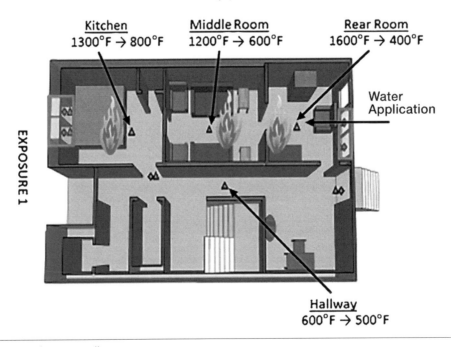

Fig. 10–12. Temperature reduction in all rooms

Rather than making conditions more hazardous, applying water directly into the fire compartment as soon as possible results in the most effective means of suppressing the fire.

Specifically, our research showed that applying a hose stream through a window or door into a room involved in a fire significantly lowered room temperatures everywhere in the home. Even a small amount of water, applied as quickly as possible regardless of where it is from, improved conditions inside the burning home. And in cases where front and rear doors were open and windows had been vented, the application of water through one of the vents enhanced conditions throughout the structure.

Our experiments showed that exterior fire attack increases the potential survival time for building occupants and provides safer conditions for firefighters performing search and rescue. In fact, our experiments demonstrated that the traditional practice of increasing ventilation to a ventilation-limited structure fire by opening doors, clearing windows or cutting the roof increased fire hazards and the potential for a rapid transition to flashover.

While the attack should be commenced from the exterior, to improve conditions for firefighters and building occupants, it must be finished inside.

Applying water to the fire as soon as possible from the outside softens the target and helps firefighters gain the upper hand, but the attack and size-up should be continued from inside the home. Once conditions inside the structure are made safer, continuing the attack from the inside increases the speed and effectiveness of fully extinguishing the fire. (UL 2013, 5–6)

Application of water to a house fire from the outside has become known as the transitional attack. In essence, this is simply quick and short application of the fire stream from the outside until the fire is significantly or mostly knocked down. This is followed up by immediate, traditional interior attack for final extinguishment. This can be done for both first- and second-floor fires if appropriate. What is appropriate? One example is a delay in forcing the door to gain access for an interior attack. Another may be a short-handed situation where the first-due engine does not have enough

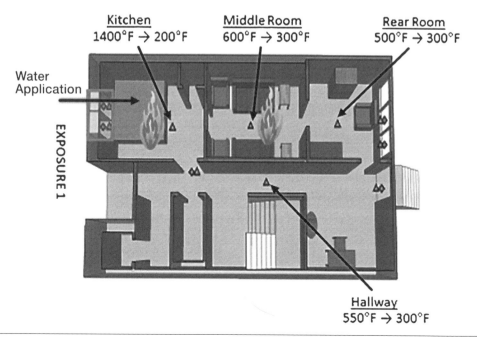

Water in a Window with No Flow Path Opposite

Kitchen
1400°F → 200°F

Middle Room
600°F → 300°F

Rear Room
500°F → 300°F

Water Application

EXPOSURE 1

Hallway
550°F → 300°F

Fig. 10–13. Temperature reduction in the rear room

manpower for an aggressive interior attack immediately. A third use of transitional attack may be if the fire is in the basement and the floor above the fire is weakened or is made of trusses or I-joists that do not withstand a fire load. Whatever the reason, if you start your aggressive attack from outside, you still need to get inside and finish off the fire. This was demonstrated in the findings of follow-on studies we will examine in the remaining pages of this chapter.

We are not suggesting that any fire department should adopt this as their procedure for *every* house fire. It is simply another option available to ICs for their strategy based on the unique circumstances of each fire. There is no substitute for good size-up and appropriate strategy and tactics. There is no one right answer for a house fire attack strategy. If there were, everyone could be a fire chief.

For complete details in the transitional attack, you should dive deep into the studies we have mentioned. Figures 10–14 and 10–15 show how you can demonstrate the effectiveness of "a dash from outside" as Bill Gustin refers to it. The nozzle must be close to the building and at a steep angle. The nozzle should not be moved in the usual directions we are used to but held at a constant position and moved only slightly when necessary to distribute the water that is bouncing off the ceiling to ensure full coverage in the room.

If you have a burn building at a fire training center or even an acquired structure, you can take trainees into the building where, in an adjacent room, they can observe the water bouncing off the ceiling and being distributed in the room. After this observation they can decide for themselves whether this may be an effective technique when applied at the right time under the right conditions. Combine this demonstration with explanation of the data from the fire attack study that shows this application of water is not detrimental to the occupants. Review of part 1 of the study, water mapping, is also helpful.

Fig. 10–14. Water distribution inside from an exterior stream

Fig. 10–15. Transitional attack

The transitional attack concept caused massive discussion and upheaval in the American fire service. There were several misunderstandings and reasons for objections to the transitional attack. First, the instrumented live burns produced massive amounts of data, which are difficult for firefighters to analyze, digest, and apply. Second, many firefighters had the false impression that this "new" technique meant that "everything we have been doing throughout my career is wrong." Third, there was the false impression that the transitional attack had to be used at every fire. Actually, the technique was presented as an option, like all other fire attack options, to be applied at the correct time, place, and fire situation. Of course the most insidious cause of opposition to the transitional attack as an option was our extreme reluctance to change.

In defense of the American fire service, we know and are comfortable with current house fire tactics. We use them every day with success. These instrumented live burns caused us to compare our fireground experience with the data acquired from the burns. Often, these seemed to conflict, but in actuality they were complementary. For example, we have all been on a search of a home, taken the window from the inside, and seen the smoke and heat lift in that room. Clearly one cannot argue with this experience.

The data show the larger effect on the overall fire situation in the house, which we were not aware of; opening that window may have provided a ventilation opening for the fire downstairs, creating a flow path and making the fire grow rapidly. Ultimately this action may trap the firefighter by a fast-moving fire. Joining and understanding both of these valid observations—data-based findings and street experience—into a new understanding of house fire dynamics remains the challenge for our fire service from 2018 and beyond.

Beyond transitional attack. The GIE dramatically demonstrated there was much more to be learned from the study of house fires than simply whether we should put water in from the outside. Many other important questions were raised that need answers to provide tactical insight on the real fireground fire attack tactics. Some of these questions include: How much water do we need for one or two rooms of fire? Where is the best place to apply it? How do we best make the push down the hallway? Can we do search operations and fire attack concurrently? And many others.

In 2014, UL was awarded $1.5 million grant for a 3-year study to conduct further research on house fires. The importance of the fire attack study cannot be overemphasized. For the first time we have measurements in real time for a variety of different fire attack methods. Think about that. Previously, all we had to evaluate a house fire were comments or observations from interior firefighters such as, "Man that was really hot!" or very localized observations such as when the smoke or heat lifted because of ventilation or when a ventilation-limited fire was driven to flashover by poorly timed ventilation.

In contrast, from the studies we have second-by-second data on temperature; heat flux; concentrations of oxygen and CO; and direction and intensity of pressure and air movement. Additionally, in each test burn there are 16 simultaneous video views of key areas in the building and several thermal imager views. Consider the value of being able to evaluate a fire from all these perspectives, not just the few feet you can see around you in the smoke or steam when you open the nozzle.

Authors of this text conducted groundbreaking research and measurements on air flow caused by fire streams and its potential effect on the fire attack. These articles were published in *Fire Engineering* in 2003. Part II of the UL study further refined the 2003 measurements and developed tactical considerations based on the data (see ch. 7). Coauthor of this book, Jerry Knapp, was selected to be one of 25 members of the technical panel for this new study of house fires. The role of the technical panel is to guide the overall direction of the work so that it is representative of what firefighters do in the field and to help write the tactical considerations.

TECHNICAL PANEL members come from a wide variety of fire service backgrounds. Some are career firefighters, some volunteer, and five on this panel were international chiefs. If you are reading this book, clearly you are a student of fire protection. You should watch the UL website and apply for a tech panel slot when they are advertised.

Working with other selectees and the staff of UL is a once-in-a-lifetime experience that will provide career-enhancing knowledge and insight into our business that you will not get anywhere else in the fire service.

You will have time to witness and understand the research and, more importantly, interact with and become friends with the researchers and other fire service leaders from around the country and the world.

The official title of the fire attack study is *Study of the Impact of Fire Attack Utilizing Interior and Exterior Streams on Firefighter Safety and Occupant Survival*. It was conducted in three parts as described below:

> Part I of the study is aimed at determining how water is distributed within a compartment, while Part II quantified the air entrainment by hose streams to provide insight into how different application methods; nozzle types and patterns; pressures/flows; and stream location and angle combinations move air inside buildings. Parts I and II were conducted without the presence of fire to gain a basic understanding of air flow and water flow before full-scale fire experiments were conducted during Part III. These full-scale fire experiments were designed based on the results from Parts I and II of the study. (UL 2018, 3)

A quote from the final report published in 2108:

> In these experiments, interior fire attack was implemented and measured for the first time and addition transitional fire attacks were conducted with more measurements than ever before. In every experiment, the fire went out and no one was injured, but that should not be a surprise to the fire service as tactics like these are executed successfully everyday. With all of the measurements made during these experiments as well as the vast experience of the technical panel, several consistent themes emerge which may be helpful to the fire service. Each of these themes is packaged as a tactical consideration with supporting text and visuals. Each one can also be traced back to the analysis section for more scientific support. Each of these considerations can have limitations so it is important that they are interpreted in the proper context.

Part III, the live burns, of this detailed study of house fires also included victim packages. These were placed where humans would likely be in the home during or trying to escape the fire. Scientists used pig skin to determine if fire suppression operations would have a positive or negative effect on their survival.

As previously stated, it is inappropriate to simply copy the study results here. What we will do is highlight some of the tactical considerations relative to first- and second-floor fires for you to consider to improve your house fire strategy and tactics. Although hope is not a plan, we hope you have already or will become very familiar with this and other house fire research done by UL, NIST, and others.

Tactical considerations. Let's look at some first- and second-floor fires with respect to just a few of the tactical considerations from this study.

This scenario (fig. 10–16) has relatively heavy smoke showing with fire in unknown location inside the home. Since this home is likely a balloon frame, immediate reconnaissance of the basement is a good idea because you don't want to put firefighters above a well-involved basement.

Assuming there is no fire in the basement, there is no substitute for getting in through the front door, locating the fire, and extinguishing it. If on-scene intelligence indicates, a search is always a priority. Ventilation should take place after water is on the fire.

In figure 10–17, fire is in a modern house. It may contain trusses supporting the roof and I-joists or parallel cord trusses supporting the second floor. The fire is well ventilated, and the open front door has created a flow path. It is difficult to tell whether fire has entered the attic, but it likely has through the soffit vent where heavy fire is impinging out the windows of failed ceilings.

Firefighters here are making an aggressive interior attack. Search operations, if indicated, should be concurrent with the fire attack, especially to areas left or toward the B side of the house. It is unlikely there are salvageable human lives in the fire rooms. If doors were closed in bedrooms in the C-D corner, there is a high probability of survival and that should be a priority for search operations.

It is worth considering applying water to this fire from the outside just prior to entry. This may make conditions better and safer for the interior attack crew should there be a collapse of lightweight support structures. It appears the crew has a 2½ in. hoseline in operation because they saw heavy fire upon arrival. The reader is directed to the fire attack study, page 172, for more detailed information.

Fig. 10–16. There is no substitute for aggressive operations at a fire like this. Courtesy of Kenny Flynn.

Fig. 10–17. Would application of water from the outside prior to entry make it safer for the interior crews? Courtesy of Kenny Flynn.

Fig. 10–18. How much time will it take to stretch to the rear of the house? Courtesy of Kenny Flynn.

When you arrive on scene and you see fire even before you get out of the engine or chief's car, it gives you a fair idea of the fire situation and a head start on your fire attack plan. But when fire is located around the rear of the house (fig. 10–18), you have to do a good size-up and evaluate the options. What is not negotiable is the fact that getting water on a room-and-contents fire at a house fire is the priority. Here again, this is where the expertise and experience of the engine officer or chief has the most impact.

Do you take the line through the front door or do you take the time to stretch around to the rear? Your size-up will help you make this decision. In the words of the fire attack study, page 176:

> For a room and contents fire, the most important timing piece is the initial application of water into the compartment, to cool the compartment and knock back the fire prior to it extending outside the compartment. If that can be achieved faster by conducting an interior attack through the front door, then that is the most effective tactical choice. If it can be achieved faster by deploying the line to side 'C', then that is the most effective tactical choice.

Fig. 10–19. Scenario for a recent fire

Figure 10–19 shows a recent scenario. We arrived to find fire blowing out the first-floor bay window on the A-B side of the front of the house. A local firefighter was standing on the porch, directing a garden hose into the pulsing flames. The entire bay window opening was showing fire and police had the sole occupant of the house across the street.

The fire was completely suppressed with the garden hose first by directing water in from the outside and then applying water from the inside. Clearly we are not advocating attempting to extinguish a fire with a garden hose flowing 5 gpm, but there are a lot of real-world lessons from this fire that are supported by the UL research. A section from the October issue of *Fire Engineering* describes a very important fact.

You don't need a lot of water for a contents fire in a home. A very limited amount of water contained this fire. For most residential fires, your 500 gallons (and often more) of onboard water will knock down a significant amount of fire when properly applied and coupled with well-coordinated ventilation. This could be especially important during operations that necessitate aggressive searches for trapped occupants. Engine companies will best support search teams with rapid and decisive amounts of water at the seat of fire to ensure improved conditions and faster advancement of search crews.

Of course, the effectiveness of tank water is not an excuse not to establish a reliable water supply. This fire and others like it (ventilation limited) provide some scale and insight as to how you may best be able to use available resources such as your first-in engine company, and situations that may further be complicated with limited staffing on your first-due engine and delayed response from second-due engine companies.

As UL reported based on 25 live fire suppression scenarios in its report *Impact of Fire Attack Utilizing Interior and Exterior Streams on Firefighter Safety and Occupant Survival*:

> When dealing with a room-and-contents fire, the energy release rate is limited by the available oxygen (ventilation limited). It *does not take a large amount* of water to absorb the energy being released and knock back the fire. Although less is not necessarily better, when a water supply has not been established, or in areas where no municipal supply exists, water application should not be delayed to establish a water supply. Even a 500-gallon supply tank can be sufficient to knock back two rooms of fire if the attack crew can get the water where it needs to go.
>
> During the 25 suppression experiments conducted, utilizing a 1¾-inch handline flowing 150 gpm to 165 gpm, the most water used for initial knockback and suppression was less than 250 gallons. When attacking a single room-and-contents fire in a residential structure, knock back and initial suppression are often possible with less than 100 gallons, in some instances less than 75 gallons. Even flowing while moving to the compartment of origin did not result in utilizing more water than available in a 500-gallon supply tank. (UL 2018, 186–87)

Oxygen was limited at this fire. The occupant was lighting a candle and dropped the match on the sofa. When she returned with a container of water the couch was involved. All windows in the house were closed and she closed the front door on her way out.

Again, it is important not to misinterpret the discussion here. Establishing a reliable water supply and delivering decisive amounts of water in your attack line remain basic plays we must be proficient at for every fire.

Speed of transition. As previously mentioned, there is no substitute for an aggressive interior attack. Knocking down the fire from outside just prior to entry may have some benefits. However, there is not time for congratulations and a victory party on the front lawn. You must continue your aggressive operation as quickly as possible if you have chosen to apply water from the outside on your way inside. A short quote from page 193 of the fire attack study explains part of the reasoning.

> During any type of suppression method, after cooling all surfaces that can be cooled from that position, the crew should rapidly relocate to a position which allows for complete extinguishment (most often the interior of the compartment). The door, which provides the fastest, most direct path to that location, should be utilized to limit the time between the last water application and final suppression.

The ability to put continuous water on this fire, as limited as it was, did not give it time to regrow. The burning couch, chair, end tables, and drapes were near the window and under constant flow from the garden hose.

Nozzle flowing. In many training situations trainers often force trainees to continue to flow the nozzle, keep the bale open, and advance the line to the seat of the fire. Intuitively this is better than shutting down, moving in and reopening the bale. In the live burns in the one-story ranch house with one or two rooms of fire, there was little difference in the cooling effect of the streams and the return of heat over the nozzle teams heads.

> When flowing and moving, there is a constant cooling effect for the hose line, both around and ahead of the advancing hose crew. When the line is shut-down to move forward the cooling effect is no longer present, and temperatures begin to rebound ahead and around the advancing crew. The temperature rebound occurs mostly over the head of the advancing crew. If the hoseline is operated again within 10–15 seconds, the rebound never reaches the temperature prior to suppression. Although structural protective gear will prevent the advancing crew from feeling a change in the environment, elevated temperatures return. If the hose line was not operated again, the temperature would rebound to the level prior to suppression. (UL 2018, 199)

The ability to constantly flow the hoseline while advancing toward the seat of the fire is likely a function of the staffing for the line and the nozzle operator's experience and muscular abilities.

We have presented only a fraction of the excellent research findings and tactical considerations developed by UL and NIST in hopes of stimulating your interest in fully understanding the findings. It is clear that this body of work will have a lasting and extremely positive impact on fire attack at modern house fires.

This chapter also reviewed traditional and modern house fire tactics. One is not right and one is not wrong. But as the world around the fire service changes and the very houses we fight fires in change, we too must consider change. Not for the sake of change, but to continually improve our operation for our customers and for firefighter safety. We urge you to understand the latest research, because frankly it is the current and next generation of firefighters' responsibility to utilize both experience and research for everyone's benefit.

References

Kerber, Steve. 2013. "What Research Tells Us about the Modern Fireground." *Fire Rescue* 8 (7). http://www.firerescuemagazine.com/articles/print/volume-8/issue-7/strategy-and-tactics/what-research-tells-us-about-the-modern-fireground.html.

Underwriters Laboratories. 2013. "Innovating Fire Attack Tactics." *New Science Fire Safety*, Summer. http://docplayer.net/6007857-New-science-fire-safety-articleinnovating-fire-attack-tactics-summer-2013-ul-comnewscience.html.

———. 2014. "Study of the Impact of Fire Attack Utilizing Interior and Exterior Streams on Firefighter Safety and Occupant Survival."

———. 2018. *Impact of Fire Attack Utilizing Interior and Exterior Streams on Firefighter Safety and Occupant Survival: Full Scale Experiments.* https://ulfirefightersafety.org/docs/DHS2013_Part_III_Full_Scale.pdf.

Attic and Exterior Fires

11

Summary

This chapter's purpose is to examine and discuss strategic and tactical considerations, options, and outcomes (positive and negative) for fire department operations at residential building attic fires. This chapter discusses the practical application of the appropriate skills and development of strategic plans based on the information from all previous chapters for both the first-due company officer and the first-due chief officers. For firefighters, this chapter provides an understanding of the larger strategies developed and directed by chief officers as well as the tactical operations you will execute for successful operations at attic fires. Techniques for tactical operations such as search, ventilation, and hoseline advancement were covered in previous chapters and will not be included in this chapter except for some very special skills directly pertinent to attic fires, special considerations, or important reminders.

Introduction

The US Fire Administration estimates that "10,000 residential building attic fires are reported to U.S. fire departments each year and cause an estimated 30 deaths, 125 injuries, and $477 million in property loss." The report also states that "electrical malfunction is the leading cause of residential building attic fires (43%)" (National Fire Data Center [NFDC] 2011, 1).

Here is an excerpt from the issue of *Topical Fire Report* that provides an overview and examples of attic fires in the United States (NFDC 2011, 10).

July 2010: An attic fire in a Fresno, CA, home started around 2 a.m. and is believed to have been caused by an electrical problem. Eight family members were home when the fire took place but no injuries or deaths occurred. The fire caused between $15,000 to $20,000 worth of damage. The firefighters were able to keep the fire contained to the attic.

June 2010: A fire that started in the attic at a family's home in East Windsor, NJ, is said to have been caused by faulty wiring in a second-story ceiling fan. The fire was brought under control in about 30 minutes and was contained to the attic and roof areas. The homeowner and a contractor who was working on the back porch were home when the fire started. Both were alerted to the fire by a smoke alarm and were able to escape safely.

June 2010: A four-alarm fire in Weatherford, TX, started when a lighting strike hit a house, causing a fire to start in its attic. The fire ended up destroying the home despite firefighter's efforts to combat the blaze. The fact that the fire department had no access to water at the home's location had a significant effect on the outcome. No deaths or injuries were reported as a result of the incident.

April 2010: A neighbor called 9-1-1 shortly after 10 p.m. when he smelled smoke coming from a neighboring house in Racine, WI. The fire is believed to have been caused by an electrical malfunction in the attic. No one was home when the fire started and no injuries were reported. It is estimated that the house sustained $30,000 worth of damage from the fire.

Storage Attic Fires

For discussion purposes we will divide attics in homes into two categories. One we will call "storage attics." The second variety (which is much less frequent) is attics that have been converted into living areas, which we will refer to as "occupied attics." Across the country, the design and construction of homes leads to various adaptations to firefighting procedures for attic fires. In this section, we will cover both traditional attic fire strategies and tactics and those new techniques based on the latest research and employed by our colleagues.

In many parts of the country, attic space is defined as the space above the top-floor living area and under the bottom of the roof. It has become—or maybe always was—a place for the occupants to accumulate and store items they no longer need but simply can't part with. Attics may contain seasonal items, like Christmas ornaments, or old yet sentimentally valuable things like children's toys. In the modern house, this space has become increasingly coveted by architects, designers, and builders for high-efficiency insulation, HVAC units, heating/cooling plants, and HVAC ductwork.

For the construction trades, it has become a place to hide mechanical equipment, particularly heating ventilation and air conditioning (HVAC) systems. Whole-house fans (or attic fans) are located either in the ceiling of the top floor or in the gable vent. The modern open floor plan (higher ceilings) has forced builders to maximize attic spaces for these systems so as not to lose valuable floor space in the living area for a mechanical equipment closet.

These storage attics, whether filled with the occupants' belongings or mechanical equipment (usually both), pose a significant risk to firefighters. The increased fuel load from the contents presents a concern for being able to deliver the proper amount of water to extinguish the volume of fire. The weight added from the water-soaked contents can also create a collapse potential for members operating below.

These spaces often have limited access and egress points, forcing firefighters to conduct operations from the living space underneath the fire. This is a dangerous place to be with a collapse and rapid fire development potential overhead.

If entry of members into the attic for extinguishment is considered, remember that the space was not designed for easy access or rapid escape. There often is just enough room for mechanics to access equipment for service. Firefighters, with bulky turnout gear and SCBA, unable to stand upright, and restrained by a hoseline, will find movement difficult at best, especially under fire conditions. Additionally, flooring may be incomplete or nonexistent in some places, creating a fall-through danger.

The HVAC equipment can have high voltages or Freon, creating an environmental or suffocation danger. The flexible ductwork can create an entanglement danger. Flex duct has spiral or helical stiffeners to keep the duct open for airflow, which is a major hazard for firefighters.

Part of what complicates an attic fire is simply getting to it (fig. 11–1). Like basement fires, attics usually have limited access. The attic is the largest void space in a home and may be packed with stored combustibles. In older homes, the attic access may be through a hatch cleverly hidden in a closet, over the bathtub, or at the top of the interior stairs, or may be through a small door hidden on the second floor. The most common access point for attics is some type of lightweight pull-down ladder. The rungs are not big enough for our boots and it is not rated for the weight of a firefighter, gear, hoseline, and other weight we routinely carry (fig. 11–2). The existing ladder is often in our way when we go to place a small fire department attic or folding ladder in the scuttle to gain access.

Fig. 11–1. Attic access is often difficult to maneuver and limited in size.

Two things further complicate any fire suppression operation from the floor below (pulling the ceilings to allow your fire stream to reach the seat of the fire). First is the amount of stored materials. Often the attic is packed with stored items, all of which are combustible. In addition to stored materials and furniture if the attic is occupied, it may contain HVAC equipment (fig. 11–3). Secondly, if the attic has a floor of any kind to support all those great things stored there, opening up from below will be very difficult (fig. 11–4). Floorboards may be nailed down, providing a complete subfloor system, or may simply be sheets of plywood to span the ceiling joists to provide a surface for all the stored material. In either case, opening up from below will be a challenge.

Fig. 11–2. You must bring your own ladder to the top floor to access the attic. Do not depend on this lightweight folding ladder to safely support your weight.

Fig. 11–3. An almost complete floor, stored materials, and an attic fan

In older homes, knob-and-tube wiring may still be present and present, posing a significant hazard to firefighters. See the article by Gregory Havel (2008) in the sidebar below.

Fig. 11–4. This attic has a ¾ in. plywood floor that is nailed down firmly. Older homes may have 1¼ in. tongue-and-groove flooring, which will be very hard to dislodge from below with a hook.

CONSTRUCTION CONCERNS: KNOB-AND-TUBE WIRING

Article and photos by Gregory Havel

In the early years of electrical service, the most common method used to install wiring inside buildings was known as "knob-and-tube" (photo 1), after the porcelain insulators (knobs) on which the wires were mounted, and after the porcelain tubes used to pass wires through wood joists and beams. This type of wiring could be exposed, as in basements, attics, and factories; or concealed inside the walls of homes, schools, and offices.

The two-piece knob and the tube were the most common insulators. The two-piece knob held the wire between the two pieces of porcelain (photo 2). This held the wire at least ½-inch away from the surface to which the insulator was attached with a nail or screw. Other types of knobs were equipped with clamps, to attach them to steel trusses and beams in factories. A porcelain tube (photo 2) supported the wire anywhere it passed through a wood joist, beam, or rafter. Another type of knob (cleat) held both wires of the circuit together and was held in place with two nails or screws (photo 3). Other styles and types of knobs for other purposes and higher voltages were also available.

The wires were single conductors, usually of copper; covered with rubber insulation and a cotton or muslin jacket; and originally intended for use in dry locations. A later development was a conductor for use in damp (not wet) locations. This was similar to the original type, except that the fabric jacket was saturated with paraffin. Wires could be spliced or tapped (the attachment of a branch line) anywhere in the run of wire. Another later development was the color-coding of each wire's fabric jacket (black, white, red, etc.), to make tracing wires easier.

Photo 1.

Photo 2.

According to the National Electrical Code (NEC; 1st edition dated 1897, decades after electrical use and electrical incidents began), supports for wires could be no more than 4½ feet apart, and wires up to 300 volts could be no closer than 2½ inches. Any wire splice or tap was required to have the insulation stripped from the conductor; the conductors twisted together to make a mechanically secure connection; soldered; insulated with rubber tape; and covered with cloth "friction tape." Each wire passing into a metal box for a receptacle, fixture, or switch passed through the steel inside a porcelain insulating bushing. When the NEC was followed, knob-and-tube wiring was safe and reliable, unless remodeling was done without attention to detail (or mice chewed up the insulation too badly).

Photo 3.

Since this method of wiring is labor-intensive and, therefore, expensive, it is no longer in common use today; it has been replaced with conduit systems and with metallic-sheathed and nonmetallic-sheathed cables, each containing one or more complete circuits. However, knob-and-tube wiring is still present in some older buildings—and this old system is often still energized and in use.

These photos were taken in August 2008 inside a 1895 building originally used to build wagons and, later, as an auto dealership, a plumbing and sheet-metal shop, a warehouse, a motorcycle dealership, and a video store. It is presently being converted into a candy store and kitchen.

Life hazard

Attic fires often threaten the occupants' lives, as evidenced by the 30 annual deaths due to attic fires (NFDC 2011, 1). Often there are no smoke detectors in the attic, so the fire can grow undetected and can extend rapidly downward into the lower floor when the ceiling (which was providing fire resistance) fails. This creates a flow path with an excellent supply of oxygen to the ventilation-limited fire. Following the path of least resistance, it may go to flashover and rapidly extend downward with deadly speed and intensity, injuring those on the floor below.

Firefighters must be aware of the hazard of fire overhead as well when we enter to pull ceilings and access the attic fire for the engine company. If you are looking for fire, be prepared to instantly deliver decisive amounts of water with a charged hoseline when you find it.

Sheetrock ceilings can be dislodged in large pieces and old plaster ceilings can loosen and fail in large pieces as well. The wood lath holding the plaster may or may not come down with it. If the lath remains, it will prevent your hose stream from getting onto the fire until truck members pull it down.

Traditional strategy and tactics

Let's use the following scenario and apply traditional methods for attic fire operations. Dispatch informs you of reported smoke coming from the eaves of 44 Shanks Road on 23 April at 0022 hours (fig. 11–5).

Fig. 11–5. A side of the building

Size-up. Even with only the A-side view, there is no question that there is a heavy body of fire in the attic of this home. Judging by the hockey net in the front driveway, the flag, the trashcans in the front, and the general well-kept appearance of this home, it is probably not a vacant structure. The front windows are open on the A-B corner and no smoke is showing from them or any other second-floor windows, which is a good indication there is not significant fire on the second floor yet. The C side shows heavy fire in the attic with no indication that it has extended downward to the second floor, though it likely will soon (fig. 11–6).

Fig. 11–6. C side of the building

Life hazard. Rescue is our first priority; therefore getting search teams inside to the most likely locations of salvageable victims is a good first step. With this large and intense attic fire, it is a sound idea to use a hoseline to protect the means of egress, which, in this case, is the stairway going to the second floor and the upstairs hallway. At this time of day, the odds are good that family members are sleeping on the second floor. Of course, this is just a general rule; occupants could be anywhere in the home.

Another option for search at this fire is VEIS with ladders to the windows where the bedrooms are. Assuming the fire has not extended downward (and there are no signs that it has), this option is not as effective as an interior search. It takes more personnel to raise the ladders and at least one firefighter to VEIS each room. The VEIS method at this fire will make searching for and finding victims who may be in hallways or trying to escape much slower. Bringing victims down ladders is slow, difficult, and time and staff consuming. VEIS in this situation does not afford interior firefighters the protection of the hoseline and the ability to rapidly search the first floor or stairs and second-floor hallways.

Hoseline positioning. Command has ordered an aggressive interior search protected by at least one hoseline, then fire attack via the interior attic stairs. When you reach the attic access, you encounter the scene in figure 11–7.

Fig. 11–7. Heavy smoke is pushing out of the cracks around the attic scuttle.

The traditional next step, to open the hatch, may introduce enough air to cause the attic fire to violently flash over and extend downward onto members operating on the floor below (fig. 11–8). If the wind is blowing, air could be driven into the attic via normal vents (soffit, ridge, or gable) or through areas in the home siding or roof where the fire has burned through. A wind-driven fire could make conditions lethal on the floor below after you open the scuttle or door. Let's assume that does not happen and that the members are safe at their location on the floor below the main body of fire in the attic.

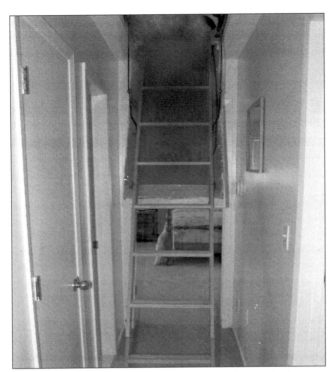

Fig. 11–8. Heavy smoke and fire follow the flow path down toward you.

Ventilation. Another common tactic employed at attic fires is vertical ventilation of the roof. If this is done before water is put on the fire, a flow path will be created. As hot gases exit the roof through the newly cut hole, fresh oxygen will be drawn up into the attic via the scuttle. Figuratively, we have just opened the air damper on the wood stove, allowing the fire to burn more freely (it was ventilation limited before) and giving it a nice supply of air. The fire will increase in the attic if ventilation is accomplished before water is applied.

Fire attack. The lightweight pull-down attic stairs usually have a load limit of 250–300 lb. The latter figure is generally assumed to be a firefighter's weight in full turnout gear and tools. Additionally, the steps are very small and the boots of even the smallest firefighter will not fit very well, causing instability issues when pushing the hoseline into the well-involved attic and fighting the nozzle reaction. It is a very good idea to use a fire department attic ladder or A-frame ladder instead of these stairs. Recall that the nozzle firefighter will have most of the nozzle reaction to deal with. The backup firefighter will be on the floor below while the nozzle person is on the ladder. If the nozzle person reaches the attic floor, the backup firefighter will be on the ladder. The nozzle firefighter alone will have to handle most of the wrestling match with the hoseline for portions of the advance.

The engine crew will encounter high heat when they penetrate the ceiling or floor and get into the attic. The plan is for the hoseline to extinguish the fire if the engine crew can remain at the attic access or make progress into the attic.

The reach of the stream should allow the engine crew to get water into all the involved areas. However, if the roof is supported by trusses you may have difficulty distributing water to all parts of the attic because there will be a lot of lumber (the webs of the trusses) in your way. Unfortunately, this wood is also small (2 in. × 4 in. truss components) dimensionally and will ignite quickly and burn intensely because there is a much larger surface-to-mass ratio than with full-cut dimensional lumber (2 in. × 8 in. or 2 in. × 10 in. rafters). Additionally, as we examined in previous chapters, the connections that hold the truss members together fail rapidly under fire. Collapse is always an issue with unprotected trusses, especially in an attic fire.

One of the downsides of this tactic is that if there is fire on all four sides of the attic access, and possibly above it, the nozzle firefighter will be surrounded by fire, and, by virtue of the air entrained in the stream, be creating massive air disturbances. We have all been there and it is a very uncomfortable position: you have crawled to the fire floor from below and are slugging it out in very close quarters.

Pulling ceilings. An alternative to entering the fire floor from the bottom (up through the scuttle) is pulling ceilings on the floor below and directing the stream upward into the burning attic. This is a time-tested method but may be hampered by low staffing levels. Additionally, as previously mentioned, if the attic has a floor and that floor is nailed down securely or there is a lot of storage on top of the floor, this tactic may not allow equal distribution of water into all the areas (fig. 11–9).

Fig. 11–9. A firefighter has pulled the ceiling only to find a solid attic floor above. Courtesy of Tom Bierds.

It is important to remember that if an exterior attack has been chosen, the gable vent or shutter for an attic fan at each end of the home may provide access to the areas your interior streams cannot reach. These gable vents are usually light plastic or aluminum and can easily be pulled by a firefighter with a hook from a ground ladder or aerial device. Although not the best place to apply water, it is an option (fig. 11–10).

Fig. 11–10. A gable vent on the side of the house can provide access for a hose stream. Courtesy of Tom Bierds.

The 2014 study by Steve Kerber and Robin Zevotek sheds some new strategic and tactical considerations for attic fires in homes. This live burn–based research and its accompanying self-paced training package are available online, and the executive summary of findings is presented in the following sidebar. If you really want to fully understand the intricacies of attic fire growth and appropriate tactics, we strongly recommend you spend a few hours using these invaluable free training resources. We will also delve into a few of the most important findings now.

Although we all "know" that adding air to the fire generally makes it worse, the study showed that opening the ceiling below the attic fire did not appreciably worsen the fire during experiments under controlled conditions. (Uncontrolled conditions such as winds or uncoordinated roof ventilation can cause rapid fire development.) There are, however, several case histories of truck company members opening ceilings with fires over their heads and having to bail out when the fire lit up and extended downward. If you are looking for hidden fire, especially in attics and void spaces, you must have a method to deal with it; bring a hoseline with you.

Additionally, opening the roof (vertical ventilation) before the engine has water on the fire is generally not a good idea, according to the study. Getting water on the fire then venting was the best strategic plan, according to the test results, and the best method for applying water was wetting the sheathing (the underside of the roof) by applying water through the soffit. This significantly slowed the fire's growth.

THIS SIDEBAR contains the Executive Summary of the UL's research on attic fires. We present it here in anticipation that this small portion of the report piques your interest to read and understand the entire report, and to compare the recommendations and findings with your current attic fire strategy to determine if modernization of your SOPs is warranted.

Excerpt of Executive Summary from *Study of Residential Attic Fire Mitigation Tactics and Exterior Fire Spread Hazards on Fire Fighter Safety* (Kerber and Zevotek 2014, 3–5)

To evaluate the exterior fire hazards of various wall construction types, medium scale testing was performed on 8 ft x 8 ft wall sections looking at ignition, flame spread, peak heat release rate and exposure potential. The results of the medium scale testing was used to establish parameters for eave experiments to further evaluate flame spread along with increasing the understanding of the dynamics of how fires transition from exterior to attic fires. Following the eave experiments full scale attics were constructed and instrumented to evaluate the effectiveness of four fire service suppression tactics on attic fires. Each tactic was evaluated both with and without vertical ventilation or simulated attic burn through to understand the fire dynamics during attic fires. Finally field experiments were conducted to investigate the fire dynamics of knee-wall fires and the effectiveness of current mitigation tactics for knee-wall and half attic space fires.

The results of the experiments were then examined with the fire service technical panel and utilized to develop 12 fire service tactical considerations for use in the mitigation of attic, knee-wall and exterior fires. An overview of these tactical considerations include...:

Increased use of plastics in exterior walls will change what you arrive to—Changes in residential wall construction methods are playing an important role in how exterior fires are initiated, as well as how they spread and extend.

If the fire starts on the outside, start fighting it from the outside—Rapid water application to knock down the exterior fire is a critical part of any attempt to control not only the fire's spread to adjacent structures but also the fire's migration into the interior of an exposed building. If the source of the fire is not suppressed, it will continue to supply heat energy to the fire developing on the interior, worsening conditions on the inside for occupants and in many cases making it impossible for the interior crews to maintain or advance their positions.

Learn to anticipate where and how an exterior fire will migrate to the interior—Exterior wall fires may easily spread to the interior at locations other than the eaves and soffits. Any penetrations—such as air vents, electrical receptacles, plumbing penetrations to faucets and drains, and especially windows—provide the opportunity for fire spread into the interior of the structure. Leaving the interior fire barrier in place until the exterior fire can be controlled will limit the extension into the structure.

Attic fires are commonly ventilation-limited fires—The openings provided for natural ventilation are not sufficient to maintain steady state burning and fuel limited fire behavior. The size of the fire is limited by the available oxygen and will nearly always become ventilation-limited. Controlled openings created below the neutral plane (such as through the ceiling below the attic space) will not cause immediate growth and can provide access for suppression operations.

Closely time or limit vertical ventilation until water is in the attic—A vented attic fire was more difficult to control with the indirect methods applied to the unvented attic test. The, "open up above and then attack it from below" tactic can and has been successfully used at attic fires. However, it can create a large amount of property damage and puts both civilians and firefighters at high risk during the initial stages of the operation if not timed properly. Once initial water absorbs some energy, a vertical vent will assist the crews with suppression and overhaul because standard fire ground ventilation tactics will be sufficient for exhausting the smoke and fire gases produced by the remaining fire. In the absence of suppression, the positive effect of a roof opening is a very short lived phenomena. The accelerating fire can overwhelm all openings and push back into the occupied space. Increased visibility does not automatically mean a reduction in the size of the fire over your head.

Plastic ridge vents can affect size-up and fire dynamics—As the vents heat, the plastic melts and collapses on the opening at the peak, creating a very effective seal. Once the ridge vent seals, the eaves will act as both the source of air as well as the exhaust and you may notice a pulsing of smoke out of the eaves.

Wetting sheathing with an eave attack slows attic fire growth—If crews wet the sheathing, either as part of an offensive fire attack or defensively to slow fire spread to uninvolved sections of the structure, the major flame spread mechanism in the attic is eliminated until the moisture evaporates. Removing the soffit and flowing water along the eave line of these structures was the most effective way to gain the upper hand on a fire that was venting through the roof.

Attic construction affects hose stream penetration—The most effective water application takes into consideration the construction of the attic, using the natural channels created by the rafters or trusses to direct the water onto the vast majority of the surfaces.

Consider flowing up instead of down with a master stream—Consider using an aerial device or portable ladders and hand lines to open up the eaves and flow water into the attic. This approach could result in controlling the fire enough to permit firefighting crews to transition back inside the structure to complete searches, suppression, and overhaul.

Knee wall fire dynamics—During a structure fire, it is possible for fire to enter void spaces and surround crews conducting interior operations. Even though there is a delay between making the breach and the change in conditions, once initiated, the transition to untenable conditions in the area of operation occurs in seconds. Knee wall construction often provides the potential for ideal fire growth, with air entering low at the eave line and combustion gases exiting the peak through mushroom vents, ridge vents or gable vents.

Apply water on a knee wall fire at the source and toward the direction of spread before committing to the attic—Applying water utilizing the same path the fire took to enter the void space may be the most effective method at slowing fire growth. Water application to the knee wall will not be effective until the source below it is controlled with direct water application.

Interior operations on knee wall fires—Tests have demonstrated that the most effective way to get a handle on knee wall fires is to control the source fire, cool the gases prior to making large breaches in the barrier, and then aggressively open the knee walls to complete extinguishment, focusing on wetting the underside of the roof decking.

Innovative fire attack

We have just reviewed how difficult, taxing, and dangerous attic fires can be. As a result of the changes in construction of modern homes, we no longer have the time to operate that we once did. Contents have much higher heat-release rates and structural systems are not designed to withstand a fire load, resulting in early collapse or floor weakness that causes firefighters to fall through floors into deadly fire conditions below. In the case of attic fires, roof collapse can trap and kill you. Let's use the previous scenario (fig. 11–11) and apply some innovative tactics for modern house fires that are recommended by the research (Kerber and Zevotek 2014).

Fig. 11–11. Fire scenario. Courtesy of Tom Bierds.

Dispatch informs you of reported smoke coming from the eaves of 44 Shanks Road on 23 April at 0022 hours.

Size-up and life hazard. Our size-up and evaluation of life hazard will remain the same: there is a heavy body of fire in the attic of this home (fig 11–12), and searching most likely locations of salvageable victims is the top priority.

Fig. 11–12. Read of the structure. Courtesy of Tom Bierds.

Hoseline positioning. We still execute an aggressive interior attack after we have finished the SAR effort. However if we can get water onto the seat of the fire before we commit members to the building for fire attack it will make it much safer for them. The studies recommend that we consider directing a hose stream from the outside through the eave or soffit overhang. The angle of the stream allows water to flow along the rafter channels, coating the underside of the sheathing of the roof and thereby limiting fire spread or extinguishing the fire. This angle of the stream also provides beneficial water distribution throughout the attic. The nozzle firefighter can move along the outside of the home directing their stream into the attic.

When considering this tactic compare how long it will take to get the line inside, pull ceilings, and find an operable area where there are no floorboards and storage. Darkening down the fire from the outside after the life hazard has been resolved is a viable option, especially in homes with lightweight construction or where there are no life hazards (from occupants, in any case). Also consider how limited your stream will be in getting water distributed to other areas if the attic is fully involved. Even if you have personnel for multiple lines, do you have personnel from multiple truck companies to pull ceiling in multiple areas of the floor below?

As we have seen in previous chapters, soffits in modern homes are covered with either perforated (for ventilation) vinyl or, in slightly older homes, thin plywood. Often if there is a large body of fire in the attic, these will be burned or melted out, providing access for your stream. If they are not already gone, you may be able to dislodge them with a powerful straight or solid stream, and as a last resort a truck company member can pull them.

Immediately after the first line darkens down the main body of fire from a safe position, a second line that was pre-positioned at the door can quickly advance for an interior attack. With the main body of fire knocked down, chances of flashover are reduced, increasing member safety while inside.

Ventilation. The aggressive exterior fire attack that darkened down the fire also produced large amounts of steam and smoke. It lowered already very poor visibility in the attic. It is time to ventilate and allow some of the smoke to exit so interior hose teams can get inside and finish the aggressive attack. The aggressive fire attack, quick water on the fire from outside, did not give the fire time to grow. It cut it off and limited its growth as soon as possible, then the attack was pressed for the knockout blow from the inside.

This tactic, like any other tactic or strategy, is not a magic bullet and cannot be used all the time and it may not be appropriate all the time. For example, in older homes with classic decorative wood trim on the eaves, this tactic may be impossible. Obviously, if you have not resolved the life safety issue, an exterior attack should not be your first choice.

The Los Angeles County Fire Department (LACoFD) has taken a modern, scientific approach to ventilating and attacking attic fires. They are much more cognizant of the condition or phase the fire is in and use that to their advantage by basing their operation on the fire conditions. Although painfully obvious, if firefighters pull ceilings and use massive amounts of water, they create sometimes unnecessary damage to the homeowners' most valuable possessions on the floors below. Of course, extinguishing the fire is a must, but limiting damage during the operation must also be a priority. Battalion Chief James Sullivan of LACoFD explains this method of extinguishing an attic fire in the following sidebar.

ATTIC FIRE ATTACK

In Los Angeles County, we developed a training platform for the suppression of attic fires based on the information that came from the UL study on attic fires and exterior water application. (The UL course is free and can be found on the UL website: http://www.firecompanies.com/modernfirebehavior/Attic-FiresOnlineCourse/story.html.)

What we have learned is that an attic fire is a ventilation-limited fire that should remain ventilation limited until water can be applied. If we open the roof before we have applied water, we are taking the fire out of its ventilation-limited state and allowing it to become free burning, thus destroying more of the structure and its contents in the process.

We train our companies to find a scuttle hole, or if necessary, make a small hole in the ceiling drywall with a pike pole or rubbish hook as near the seat of the fire as possible. We then apply a fog pattern into the attic space, allowing the water to expand as it turns to steam and is forced out of the vents in the roof. By doing this, we keep the fire ventilation limited and suppress the fire by cooling the gases and lowering the heat in the attic space.

In some cases, there is no need to cut any holes in the roof during the operation. The fire is taken into its decay phase due to the lack of heat and oxygen caused by the application of the fog stream. In other cases, it will be necessary to open the roof to ensure that full extinguishment has occurred, but only after the application of water, keeping the fire from entering a free-burning state.

By keeping an attic fire in the ventilation-limited state and applying a fog stream from the bottom up to suppress the fire, we do less damage to the structure and have the ability to do more salvage work prior to pulling ceilings and ensuring full extinguishment.

Fire conditions

LACoFD and other fire departments across our nation are embracing modern fire attack methods. These methods and procedures are based on new research that gives us the ability to understand the fire conditions in attics and all kinds of house fire situations: basement, first- and second-floor, and attic and exterior fires.

The fire researchers at UL have been providing us with exceptional amounts of data to analyze and improve our house fire operations and firefighter safety. It is impossible to provide it all in this book but the sidebar on pages 237–38 contains a small synopsis of critical attic-fire suppression tactics. We urge you to read and study this research, which will provide you with an understanding of the fire conditions you are trying to suppress.

Additional scenario

Fig. 11–13. A side of the building. Courtesy of Tom Bierds.

For the purpose of this scenario we will assume the life hazard for occupants has been resolved and you have a reliable water supply. Apparatus and personnel available to you is whatever is typical for your department.

Attic fires are one of the costliest types of house fires due to the damage we need to do to gain access to the fire, usually by pulling ceilings. Additionally, water damage from our hose streams and wet insulation, sheetrock, and stored materials falling from the attic to lower floors is always extreme and damaging. It is very frustrating to see a family's home and possessions destroyed by water and other damage, not by fire. Battalion Chief (retired) Rick Fritz, High Point FD (NC), summarizes the necessary actions of the fire department and judicious use of water at attic fires this way, "I have seen a lot of things dry out. I have never seen anything un-burn." Nobody wants to see property get damaged, but that is why we sometimes have to do minimal damage to save maximum amount of property.

Salvage is the prevention of damage to property and possessions, and should be one of your major concerns and efforts before the fire attack operation begins, assuming you have the staff to do this. For example, removing priceless family photos from the wall to a safe location will help the family recover from the trauma of having their home destroyed. Throwing a few well-placed salvage tarps also will be greatly appreciated and will result in saved property, our second-highest priority.

Size-up. Size-up of this fire reveals a well-involved attic fire. It appears that it has not extended down to the second floor yet. There is or was heavy fire on the B side. The smoke tells you the fire has extended or soon will throughout the entire attic (fig. 11–14).

Fig. 11–14. B side of the building. Courtesy of Tom Bierds.

Your view of the B and C side confirms that the fire has not extended down to the second floor but the soffit has burned through on the B-C corner. The sheathing (if it was there before the fire) on the B side has burned through, giving you great access to the seat of the fire. A strategic option is to pull members from inside (since they are now the only life hazard) or hold them at the front door, and apply water directly to the seat of the fire from outside with a second line.

Looking at the fire from this side, it seems that the main body of fire is on the B end of the house. The firefighter on the deck can easily pull the plastic soffit covers, allowing your hose stream complete access to the attic fire from a position of safety. Then, once the fire is darkened down, members can get inside for an aggressive attack using traditional methods.

We have been taught not to put water on from the outside because it pushes fire. In this case, if the stream operated from the outside did push fire, where would it push it to? With a direct assault on the seat of the fire with decisive amounts of water (150–180 gpm), it is reasonable to assume this action would eliminate most of main body of fire.

After the main body of fire is knocked down, any firefighters who climbed into the attic through the scuttle will now be surrounded by steam, not raging fire. Should the roof collapse, these firefighters' chances of survival with minimal injuries are vastly increased (figs. 11–15, 11–16, and 11–17).

Fig. 11–15. Easy access to the main body of fire. Courtesy of Tom Bierds.

Fig. 11–16. Getting water on the fire to slow its extension and growth. Courtesy of Tom Bierds.

Fig. 11–17. Main body of fire still burning as firefighters conduct vertical ventilation. Courtesy of Tom Bierds.

The main body of fire has been extinguished and fire now vents out of the roof cut (fig. 11–18). Note that the roof on the B side has burned away.

Fig. 11–18. A side after water application. Courtesy of Tom Bierds.

As stated previously, salvage is of prime importance at attic fires. No matter what strategy or tactics you apply, discharging large volumes of water on the top floor of an occupied building will cause a lot of damage.

Occupied Attics

Attics, although normally not occupied, may be used as living space in high-cost real estate areas, college towns, or where there is a migrant population during certain seasons, as in some farming communities. Usually, if the attic is occupied, there is at least a semi-full set of stairs that provides relatively easy, routine access from the top floor. In some houses, these stairs wind up near the outside of the home and when the attic floor level is reached, access is compromised by the steep pitch of the roof (fig. 11–19). This compromises the egress for occupants in case of emergency and makes access for firefighters, with turnout gear and SCBA on, difficult and dangerous.

Knee walls

A knee wall is a short wall, usually around 3 feet in height, about the same height as a human knee. They are used to support the rafters in roof construction. They are also used to create an aesthetic living space directly under a pitched roof in an attic. Cape Cods have knee walls on the second floor. Finished attics, legal or illegal, will have knee walls (fig. 11–20).

Fig. 11–19. A door of nonstandard size and shape provides access to this illegal attic living space.

Fig. 11–20. The knee wall is the vertical portion of the wall behind the television that creates a void space and an aesthetically pleasing room design.

If an attic is a void space, knee walls are a void space (subdivided attic space) within a void space (attic) on top of multiple void spaces (the walls). All these hidden areas should make you apprehensive simply because you can't see where the fire and smoke is, how intensely it is growing, and whether it's under control or not. Simply, it is all hidden from your view. Fire concealed by the knee walls has the ability to grow undetected to killer proportions and is beyond our ability to visualize or gauge how dangerous it is. Don't forget, if you are operating in the attic surrounded by a knee wall that is loaded with flammable smoke and maybe intense fire, you only have one way out when the fire breaks out of its temporary cage.

Attacking a knee-wall fire from inside the attic necessitates entering the attic and truck company members opening the knee wall followed by engine company members applying water. Due to the intense fire that may be hidden behind the knee wall, it is possible for conditions in the attic to worsen in seconds, trapping, injuring, or killing members operating there.

Recent research on fires in knee walls from live burns in acquired structures by the Milwaukee Fire Department shows that knee walls, by virtue of their location and construction, are excellent areas for concealed fire growth. Air can enter the eaves and the void spaces, allowing the fire to grow. Products of combustion continue to move vertically and exit through normal vents, including ridge, gable, and soffit vents (figs. 11–21 and 11–22).

Fig. 11–21. Attic of a balloon-frame home. The tape measure shows where the top of the wall is open to the concealed knee-wall area. Fire can enter the knee-wall area from the outside eave or from a basement fire or fire on lower floors that enters the walls.

Experimental application of water (supported by thorough instrumentation and data analysis) at various areas of the knee wall fires showed that water application from below and outside was the best option. This water controlled the fire below the knee wall and limited fire growth in the concealed space. (Consult the free online training program at https://ulfirefightersafety.org/ for the complete information.)

Fig. 11–22. This knee wall has a large volume to build up flammable smoke and conceal fire. Note also the electric wiring and insulation and flexible ductwork for heating and cooling systems.

Exterior Fires

Exterior fires that lead to residential fire envelopment were covered extensively in chapter 3. It is important to note that exterior fires often result in attic fires. An exterior fire can spread quickly up the exterior of the home and enter the attic through the soffits.

At these fires you will be faced with both an exterior fire and an attic fire. The exterior fire—caused by burning vinyl siding, foam insulation, and sheathing—must be extinguished first. The good news is that it is easily accessible to one firefighter with a mobile exterior hose stream. Strategically, this stream should continue its tactical advantage after extinguishing the exterior fire: it can continue to apply water into the soffits as described previously to suppress the attic fire. Although you were faced with two main bodies of fire, the exterior application of water can have a very effective result on suppressing both.

Your size-up should determine, as quickly as possible, if the fire has entered the living spaces on the first or second floors. If the fire has enveloped a significant portion of the exterior of the home and has entered open windows (or burned through vinyl-framed windows) and doors, the safety of everyone inside is in serious danger, a situation that requires your immediate attention. As we noted in chapter 3, the correct action may be to be a bit less aggressive in assigning members to inside operations in traditional ways.

Further complicating these types of fires is the fact that many modern homes are built with truss construction, wooden I-joists, and other void-producing methods. Once the exterior fire gains access to these places, you should anticipate these structural members to weaken throughout the home. The void spaces, especially in modular homes, often run the length and width of the house. (See chapter 3 for full details and fireground considerations.)

References

Berry, Joseph. 2016. "Outside/In Fires: Two Incidents, Two Outcomes." *Fire Engineering* 169 (6). http://www. fireengineering.com/articles/print/volume-169/issue-6/features/outside-in-fires-two-incidents-two-outcomes. html.

Havel, Gregory. 2008. "Construction Concerns: Knob and Tube Wiring." FireEngineering.com (blog), 13 November. http://www.fireengineering.com/articles/2008/11/construction-concerns-knob-and-tube-wiring.html.

Kerber, Stephen, and Robin Zevotek. 2014. *Study of Residential Attic Fire Mitigation Tactics and Exterior Fire Spread Hazards on Fire Fighter Safety*. Northbrook, IL: Underwriters Laboratories. https://firenotes.ca/files/UL_FSRI_Attic_Final_Report.pdf.

Knapp, Jerry. 2013. "Rapid Fire Spread at Private Dwelling Fires." *Fire Engineering* 166 (10). http://www. fireengineering.com/articles/print/volume-166/issue-10/features/rapid-fire-spread-at-private-dwelling -fires.html.

———. 2015. "Modern House Fires Warrant Tactical Agility." *Fire Engineering* 168 (10). http://www.fireengineering. com/articles/print/volume-168/issue-10/features/modern-house-fires-warrant-tactical-agility.html.

National Fire Data Center. 2011. "Attic Fires in Residential Buildings." *Topical Fire Report* 11 (6). https://www.usfa. fema.gov/downloads/pdf/statistics/v11i6.pdf.

12

Common Nonfire Calls to Homes

Summary

This chapter's purpose is to highlight some of the more common but nonfire, nonmedical calls to homes.

Introduction

Maybe the fire department should just be called the "everything department." It seems we get called every time Joe or Jane Citizen gets into trouble with anything. This chapter will cover some of the most common and most dangerous nonfire calls. We're sure you have been to one or more crazy calls that are far beyond what follows. The focus here is that we need an understanding of the situation to determine if it is an emergency or simply a repair to be handled by others.

Carbon Monoxide

Carbon monoxide (CO) investigations may not be routine calls, though they happen frequently. As a company officer, one of the more common alarms you will respond to is going to be a CO alarm investigation. This may seem like a routine call but it is not the time to be complacent. According to US Product Safety Commission statistics, on average 250 people die and 10,000 more are seriously injured from CO poisoning each year. These high numbers have prompted many jurisdictions, like New York City, to adopt carbon monoxide detector legislation requiring detectors in dwelling units.

In New York City, the FDNY alarm assignment for a CO device activation is one engine and one ladder company, no chief. This puts all of the initial decision-making in the hands of company officers, and it doesn't matter if you've been promoted for ten years or one day, you have to make the calls. To make these critical decisions, you must know how to interview occupants, properly investigate the alarm, and determine the source of the CO. You must know the department's operational guidelines and be able to take appropriate actions.

In other jurisdictions, the fire department or EMS is only notified if the occupants are exhibiting symptoms. The alarm investigations are handled by the local public utility. If this is how it is done in your area, you know you're going into a more complex situation if you're called.

History

Home CO detectors are much different from their industrial counterparts. Industrial units are complicated, expensive, and lack the sensitivity for the home environment. As detectors for home use have gained popularity, their level of sophistication has improved. The earliest home detectors were nothing more than a fiberboard card treated with a chemical that would change color if exposed to carbon monoxide. This was not acceptable because the card had to be visually inspected and was not able to provide an audible alert to occupants.

The next generation of home detectors employed a biomimetic technology. Biomimetic detectors are designed to mimic the human body response to exposure to CO. This gel sensor was an improvement over the fiberboard cards but had its own drawbacks. It was designed to continually absorb air samples, so even low levels of CO could build up in the unit and cause false alarms. To reduce the false alarms, the detector was designed not to go into alarm until a time-weighted average was reached. The time-weighted average would put the detector in alarm more quickly at higher concentrations while extending the time at low levels. It functioned like a rate of heat rise detector. Once a sensor had been exposed to CO, the biomimetic gel would require many hours of exposure to clean air to reset. Basically, the sensor would have to "off-gas" to reset. These detectors had wording to the effect of "remove to fresh air to reset."

Table 12–1 is based on the American National Standards Institute (ANSI)/UL 2034–2005 standard for sensitivity (the parameters of the time-weighted average; ANSI and UL 2005).

Concentration (ppm)	Response time in minutes
70 ±5	60–240
150 ±5	10–50
400 ±10	4–15
FALSE ALARM—CARBON MONOXIDE CONCENTRATION RESISTANCE	
Concentration (ppm)	Exposure time (no alarm)
30 ±3	30 days
70 ±5	60 minutes

Table 12–1. Carbon monoxide concentration and response times

The next type of sensor designed for the home market was the metal oxide sensor. The tin-dioxide circuits in these units are more accurate and are the most common in use today. They continually monitor the air and provide a digital readout of CO concentrations. The alarm levels are based on the LCD readout. They are, however, susceptible to false positives in that they can be sent into alarm by exposure to carbon monoxide found in other gases such as methane (natural gas) and propellants in household items like hair spray. The cumulative effects of these false positives can cause a detector's accuracy to stray up to 40% in six months.

The newest sensor technology is instant detection and response (IDR). It is an electrochemical sensor and is becoming the industry standard for detection equipment. It has an accuracy of ±3% and is not cross-sensitive to other gases, eliminating false positives. This is a new technology that has an increasing market share in home detection equipment. It is similar to the processor found in the Gasalert personal CO monitors carried by FDNY members. However, this technology is not without drawbacks. One of the concerns is about the detector's ability to monitor air in high humidity. The detector has a filter medium that covers the sensor. When the filter is soaked with moisture it may not sense accurately. This is akin to protecting yourself from smoke by breathing through a damp cloth. The moisture in a home can come from rainfall or it could come from moist air generated by a shower or boiling water during cooking. The dampening of the filter medium can cause differences in the reading on both the home detector and FDNY meters. This can occur if the assigned member is wearing the Gasalert meter under their bunker coat during hot weather. The member's own perspiration can create humidity that is trapped under the coat and can cause the filter medium to become soaked to the point that accuracy is affected. The operating range for the Gasalert CO detector is 15%–90% relative humidity.

Investigation

Now that you know more about the types of detectors you may come across in the field, we can conduct the investigation. Since life and health are our first priorities, begin the investigation by interviewing the occupants. Are there any signs or symptoms? Were they home when the detector went off? If they were home all day and the detector just went off, ask what were they doing during the day. This can provide clues. If they just arrived home and found the detector going off, it is reasonable to assume that they will not show signs or symptoms. Remember the signs of CO exposure: confusion is one of the early signs. If, during your interview, the occupants are confused or unsure of answers, they may be answering your question by not being able to answer.

During the interview, do not ask leading questions. Do not ask, "Do you have a headache?" or "Do you feel lightheaded?" because if they think they were exposed, they will tell you they have those symptoms. Instead ask, "How do you feel?" and let them explain. If you ask leading questions, everyone in the building will have some combination of all the symptoms by the time you are done.

Look for signs in the high-risk populations: small children and older adults may exhibit signs of CO exposure more quickly than healthy, middle-aged adults. Pregnant women should be carefully evaluated for CO exposure since CO can pass from mother to fetus.

CO will also affect pets. If the family has been out all day and the pets were home, they may provide the best clues. Is the dog vomiting or listless?

Have contractors been doing any work recently in the building? Repairs and replacement of fuel-burning appliances can have problems during initial startup, which may produce CO until final adjustments are complete.

Ask the occupants if they opened or closed any windows since they called for assistance. If the area has been vented before your arrival, the source of the CO may not be evident. You may need to recreate the conditions, as they were when the detector went off, by closing windows. If the room has been vented it may take some time before the CO level rises to detectable levels. Readings on any instrument should be verified by a second device to ensure accuracy.

While conducting your investigation, never forget your senses. Though CO is odorless, the fumes that produce it may not be. For example, if the CO alarm is going off in a first-floor apartment and you enter the front door and smell an oil burner, it may indicate that the heating equipment is malfunctioning. Odors of food might indicate that someone was cooking on a defective stove. A properly functioning stove has bright blue flames; any other color toward gold or yellow indicates that the stove is not burning with the proper air mixture, and thus is combusting incompletely, which produces CO. A smell of gas can indicate a leak in the gas piping to the stove. Remember false positives. Check for a natural gas leak behind the stove with the combustible gas detector (such as TIF 8800). A soap and water mixture applied to the piping will cause large bubbles to form at the point of the leak.

The last and most important point in conducting a CO investigation probably should have been the first point: know how to protect your members. Know the operational limits of your PPE and when you *must* wear your SCBA.

At some point, you must examine the detector. Do you know what the chirps mean? It would be impossible to explain every brand and model on the market. The detector itself will have a label on the back with a sticker to decode the chirps. This will allow you to determine if the device is in alarm, just in need of a replacement battery, or chirping about some other condition.

On the device, you will also find a date of manufacture for the detector. Most brands now are guaranteeing their devices for five years (some longer), but recommending replacement at seven years. The older detectors may still be found in the field. They will still have all the drawbacks previously stated and may provide the homeowner with a false sense of security.

Decision time

What do you do when our fire department meters verify the activation of the homeowner's device? What is the reading? How many people are involved? Are there signs or symptoms? Is the CO is affecting a large nursing home? Remember we spoke about high-risk populations earlier.

Any incident that has persons displaying signs or symptoms requires a notification for EMS and a chief officer to respond. Advise the dispatcher of the CO readings and request a response from the local utility. The Public Service Commission requires utility companies to conduct an investigation when injuries are sustained from carbon monoxide where natural gas may have contributed to those injuries. This can also be a proactive measure since if members turn off the gas to any fuel-burning appliance, the utility must be notified anyway, and this can expedite their response.

Provide first aid

Remove to a clean environment any individuals suspected of being exposed. Administer oxygen and monitor vital signs. Administering a high flow of oxygen through a non-rebreather mask will reduce the time it takes to remove the CO from the blood stream by a factor of five. It normally takes five hours to reduce the level of CO in the body by half, when the person is breathing ambient air (21% oxygen). Breathing oxygen at 100% (about five times the percentage in ambient air) will reduce that time to one hour.

In the prehospital setting, it is difficult to determine the CO level in a person's blood from symptoms alone. Individual tolerance can vary and assessment can be subjective. Commercial monitors are available, such as a Massimo RAD 57, a CO monitor that clamps onto the finger similar to a pulse oximeter. The device will digitally display the percent of CO in the patient's blood.

Most decisions you will make at these alarms will not be dramatic, but they are not without consequence. Consider the young family who calls you late at night when the detector wakes them up. You are unable to clearly determine the source of the CO, even though you suspect the boiler. Do you shut down the boiler, turn off the oil or gas, notify the utility, and make them get a licensed plumber to service the boiler before they restart the unit? Or do you do the "right thing" and tell them to call back if the detector goes off again?

This is probably closer to the type of decision you will have to make. What if they never wake up to call you back? What would have been the cost of a hotel room for the night? The service call for the plumber? The inconvenience? What did it ultimately cost them and their family in the long run?

As fire officers, we hold people's lives in our hands every day. This is just one more example of why we need to be professional, know our job, and have the expert judgment to handle crisis. You may not be able to handle all routine calls routinely.

Helpful hints for carbon monoxide investigations

Properties:

- Colorless, odorless, flammable, and toxic
- Vapor density slightly lighter than air; carried on convected air currents
- Reduces the blood's ability to carry oxygen
- Flammability range: 12%–74%

Symptoms:

- Low level exposure: headache, dizziness, confusion, fatigue
- Medium level exposure: vomiting, loss of consciousness
- High level exposure: seizure, coma, brain damage

Operational guidelines:

- Begin metering before you enter the building.
- Locate the source and shut down the appliance.
- Ventilate.

PPE:

- SCBA shall be worn at all CO investigations and used during CO emergencies.

By the numbers

Incidents:

- Readings less than 9 ppm: evacuation NOT required unless symptoms

Emergencies:

- Readings over 9 ppm or symptoms
  - Recommend occupants leave the area; potentially dangerous CO reading
  - Ventilate to reduce CO before reoccupying
- Greater than 100 ppm or symptoms
  - Mandatory evacuation; potentially lethal levels
  - Ventilate to reduce CO before reoccupying

Natural Gas Emergencies

Gas leaks or odors of gas are like Rodney Dangerfield: They don't get no respect. This is because most people can smell the odorant put in natural gas long before it is dangerous. So we get the call and become conditioned, response after response, when nothing bad happens and begin to think nothing bad will ever happen.

As trained professionals, we must remember that we are dealing with a highly flammable invisible gas that flows undetected underground, taking the path of least resistance into a nearby building; is supplied by pipeline with an unlimited source; and, if contained under the right conditions, is explosive and will kill us and the people we are sworn to protect. In a microsecond, you can go from well-trained firefighter to a badly injured firefighter-victim whose survival often depends on dumb luck.

We should always work from the mindset that this call could develop into a worst-case scenario until we prove otherwise. This is exactly what we do for search-and-rescue (SAR) situations. Some of my coworkers are fond of saying, "The building is clear only after we conduct our primary and secondary searches!" Often, these searches include great personal risk. Instead, we need to think, "The building is going to blow unless we prove otherwise" (fig. 12–1).

Fig. 12–1. This is what remained after a gas explosion in the end unit of a townhome. Courtesy of Tom Bierds.

Procedures

To ensure a safe and efficient response to natural gas explosions, gas emergencies, gas mains struck or damaged, and odors of gas, it is critical to have procedures that are based on the industry best practices. Surely, you regard your local gas supplier and the service technicians who respond with you to gas calls as the subject matter experts. They are; they do it every day and their procedures are based on years of experience in the gas industry. Fire department procedures should be written in close coordination with utilities and look a lot like gas companies' procedures.

The one main and critically important difference to our procedures is that the fire department mission is life safety. The gas company mission is to supply gas and keep that supply uninterrupted. Human nature being what it is, sometimes a for-profit business may take some liberties in safety to keep their mission intact—which in this case means keeping their customers happy and their profits high. Sound, step-by-step procedures are your key to success and firefighter safety at gas leak emergencies. The utility technicians on the scene have well-written procedures and will be following them. We should be no different.

The utility company that provides natural gas to your area is required to provide emergency responder training. In New York, the regulation on training is taken from the Code of Federal Regulations (2009):

(c) Each operator shall establish and maintain liaison with appropriate fire, police, and other public officials to:

(1) learn the responsibility and resources of each government organization that may respond to a gas pipeline emergency;

(2) acquaint the officials with the operator's ability in responding to a gas pipeline emergency;

(3) identify the types of gas pipeline emergencies of which the operator notifies the officials;

(4) plan how the operator and officials can engage in mutual assistance to minimize hazards to life or property; and

(5) offer annual training, at mutually acceptable locations, to volunteer fire departments regarding the appropriate response to gas related emergencies and to police departments regarding the recognition of gas related emergencies. For non-volunteer fire departments, annually offer to assist the training coordinator in developing training programs for gas safety related matters.

Other states have similar rules that require gas utilities to supply appropriate training to fire departments. Seek out this training and procedures.

Consider this: Imagine one of your loved ones is having a severe heart attack. Like a gas leak, this can be a life-or-death event. You call the paramedics; they quickly arrive at your home and come into your living room with the usual complement of equipment. You expect them to have a clear and appropriate procedure, based on sound medical best practices, to evaluate and correctly treat your loved one. But then you hear this conversation.

Paramedic Billy Bob says, "You know Chris, on the last call we shocked the guy because we thought we would try it."

"Jeez, Billy Bob, I think we should try something different, that one did not go so well. What do you want to do this time?"

Billy Bob responds, "Aw, nothing is gonna happen before we get to the hospital so let's try some epinephrine, maybe that will wake him up."

You would be outraged to hear these paramedics playing fast and loose with your loved one's life, but this is exactly the same as responding to a natural gas incident without any standard operating procedures or even a plan in mind.

No Plan B

Another key factor in gas leak response that makes it unique from most of our other responses is that there is no plan B (fig. 12–2). Let's say you are caught by a fast-moving fire on the second floor of a home while searching. You have at least three plan Bs. First, you can shut the door or duck into an uninvolved room or, if things are really bad, bail out the window using your bailout rope or head-first ladder slide. Worst case you can hang out the window and drop to the ground to minimize your injuries. You have options.

Fig. 12–2. Plan A, the main chute, is on the jumper's back; plan B, his reserve chute, is on the front.

When gas is contained in a building and it explodes, the building is torn apart. It is instantaneous and you have no time to react. If you are too close to the building or simply standing in the wrong place, your survival will depend on luck. What is the "wrong place"? As the building comes apart, large pieces of it will become deadly shrapnel. Doors, windows, whole section of walls and floors, and individual deadly projectiles like 2 × 4 boards and header boards. It is also critical to remember that our turnout gear is not rated for explosions or flying metal, wood, or glass that goes with it. Obviously, our gear does nothing for the overpressure of the blast wave either. Although this blast wave may be weak compared to the high-power crack of an explosive like C4 or dynamite, it will be strong enough to lift you up and toss you into nearby cars, lawns, and buildings. As the old saying goes, it is not the flight, but the sudden stop at the end that hurts (figs. 12–3 and 12–4).

Fig. 12–3. A gas explosion completely leveled this home and sent large, firefighter-killing debris up to 150 ft.

Fig. 12–4. Very heavy parts of the building were rocketed across the lawn and landed in the street and beyond.

Procedures

The photo of the skydiver above contains an important lesson about procedures. Not only do they have a plan B (a reserve parachute), they have also had *procedures* drilled into their head for how to use it. The reserve is only used when there is some type of failure of the main canopy. For example, the main comes out but does not inflate properly to slow the descent. This is called a streamer and is a life-threatening

malfunction with deadly consequences. The skydiver must know and be able to implement the procedure that will keep them alive. This life-saving procedure is not just common sense, but is based on state-of-the-art tactics and an understanding of parachute science. We should be using the same mentality to respond to gas leaks. We must have procedures based on the experience and knowledge the gas industry has to offer, procedures very similar to what their employees utilize.

As you can see, having, understanding, and training to execute your procedures at a natural gas emergency are key to firefighter safety at gas emergencies. Simply, it is the responsibility of senior leadership to provide drivers, subordinate officers, and firefighters with solid procedures for responding to gas emergencies.

For discussion purposes, in this section, we will refer only to natural gas. Natural gas, mostly methane, is lighter than air and has an explosive range of 4%–15% gas in air. It has an ignition temperature of 900°F and is odorless. Ethyl mercaptan is added as an odorant before it reaches the consumer to aid in leak detection.

Policies

Standard operating procedures, guidelines, protocols, field guides—whatever you call them, they are made of two parts: policy and procedures. Let's look at policy first.

Policy is the general guidance for officers and members to follow. For example, a good gas response policy should say that the mission of your fire department is life safety. This provides the mindset for members that all on-scene actions will be in support of life safety.

If your SOP says you will also be expected to find the leak and secure it, causing the least service outage possible, you have other responsibilities in addition to life safety. The general OSHA standard clearly says you have to be trained to the level you are expected to perform at; therefore, training on finding and securing a gas leak must be provided for the department to be in compliance. To meet the goal of limited service interruption, you may have to take more time to figure out which meter serves which apartment before shutting it down. This time could result in further accumulation of gas and a reduction in life safety of both members and civilians.

Especially for single-family homes and small apartment buildings, shutting the gas off as soon as practical if there is a confirmed leak inside the building increases firefighter safety by eliminating sources of ignition (pilot lights) and stopping the accumulation of gas. This does not guarantee safety but the increases the odds in everyone's favor and helps get you closer to your mission of protecting lives, firefighters' lives included.

"The building will freeze, don't shut off the gas!" is what we hear so often. It will take a considerable amount of time for the building to cool down and freeze pipes. Even in the coldest climates and seasons that is better than getting blown up and if the gas company technician is responding, it is very likely they will be able to fix the problem in short order. If not, the situation was serious enough to warrant shutting the gas off to fulfil your life safety mission.

Safety policy. Your SOP should have a safety policy that goes something like this: if circumstances warrant or the officer in charge (OIC) has a suspicion of a hazardous condition, they are authorized to take whatever actions are necessary to protect civilian and firefighter lives and safety. Essentially this means if you think we need to evacuate or shut off the gas, you are authorized to do it. Another good choice of words for your SOP is: Any level or odor that is situationally concerning is grounds for shutting off the gas supply, tactical withdrawal of members, and evacuation of civilians. These statements are good, general policy guidance.

Kill box. Another important policy is to establish the concept of the kill box. This is similar to a hot zone in hazmat lingo. This is your estimate of the area around the building where members or civilians will be killed or injured by flying debris, glass, or significant parts of the building (fig. 12–5). Also consider the collapse zone, should there be a collapse (involved and nearby buildings) secondary to a gas explosion.

Fig. 12–5. Note the header board that flew across the lawn and into the street. If these firefighters had been in the kill box when the home exploded, they would have suffered serious injuries or worse from the heavy flying debris.

In your SOP, you can generally define the kill box as one to two houses away from the gas leak (or home reporting the leak) if you are in a suburban response area. In an urban setting, you may use half a block or another suitable distance. Rural fire departments may want to use a distance of 150–200 ft as a kill box estimate.

What should we do about the kill box? Be sure all firefighters are doing their size-up, estimating the kill box and keeping themselves out of it on every call. Minimizing the members in the kill box is important, at least until you have proven that there is not an improvised explosive building (a building that contains gas and ignition sources) on that street.

Many departments have different codes for leaks that trigger automatic actions by the dispatcher or responding chief officers. These may include setting up a command post, requesting additional units, requesting EMS to stand by, and other actions. These should be spelled out in your policy section. Mandatory action levels for meter readings can also be included.

Meters

Meter use at every gas call is essential for safe response to gas emergencies. Although humans can smell gas at very low levels, some of us cannot smell the odorant, some civilians confuse the odor with that of gasoline (or don't know the difference), and all of us will get used to it and have our olfactory sense overwhelmed to the point of not functioning.

Another critical reason to use a combustible gas indicator is that natural gas has no odor. The ethyl mercaptan that has been added to it may be scrubbed out as it passes through the soil from an underground leak. No matter how good your nose is, you will never smell it in this case.

Although we cannot review all the important points necessary to be fully versed in gas meter use, here are a few important tips:

1. Your department must establish an action level. If you are metering for any gas, you are going to get a value from your meter. It should not be up to the officer on the scene to determine what to do at certain levels. For gas leaks, one of the most frequent actions is to evacuate civilians. Check with your utility but a common action level for *mandatory* evacuation is 10% lower explosive limit (LEL). This is usually the low alarm setting for your meter. It may also be the action level used by your utility.

   a. It is critical to recall that this is a *mandatory* evacuation level. Obviously, good sense dictates that if there is an imminent or perceived danger, you and your members don't have to wait around for 10% LEL. Mandatory action levels provide firefighters who may not have a thorough understanding of gas emergencies a safety factor. They may not know why they are doing something but understand the actions required.

2. Low alarm, high alarm. Usually combustible gas indicators are set to both low and high alarm. Members must be familiar with these alarms and understand the appropriate action for each as laid out in your SOP.

3. LEL versus percent gas. Most gas company employees will speak in terms of percent gas in air. Most fire departments meters only measure up to 100% LEL. For natural gas (depending on the exact mix), the LEL is 4% gas in air. So if the gas technician says they have 4%, that means they measured 100% LEL and all that is needed is an ignition source for an explosion or fire. Ask the gas technician if they are reading percent gas in air or percent LEL!

4. Meter over 100% LEL reading. Since most of our meters cannot read over 100% LEL and most firefighters have never seen it at this dangerous level, most firefighters do not know how their meter indicates this very dangerous condition. Working with different instruments, we have learned they may read any of the following: O/L for over limit, OFF for off scale, 999 for 100% LEL and something as unhelpful as *** to tell the firefighter they are in a flammable or explosive atmosphere. None of these intuitively indicate the extreme danger.

Step by step

Paramedics who respond to life-and-death calls have step-by-step procedures. Gas utility workers who respond to gas leaks have clearly defined procedures. Fire departments that respond to gas emergencies that can quickly turn into life-and-death calls must have step-by-step procedures supported by sound policies.

Many SOPs say what not to do: don't ring the doorbell, don't turn on a light switch. It is much more important to know what to do! The paramedic protocols mentioned above tell what specific steps the medic is supposed to take for a particular diagnosis. The protocol is a process that guides the medic at this critical time. A gas emergency is a very critical time. It is the procedure that will determine if members minimize the threats to life and property, ensure success, and maximize member safety.

Those fire officers and firefighters who do not want to spend the mental effort to develop or train on these procedures will say there are too many variables, that we go to gas leaks all the time and nothing ever happens. An alternate description for this dangerous mindset is: are we experienced

and good at what we do, or are we just lucky at those calls and got away with poor procedures?

There are five major types of gas emergencies you will be called to:

1. Inside odor of gas
2. Outside odor of gas
3. Gas explosion
4. Locked building with gas leak
5. Gas main struck (with or without fire)

Let's take these scenarios one at a time and look at possible procedures for a safe response.

Scenarios

Inside odor of gas. What is your first thought when you hear dispatch giving you information about an inside odor of gas? "Here we go again, an unlit pilot light, broken flex line, it is always BS!" We have all been on far too many of these calls. What you should be thinking is this: What is the worst case scenario? Is there an outside, underground gas leak that has migrated into the house, causing the odor? If there is, this house could be loaded with gas in the basement where water and sewer pipes, as well as electric and communication lines, enter the house. All these are conduits for gas migration into the house.

Planning for the worst, you park your rigs one to two houses away from the address of the caller, just in case things go bad. You want to be part of the solution, not part of the problem or list of victims. An officer and one firefighter with your combustible gas indicator go to the building. As the officer interviews the caller for additional information, the firefighter takes readings at or near the front door of the house. If conditions warrant it or the action level has been reached by your meter, you can evacuate both civilians and firefighters right now. If you get no or very low readings and your department polices state that you can enter the building to find the source of the leak, do so. Many departments shut off the gas at the meter and leave the detective work to the utility technician who is responding.

The questions the officer asks the occupant will help you focus on where the leak is. The answers the officer gets will have a huge bearing on your actions. For example, if the occupant says the odor is only in the kitchen, the leak is likely the flex line to the stove or an unignited pilot light or partially open gas valve. If the answer is something like, "It's all over the house" or "I really can't pinpoint it," your suspicions should be that you have a bigger problem and hazard. One of your biggest concerns at this call is whether the leaking gas from outside made it into the basement or utility room of this building, or if the leak is in the basement permeating the house with the gas and its odor. A few meter readings in the basement will help rule out this hazard.

Of course the gas odor could be coming from another apartment, another appliance, or even outside. Be thorough when you are determining the source.

Outside odor of gas. This is another of the calls that we have all been to a thousand times. We commonly don't see this as a hazard, because we erroneously think all the gas is going up harmlessly into the air. Sure, some of it is going up, but what if it is coming from a cracked or broken 2 in. gas main operating at about 60 psi gas pressure? All that gas will take the path of least resistance, which may lead into nearby buildings. So here are some considerations for your response procedures for outside odors of gas.

As always, park your apparatus a safe distance from the reported site. The first-in engine should secure and test a hydrant. Usually a house or two away is fine for your initial kill box estimate, or your SOP may state a certain number of feet.

Now it is time to be a detective. If possible, interview the caller and determine where they smell it. Air monitor the nearest building to see if migrating gas entered this building. Also monitor the next nearest buildings on all four sides of the odor source. If gas has entered the buildings, evacuate them and monitor for gas using the indoor protocols: outside the door first, then slowly making your way inside and to the basement or where utilities enter the building.

Since this leak is underground, you can't see its path or directly monitor its location, but you still have to determine how far it has gone. Subsurface structures such as storm drains, sanitary sewers, and electric and communication manholes are ready-made openings into the street where gas may be accumulating. Openings in the covers of these manholes will allow gas to escape and are a good place for you to put your meter probe. It is vital to increase the size of your kill box if your meter readings indicate a hazard.

Another helpful tool for finding the underground leak source is to ask the utility (through your dispatcher) if they are aware of any existing leaks in the area of the complaint. You still need to go through your procedures but with this information you will have a better idea of where to start.

Utilities often let gas leak underground for months or years. Underground gas leaks are rated by class. Class 1 leaks are near buildings, present an immediate hazard, and must be repaired immediately for obvious reasons. Class 2 and 3 leaks are away from structures and deemed a minimal hazard. They may last for many months, until the utility has time, money, and manpower to repair them.

Gas explosion. You are dispatched and respond to a reported explosion. As you turn down the street, you see debris strewn wildly and widely around the reported address. It is immediately clear that at least one building has indeed suffered an explosion (fig. 12–6).

Fig. 12–6. Note the wide dispersion of large debris. Gas seeping from underground continues to burn as firefighters extinguish the debris fire.

One of your first actions is to confirm that the explosion was caused by natural gas. In this case, a utility worker meets you and claims they were investigating a gas leak. If you cannot confirm with a reasonable degree of certainty that natural gas was the cause, keep an open mind and watch for and actively seek out clues related to terrorism, hate crimes, suicide, or criminal actions. Assuming you rule out nefarious activity soon after arrival, there are several concerns that should enter your mind. First, how big is the kill box? Of course you need to consider civilian life safety but if you send in your members immediately and another building explodes or collapses, dead rescuers will save no one.

As in earlier examples, your park apparatus a safe distance away if possible. Conduct your size-up and get any walking wounded to medical attention. Meet with the gas technicians as soon as possible at the scene and ask their help in the initial estimation of the kill box.

Gathering information about any civilians injured or trapped in the rubble is a priority so you can set up SAR and EMS sectors. Life safety is our top priority. It may be necessary to risk firefighters (within reason) for SAR activities early on.

As in the other scenarios above, one of your main concerns is: If there is an underground pipe leak, where is the gas going? This is a critical concern and you will need to call enough personnel and equipment to conduct the SAR operation and, as soon as possible, check surrounding buildings for infiltrating gas. Utility technicians are well trained at this task and will be a tremendous help to you. Be sure you have a unified command post with utility supervisors. The results of these investigations, interpreted with the help and expertise of utility personnel, will determine if you need to increase or can decrease the size of your initial kill box.

There are a lot of command and control concerns after a building explodes, too many to mention here. The key factor to remember when dealing with gas explosions is that you should be closely monitoring the nearby buildings to make sure they do not join the fray and compound your problems.

Locked building. We call these improvised explosive buildings (IEBs). If they contain gas, they also contain ignition sources such as pilot lights, thermostats, telephones, or other electric items that could make a spark and set off an explosion. They are a bomb prepared to go off if there is enough gas inside, and you have no way of knowing how much gas is inside.

If a police officer came up to you and said, "I looked in the window and I think I saw a bomb on the living room floor, would you go in and look," what would you tell them? Hopefully a very emphatic no! A building that contains gas and ignition sources is a exactly that: a big bomb with a large kill box.

Often, the utility will ask the fire department to force open locked doors to buildings that may contain dangerous amounts of gas. The utility technicians want to get into the locked buildings both to check for gas migration from an underground leak and to vent the structure before it builds up enough gas to be a hazard and fill the kill box with dangerous firefighter-killing debris. This is a noble but very dangerous mission. Although you get paid to take manageable risks, this is a loser. Your mission is life safety. Before you commit firefighters' lives to this building, you need to determine that there is a reasonable likelihood of salvageable human life inside. Question neighbors, look around the home for signs of nonoccupancy, such as a for-sale sign or no curtains on the windows. We are not saying *never* take a risk, especially for a life hazard; just take a *sensible* risk. Never forget that the fire officer you will send in does not have a plan B and has no personal protective equipment rated for an exploding building.

Mitigate the risk. One tactic you can do to determine the risk level is to monitor the air outside the locked building at a window or door (fig. 12–7). Of course this assumes some risk as well since firefighters are still inside the kill box, but sometimes it is a reasonable risk in order to better determine safety. Place your air monitor at the crack in a door or window from the outside (fig. 12–8). If you get any reading of gas you should suspect an IEB and retreat immediately, especially if no civilian lives are involved. For this particular situation you should check the safety policy and procedures for the utility that services your area.

Fig. 12–7. A firefighter samples for natural gas seeping out of a window of a locked building. The tool helps open the window so the gas monitor can get a good sample and reading.

Fig. 12–8. A firefighter demonstrates the correct placement of a meter to monitor for gas behind a locked door.

There is another important preresponse phase to this operation. Get a copy of your utility companies' procedures for locked buildings that contain gas. You may be surprised to find that they rate this situation as very dangerous and mandate several actions to make the building safe before anyone enters. These actions include shutting off the gas from the outside meter and shutting down the electric service if possible. Some utility company procedures state that no one should enter the building until the fire department determines it is safe. If your utility is uncooperative and will not share their procedures, use your local freedom of information laws to obtain it. The agency that regulates the utility in your state will provide it to you because it is a public document.

Firefighter safety should be considered if there is a life hazard and you have determined you need to send members inside. Since forcing the front door may cause sparks, consider sending one of your officers to pry up or break the glass in a window. This will not cause a spark and may be considerably safer, though still a very dangerous operation.

If there is no life hazard to civilians, don't reinsert the life hazard by putting firefighters in harm's way. Withdraw well out of the kill box and stand by while the utility makes the situation safe by shutting off the gas, reducing ignition sources, and waiting until the building vents down on its own. That is their mission. This is not an emergency; time is on your side. If the building blows up and no one is injured or killed, it was still a successful operation. Consider the alternative: you risk firefighters, they are not lucky and the building explodes, and we suffer injuries to our members or worse, an LODD.

We don't have to fix every situation we are called to! It is important to remember that there are some situations that are not the fire department's responsibility to fix. We do not and should not become involved in disarming improvised explosive devices (IEDs) such as pipe bombs. Why would we become involved in disarming IEBs?

Gas main struck. The utilities call this third-party damage. Nearby homes can become easily ignitable exposures at this type of call. It is the most common type of gas leak and damage to gas mains and service lines (gas pipes from street mains to buildings). Gas mains are struck by contractors of various natures. It may be a contractor repairing a water or sewer pipe, a communications company pulling new lines into buildings, or any contractor with a reason for excavating. Sometimes it is a homeowner putting in a fence or doing some landscaping.

Gas mains and service lines, especially at or near homes, that were originally buried in compliance with the code in your area may become more susceptible to damage when landscaping, driveway installation, drainage work, or other construction removes the soil covering the gas main. It is not unusual to find gas mains at relatively shallower depths than we would expect. Of course, there is always the case of the shoddy contractor on a new installation who hits bedrock and does not bury the main to a safe depth then covers it before enforcement officials see it. Chipping out solid rock is costly.

Typically what you will see upon arrival at the scene is some type of excavation equipment and gas leaking out of the pipe straight up into the air. To the layman, this doesn't appear to be a problem, as all the gas is being vented harmlessly into the air. But we know that there are two situations that pose a significant hazard at this call.

First, if soil, boulders, or other debris are covering the leaking gas, some of it could be following the path of least resistance and flowing back into nearby basements. As we

discussed above, it is important to check nearby structures for gas accumulation. Second, if the excavator's bucket got under the gas pipe and pulled it, the pipe connections in nearby buildings could be broken and gas could be leaking into the buildings and into the excavation hole.

The excavator, nearby vehicles, or even a static spark could ignite the gas flowing outside, which will cause exposure problems. Use your normal exposure-protection methods and cover exposed buildings with water to prevent their ignition due to radiant heat. If possible, try to keep water out of the excavation hole since gas technicians may have to enter the hole to pinch off the pipe and stop the leak.

As with all types of flammable gases we do not want to extinguish the fire unless we can shut off the flow of gas.

The following article summarizes real-world experiences from a fire department that has learned valuable lessons from deadly gas calls and implemented excellent procedures.

FIREFIGHTER RESPONSE TO NATURAL GAS LEAK

By James R. Kutz

Photos by Dennis Wetherhold

Photo 1. The scene of a February 2011 natural gas explosion at 13th and Allen streets in Allentown, Pennsylvania. Five people were killed.

Natural gas leaks and explosions are growing problems across the country; aging infrastructure is one of the reasons for this growth. The Northeast region of the United States has the oldest infrastructure in the country. These gas pipelines have naturally degraded over a long time and desperately need to be replaced. In light of the current recession and crippled economy, it is unlikely this problem will be fixed in the near future. Even if the money were available, it likely would take many years to dig up and replace all the old gas lines across the nation. There are 2.5 million miles of pipelines nationwide; in

Pennsylvania alone, there are 46,000 miles of natural gas pipelines, which would go around the circumference of the earth twice.

The city of Allentown, Pennsylvania, is no stranger to gas leaks and explosions. Allentown, 60 miles north of Philadelphia, is a third-class city with 118,000 residents and the third largest city in the commonwealth. It is in the center of the Lehigh Valley, home to more than 821,000 people. The Allentown Fire Department has six stations that include seven engine companies, two ladder companies, and a battalion chief.

I have worked for the Allentown Fire Department for 10 years and am assigned as the lieutenant of Engine Company 6. Previously, I was a volunteer at a neighboring township and worked part time for another combination fire department in the Lehigh Valley. As a volunteer and part-time paid firefighter, we responded frequently to gas odor calls; they generally were not taken very seriously and were considered to be nuisance calls.

When I was hired for the city in 2001, I immediately noticed a tremendous difference in the way gas calls were handled here. As a probationary firefighter, I wanted to know why. I started asking my instructors and some of the senior officers why we handled gas calls so differently from the way I saw them handled elsewhere. It seemed each of them had a personal story about a structure or multiple structures blowing up from natural gas. Each story they shared resulted either in complete destruction of the structure, someone being blown across the street, people dying, or even the deaths of two of our brother firefighters.

AREA EXPLOSIONS

On August 8, 1976, in the 1100 block of Oak Street, a home exploded. When engine companies arrived, they found one house completely leveled and began to fight the resulting fire. While operating on the scene, the home across the street exploded, trapping two of our firefighters in the debris and resulting fire; Lt. John McGinley and Firefighter William Berger succumbed to their injuries. A memorial plaque for these brothers hangs by my office and is a constant reminder of the dangers we face every day specifically from gas-related disasters.

Between the years of 1925 and 1976, there were two significant gas explosions in Allentown that killed 10 people—including the two Allentown firefighters previously mentioned—and injured 24 other people. Since 1976, there have been nine gas explosions in

Allentown, killing 10 people and injuring 110 more. In fact, the most recent explosion occurred while I was in the middle of writing this article.

Shortly after this explosion, my crew and I responded to a gas main that had been struck directly in front of a high-rise that housed the elderly and a daycare center. It clearly would have been a disaster if it had exploded. Our meters were detecting 50 percent of the lower explosive limit (LEL) standing across the street, and an odor of natural gas was apparent from blocks away. We safely evacuated the day care and moved 150 residents of the high-rise to the far side of the building.

Since graduating from the Allentown Fire Academy in 2001, I have responded to two gas explosions, both of which were significant and made national news. The first was on December 9, 2006; it left three homes leveled and a fourth house burning because of the large fire from the explosion. The second took place on February 9, 2011, resulting in the deaths of five civilians, six injuries, the evacuation of 500 people, the instant demolition of two houses from the blast, the destruction of six neighboring houses by a fire directly fed by the gas main that couldn't be shut off, and secondary damage to 47 nearby structures. During this incident, numerous pockets of gas-fed fire broke through the ground where the mains had been ruptured.

Photo 2. A photo of the Mohawk Street gas explosion in December 2006.

As horrible as these incidents are, I realize how fortunate I have been at the countless calls where natural gas readings were taken and were within the explosive range. Thankfully, none of those incidents resulted in an explosion after our arrival, resolutions I attribute to the quick response time of the fire department and the utility company, the serious treatment of every call, and the protocols and procedures in place that were precisely followed.

THE DANGER OF COMPLACENCY

I do not consider myself an expert on natural gas emergencies, but many of you may be like I once was—naïve and complacent in my approach to gas odor and gas leak calls. I'm compelled to share my experiences since the rest of the country will soon be where we are now. The problems our region is experiencing are the problems other areas will face; unfortunately, more people will probably die, be injured, and lose their homes in the process. I hope that more of us are prepared and preparing for how to respond to gas explosions and, more importantly, how to prevent a great deal more of them from occurring.

Allentown is relatively small compared to cities like Philadelphia, New York, and Boston, but to put the growing risk from natural gas into perspective, in 2009 AFD responded to 12,206 calls of service: 76 of them were for gas odors, 68 were for leaks, 14 of them were significant enough that people had to be evacuated or the readings were in the explosive range. In 2010, we responded to 11,394 calls: 206 of them for gas odors, 44 for leaks, nine were significant. In 2011, there were 11,206 calls: 292 for gas odors, 125 for leaks, and 49 of them significant. This is an astounding 250 percent increase from 2009 to 2011, and a 444 percent increase in significant leaks just from 2010 to 2011.

If you were like I once was, unaware of or inexperienced with natural gas emergencies, would you know how to handle these incidents properly? The first thing is to have a proper attitude. Don't be complacent. I am more on edge on a gas call than just about any other type of incident. The second would be to read up on natural gas fires and emergencies. The third would be to contact your local gas utility company. Many of our standard operating procedures have come from the proactive training they offer us. The fourth would be to contact other fire departments in your region that have more experience with these types of incidents and see what they are doing.

AN EXAMPLE OF A STANDARD NATURAL GAS RESPONSE

I will share briefly some of the things our department does routinely on gas odor calls:

When our 911 communication center receives a call for a gas odor or leak, inside or outside, it dispatches a full box, consisting of a battalion chief, one truck company, and three engines—bringing 13–15 members, depending on staffing that day. The dispatcher then notifies the utility company to respond.

The first-arriving units shut off traffic flow by blocking all intersections and lanes of traffic to the entire square block surrounding the reported leak. This is to protect the apparatus and personnel, as well as civilians traveling on those roads from being in front of or near a potential explosion, or near a hydrant in case an explosion does occur. At a minimum, the apparatus should be clear of the collapse zone and far enough away so that it doesn't become an ignition source to any leaking gas. Also, be sure not to park over manhole covers because if the leak made its way into the sewer system and an explosion occurs, this manhole could damage the underside of the apparatus or injure the operator.

Next, the responding firefighters walk to the reported incident in full turnout gear and self-contained breathing apparatus (SCBA), carrying natural gas meters and forcible entry tools, at a minimum. The meters we use as those used by the gas company, which calibrates them for us once a month to keep them in working order. These meters first read in percent-LEL; once the readings reach 100 percent LEL, they switch to percent gas in the air—the explosive range being 5 percent to 15 percent gas in the air.

We approach on the same side of the street as the reported address. If an explosion occurs, you tend to be more protected if you are on the same side of the street, since the explosion normally pushes straight out, across the street.

The firefighters begin taking samples as soon as they get off the apparatus, working their way toward the reported leak, trying to locate the source of the leak.

While monitoring for gas, we give very specific directives to bystanders and occupants: We direct them to extinguish any possible ignition sources such as cigarettes, running cars, an appliance that has an open flame, or anything capable of producing enough heat to ignite natural gas. We also direct them not to touch anything such as light switches, plastic supply pipe or door bells. Sometimes, we even ask them to put their hands in their pockets to keep them from forgetting. For ourselves, we try to inhibit static electricity by not dragging our feet and using all intrinsically safe equipment.

Readings are taken in and around the originating structure and exposures on either side of the structure, if they are attached. We also check both sides of the street to make sure the leaking gas doesn't follow water or sewer lines into the other homes.

If we detect any natural gas with our meters, we continue to check and evacuate until we have two clean buildings. Our gas utility company asked us to do this because in the event of an explosion, it has the potential to take the two adjacent structures with it. I have seen this happen at both explosions to which I responded.

Meanwhile, we evacuate the buildings and shut off the gas at the appliances, delivery meters, or curb shut-offs; our utility company allows us to accomplish this by supplying every company with a curb key.

Finally, we secure the scene, ventilate if safe to do so, and wait for the arrival of the gas utility company to fix the problem.

Although I hope this article serves as a reminder or a wake-up call to some of you, it is also important to keep in mind that even though we may do our best to prevent occupants and ourselves from creating and extinguishing ignition sources, we can't control everything all at once. Don't forget to operate safely, don't become complacent, prepare for the worst, and hope for the best—and as my utility company instructor told me, "If you see the gas man running, RUN FASTER!"

This article is dedicated to Lt. John McGinley and Firefighter William Berger who died in the line of duty August 8, 1976, while operating at a gas explosion.

James Kutz is a 10-year career firefighter and lieutenant with city of Allentown, Pennsylvania. He is assigned to Engine 6 at the Hibernia Fire Station. James is also a suppression instructor for the Commonwealth of Pennsylvania and assists with the Allentown Fire Academy.

Lightning Strikes

We are including in this book on house fires a discussion of lightning strikes. The increasing amount of electronics makes the modern home more vulnerable to disaster from lightning and other electrical surges. Electrical circuit boards are not limited to computers and televisions; they are in most major appliances.

We are going to provide some background as well as some protective measures that can be put into place to reduce the effects of an electrical incident. We will be looking at the standard electrical system for a residential building: a two-phase, 240/120 volt alternating current (AC) system.

There are several codes and standards that electricians must comply with, and local ordinances may supersede national code. Firefighters and fire officers should be familiar with these regulations. Common codes include the National Electrical Code (NEC) for the United States or the Canadian Electrical Code (CEC) for Canada; the lightning protection code, NFPA 780, for installing lightning protection systems; and eight Underwriters Laboratories (UL) standards.

Data from State Farm Insurance in 2013 show that there were 151,000 claims related to lightning strikes (NLSI 2014). Lightning causes more than 26,000 fires, with damage to property in excess of $8–10 billion annually in the United States (NLSI 2014). According to the executive summary of study from the NFPA,

> During 2007–2011, U.S. local fire departments responded to an estimated average of 22,600 fires per year that were started by lightning. These fires caused an average of nine civilian deaths, 53 civilian injuries, and $451 million in direct property damage per year. Most of these fires occurred outdoors, but most associated deaths, injuries, and property damage were associated with home fires. Fires started by lightning peak in the summer months and in the later afternoon and early evening....

> Over the 10 years from 2003–2012, 42 U.S. firefighters were killed as a result of lightning-caused fires. These deaths include fatalities during fireground activities, as well as responding or returning to fires. Four of these deaths occurred at structure fires, and the remaining 38 were killed as the result of wildland fires. Eleven of these deaths occurred in helicopter crashes.

> In addition to causing fires, lightning is dangerous on its own. Data from the National Weather Service show that in 2008–2012, an average of 29 people per year died as a result of lightning strikes. The most common location for these deaths was outside or in an open area. The average number of lightning flashes per square mile varies considerably by state, as does the death rate from lightning incidents. (Ahrens 2013, i)

There is a misconception about lightning that is important to clarify here. The current induced by a lightning strike does *not* take solely "the path of least resistance," as popularly believed. Lightning currents take *all* paths to ground. This includes all metallic piping whether it is copper waterlines, cast-iron sewer and drain lines, or black-iron gas piping.

Leaking gas from an electrically charged pipe can be the cause of a fire.

To get an understanding of the number of paths to ground, we need to understand the amount of copper wire found in a typical home today. The industry estimates that there is one linear foot of general purpose wiring (14-gauge, two-conductor wire) for every square foot of floor area. Then add another 25% of the linear figure for specialty wiring the heavy duty (appliance) circuits that require 12-gauge and three-conductor wire for smoke detectors or the like. This figure will of course change based on where in the country you live and whether appliances like cooktops, ovens, and clothes dryers are electric. Add again to that figure the amount of low-voltage wire in a standard home: RG6 coaxial cable for television; Cat5 for computers, home entertainment systems, and other networked devices; landline telephone wiring; and security system control wiring. The constantly developing technology of home automation systems requires devices and interfaces that are susceptible to electronic pulses. An understanding of these systems and their "failsafe" features—do they fail open or closed—can affect security and life safety. As the technology improves, more of these systems are becoming wireless. However, the existing wire is not removed from inside the wall.

The concern with lightning strikes is that as each length of wire is energized with current levels in excess of 400 kilovolt-amperes (kVA) and temperatures of 50,000°F, each section of wire becomes a burning fuse throughout the building. Controlling a fire under these conditions is a very real challenge. The craftsmanship of the electricians and the diligence of the inspectors to ensure that floor-to-floor penetrations are properly sealed will be a key to a successful outcome.

Though lightning most often hits overhead or elevated structures, trees, flagpoles, and signs, these are not the only areas threatened by lightning. The building does not have to be struck directly. A ground strike of lightning can affect underground conduits and feed the electric pulse into a structure through underground utilities. Electronic dog fences use a low-power signal wire buried in the ground surrounding the containment area. This wire can become a lightning antenna and transmit the surge to the AC-powered transmitter. The electromagnetic pulse from a strike near a structure can be picked up by the wiring system and cause damage to electronic equipment.

There are several commercially available surge protectors for residential use. They are grouped based on the amount of energy, or surge, you wish to protect against. These devices are only designed to prevent damage to electronic equipment, not to protect against fires. That being said, properly grounding the surge does indirectly protect against fires. These protection systems fall into four broad categories:

1. Lightning protection systems (LPSs) are the most complex installations. They use lightning rods, conductors that bond to the metallic parts of the building, such as gutter or metal roof panels, and connect to ground electrodes.

2. Surge protectors for AC wiring. These are commonly tied into the circuit breaker panel and provide a level of protection for the whole house.

3. Low-voltage systems, also referred to as signal wiring, include cable TV, phone, and central station monitors for alarm systems. These are required to separately protect their proprietary equipment but may be tied in to other systems as well.

4. Point of use (POU) devices. These are installed in the power supply at the point closest to the device they are intended to protect. They may be in the form of multi-outlet plug strips or plug-in models that connect directly to the outlet. These surge protectors can be inexpensive devices with lower ratings based on the assumption that the energy will be dissipated through other paths before reaching the protected device. This could create a false sense of security about the level of protection provided.

We have included all the details on the lightning protection systems to provide you with a basis for determining if a building has been struck by lightning or suffered an electronic pulse event. This type of call may be dispatched as "an odor of something burning" following a thunderstorm or on a hot humid day when "dry lightning" (also known as heat lightning) is predicted.

We spoke of circuit boards and their vulnerability to electrical spikes. Fan motors can be vulnerable as well, especially if they were operating when the spike occurred. Fan motors on ventilation systems, whether for heat or cooling, can spread an odor throughout the building. An understanding of HVAC systems would serve you well in tracking down the sources of "smells-and-bells" calls. This is not to suggest that the firefighter's role at these nonfire emergencies is that of service technician; however, if you need to turn a circuit breaker off, it is important to know that you have identified and de-energized the proper device.

Case History

Let's look at a case study of a home struck by lightning. The experience is relayed to us by a good friend, Chief Tim Pillsworth. Tim is a past chief at Winona Lake Engine Company #2 in Orange County (NY), is a frequent contributor to *Fire Engineering*, and teaches with us throughout the country.

Over the years, all fire departments have become trained in responding to general hazard alarms. In most cases, these are weather-related emergencies ranging from blizzards to tropical storms and everything else in between. This does not take away from our primary response (structural fire-fighting); in fact, in many cases, it pushes us right into it.

We have all responded to a house struck by lightning call and found a home that has not taken a direct hit but has damage to its electrical system and sometimes even fire. My experience was a little more interesting.

I was home on the phone with one of the other chiefs talking about the family picnic and whether to delay or cancel it due to the weather. We hung up and I went to check the radar on the computer to see how long the storms were supposed to last, and that's when it happened: a very large flash of light and at the same time a boom like I have never felt. Sparks shot out of the outlet 18 in. to my arm and hit my hand and watch, stopping the watch. Everything in the house was popping and buzzing, light bulbs blew out, and light smoke was coming from some of the kitchen appliances that were plugged in. Right then I knew we'd been hit by lightning. I was thankful that my wife and son were out running some errands.

I went downstairs to kill the main electric breaker and that was when the real problem showed: heavy smoke was coming from the laundry room. My house was on fire. Running through the garage to my chief's truck, I called it in, and luckily was able to knock down the bulk of the fire with an extinguisher.

After the fire was knocked down, I went outside to inspect my home and expected to see a hole in the roof, the wall, somewhere. But there wasn't one. My house had not taken a direct hit. We found the strike location later, about 30 ft behind the house. The electrical charge traveled through the ground and up into the house through the dryer duct. This ignited the duct and everything around it in the ceiling.

Through the investigation from the fire inspector, from the electrician doing the repairs, and from my own searching, I learned that there was damage to many electric, telephone, and cable-operated devices within the home. The current from the lightning entered my home in an area that had power, telephone, and cable wires, and traveled throughout the entire home. Later in the day, as we were cleaning up and waiting for the insurance agent to show, I noticed that a phone in the bedroom on the other end of the house had some smoke damage on the body of it. The electricity had been strong enough to burn up a phone in the other end of the house.

What can we learn from this? Many general hazard alarms turn into fires. We are all trained to go to the seat of the fire and, because of the home layout, smoke conditions, and so on, we might not check other areas of the home as closely or often. But in the case of a lighting strike, there could be issues in more than one location in the home. As soon as the main fire is contained, the entire home must be checked. A room-by-room check with a thermal imaging camera can locate possible hidden fires and problems that, if not found, will bring us right back in the near future. While the small burned up phone in our bedroom did no damage to anything, it could have.

Don't let your guard down when responding to general hazard alarms. Many times they can be more challenging than expected. Ask the homeowner what works and does not work in the home, size up conditions both inside and outside (maybe the home or yard is always that messy), and take the time to check all areas. You do not want to miss something and have the dreaded call back.

Leadership: The Most Common Call

If you have reached this point in the book, you have learned many things and gained insight into many problems encountered at modern house fires. These concepts were presented to inspire a thoughtful discussion and robust debate to develop creative solutions to the problems in your own department.

We would be remiss if at this point we did not discuss what you should do with this knowledge. The fire service is built on sharing information and handing down best practices. That was part of the motivation behind this book. We have learned many things in many ways, and some of what we learned was painful both physically and emotionally.

So who do you share this information with? Our answer is anyone who will listen. There are those few in the fire service who can't be told anything; because they are the chief they must be the smartest person on the planet! We have always professed that we will never say, "I told you so," but should always be able to say, with a clear conscience, "We did tell you."

Getting people to embrace a new concept can be challenging in a culture steeped in tradition, but the fire service has to evolve. A key to creating change is identifying those personnel who have vision and are willing to hear the evidence and make an informed decision.

Don't be hamstrung by a person's rank. There are many brilliant firefighters and company officers. Everyone did something before they became a firefighter; that gives them very diverse skillsets that can be tapped and used to better the service. Find individuals with those qualities who can support your cause: the cause of finding better solutions to the issues facing your department.

Leadership is personal. An experienced leader knows to be surrounded by talent. Don't surround yourself with boot-lickers who agree with everything you say; you will lose sight of the big picture. Dissent generates debate and can allow you to identify and solve problems early in the process.

If you are a subordinate identified by a superior, embrace the position. Whoever has identified you has done so for a reason. Maybe you are getting the attention because you are a motivated individual who has the ability to challenge the status quo, or maybe you possess the unique skillset needed for a particular project. The best advice we can give to you is never say no to an opportunity. There are a few reasons you should strongly consider agreeing to the request. One, you will have an experience that potentially no one else in your organization will have, which makes you more valuable to the organization as a whole. Second, as an agent of change in your organization, having the ear of the leadership is valuable and being acknowledged for your talent is priceless.

If your organization is like ours, 10% of the members do 90% of the work. So don't be surprised if you are asked to work on multiple programs and projects once the administration sees your work.

Get involved

Find your niche in the department; there are many jobs filled by unsung heroes who make the department what it is. Offline and administrative support functions are areas where expertise is valuable. FDNY's Bureau of Training functions with a small cadre of offline staff for program administration. Many of the instructors come from field units and are seasoned fire and EMS officers with the reputation to speak with authority.

Research and development is another offline position that has a great impact on your department. Have you ever said, "Who bought this piece of junk?" Without input from the end users in research and development, the department is most likely going to purchase what *they* think you need.

The FDNY has a family assistance unit, which has an impact on how the public as well as the job is perceived. After all, doesn't the fire service have a responsibly to care for its own?

Everyone enters the fire service for their own personal reasons, but most want to make a contribution and be remembered for it. It's your legacy—what will it say?

Train your replacement

Another key function of leadership is planning for the future. What will your department look like in 20 or 30 years?

Who will be in charge? What experiences will they have? Experience is the hardest thing to pass on in the fire service, but is your experience going to serve your replacement well? Look back at fire service magazines from years ago: Where are the bunker gear and SCBAs? If you told those guys that one day we would have a thermal imaging camera that can "see" through smoke, they would have said you were crazy.

Basic skills remain unchanged, but knowing how to apply them to changing circumstances should be part of training. Teach trainees to think outside the box. Don't give them a fish; teach them to fish.

Be ready to step into a new role within the department

From the day you walk into the firehouse you should be ready to accept new challenges. You already have—you're a firefighter. Those challenges will come as you gain experience, seniority, or promotion. You should want to be the best you can be, to know your job and the jobs of those around you. Someday you may need to step into a new role, either through battlefield commission or some other mechanism, so be ready. Keep a list of contacts close in case you need to make a call and ask for some advice. Everyone has a first day in a new assignment, but with a good foundation of knowledge, you will get through it.

References

Ahrens, Marty. 2013. *Lightning Fires and Lightning Strikes*. Quincy, MA: National Fire Protection Association.

American National Standards Institute and Underwriters Laboratories. 2005. *ANSI/UL 2034-2005: Standard for Single and Multiple Station Carbon Monoxide Alarms*. Northbrook, IL: Underwriters Laboratories.

Code of Federal Regulations, title 49, part 192, subpart L, section 192.615: "Emergency plans" (2009). https://www.law.cornell.edu/cfr/text/49/192.615.

Kutz, James R. 2012. "Firefighter Response to Natural Gas Leaks and Emergencies." FireEngineering.com (blog), May 15. http://www.fireengineering.com/articles/2012/05/firefighter-response-to-natural-gas-leaks-and-emergencies.html.

National Lightning Safety Institute. 2014. "Lightning Costs and Losses from Attributed Sources." http://lightningsafety.com/nlsi_lls/ListofLosses14.pdf.

Bibliography

Ahrens, Marty. 2010. *Home Structure Fires*. Quincy, MA: National Fire Protection Association.

———. 2013. *Lightning Fires and Lightning Strikes*. Quincy, MA: National Fire Protection Association.

———. 2014. *Characteristics of Home Fire Victims*. NFPA No. USS01. Quincy, MA: National Fire Protection Association. http://www.nfpa.org/news-and-research/fire-statistics-and-reports/ fire-statistics/demographics-and-victim-patterns/characteristics-of-home-fire-victims.

———. 2016. *Home Structure Fires*. Quincy, MA: National Fire Protection Association. http://www.nfpa. org/news-and-research/fire-statistics-and-reports/fire-statistics/fires-by-property-type/ residential/home-structure-fires.

American National Standards Institute and Underwriters Laboratories. 2005. *ANSI/UL 2034-2005: Standard for Single and Multiple Station Carbon Monoxide Alarms*. Northbrook, IL: Underwriters Laboratories.

Anderson, Leroy. 1975. *Wood-Frame House Construction*. 2nd rev. ed. Washington, DC: Government Printing Office.

Avillo, Anthony, 2008. *Fireground Strategies*. Tulsa: PennWell.

———. 2009. *Fireground Strategies*. 2nd ed. Tulsa: PennWell Corporation.

———. 2015. *Fireground Strategies*. 3rd ed. Tulsa: Pennwell Corporation.

Berry, Joseph. 2016. "Outside/In Fires: Two Incidents, Two Outcomes." *Fire Engineering* 169 (6). http:// www.fireengineering.com/articles/print/volume-169/issue-6/features/outside-in-fires-two -incidents-two-outcomes.html.

Bossert, Lisa. 2007. *Fire Protection Study: Pine Knoll Townhome Fire*. Presented to the City of Raleigh. http://raleighnc.gov/content/Fire/Documents/Fire%20Prevention/Townhouse_Fire_Full_Report.pdf.

Bowers, Richie, Corey Parker, Jennie Collins, Bill McGann, Justin Green, and Greg Moore. 2008. *Significant Injury Investigative Report: 43238 Meadowood Court, May 25, 2008*. Leesburg, VA: Loudoun County Department of Fire, Rescue, and Emergency Management.

Brannigan, Francis. 1997. *Building Construction for the Fire Service*. 4th ed. Quincy, MA: National Fire Protection Association.

Bricault, Michael. 2006. "Rescue Profiles for Residential Occupancies." *Fire Engineering* 159 (9). http:// www.fireengineering.com/articles/print/volume-159/issue-9/features/rescue-profiles-for -residential-occupancies.html.

Buchanan, Eddie. 2013. "Rethinking RECEO VS: Breaking Up with an Old Friend." FireEngineering.com (blog), 5 November. http://www.fireengineering.com/articles/2013/11/rethinking-receo-vs -breaking-up-with-an-old-friend.html.

Bukowski, Richard. 1995. "Modeling a Backdraft: The Fire at 62 Watts Street." *National Fire Protection Association Journal* 89 (6): 85–89.

Code of Federal Regulations, title 49, part 192, subpart L, section 192.615: "Emergency plans" (2009). https://www.law.cornell.edu/cfr/text/49/192.615.

Coleman, John "Skip." 2011. *Searching Smarter*. Tulsa: PennWell Corporation.

Comella, Jay. 2003. "Planning a Hose and Nozzle System for Effective Operations." *Fire Engineering* 156 (4). http://www.fireengineering.com/articles/print/volume-156/issue-4/features/planning-a -hose-and-nozzle-system-for-effective-operations.html.

Corbett, Glenn, ed. 2013. *Fire Engineering's Handbook for Firefighter I and II*. Updated ed. Tulsa: PennWell Corporation.

Cote, Arthur, ed. 2008. *Fire Protection Handbook.* 2 vols. 20th ed. Quincy, MA: National Fire Protection Association.

Crapo, Robert, and Noel Nellis, eds. 1981. *Management of Smoke-Inhalation Injuries.* Salt Lake City: Intermountain Thoracic Society.

Dalton, James, Robert Backstrom, and Steve Kerber. 2009. "Structural Collapse: The Hidden Dangers of Residential Fires." *Fire Engineering* 162 (10). http://www.fireengineering.com/articles/print/volume-162/issue-10/features/structural-collapse.html.

doityourself.com. 2016. "7 Types of Sheetrock Explained." http://www.doityourself.com/stry/7-types-of-sheetrock-explained.

Dunn, Vincent. 2015. *Safety and Survival on the Fireground.* 2nd ed. Tulsa: PennWell.

Earls, Alan. 2009. "It's Not Lightweight Construction. It's What Happens When Lightweight Construction Meets Fire." *National Fire Protection Association Journal*, July/August. http://www.nfpa.org/news-and-research/publications/nfpa-journal/2009/july-august-2009/features/lightweight-construction.

Fahy, Rita, Paul LeBlanc, and Arthur Washburn. 1990. "Fire Fighter Deaths As a Result of Rapid Fire Progress in Structures: 1980–1989." In *Analysis Report on Fire Fighter Fatalities*, 42–45. Quincy, MA: National Fire Protection Association.

Fang, J., and J. Breese. 1980. *Fire Development in Residential Basement Rooms.* NBSIR 80-2120. Washington, DC: Department of Commerce. http://nvlpubs.nist.gov/nistpubs/Legacy/IR/nbsir80-2120.pdf.

Fire Analysis and Research Division. 2006. "U.S. Home Structure Fires." Fact sheet. Quincy, MA: One-Stop Data Shop.

Fire Department of the City of New York. 1997. *Ladder Company Operations: Private Dwellings.* Vol. 3, book 4 of *FDNY Firefighting Procedures.* N.p.: Fire Department of the City of New York.

———. 2013. *Ventilation.* Vol. 1, book 10 of *FDNY Firefighting Procedures.* http://www.firecompanies.com/modernfirebehavior/governors%20island%20online%20course/story_content/external_files/ffp_b10.pdf.

Gallagher, Kevin. 2009. "The Dangers of Modular Construction." *Fire Engineering* 162 (5). http://www.fireengineering.com/articles/print/volume-162/issue-5/features/the-dangers-of-modular-construction.html.

Grosshandler, William, Nelson Bryner, Daniel Madrzykowski, and Kenneth Kuntz. 2005. *Draft Report of the Technical Investigation of the Station Nightclub Fire.* NIST NCSTAR 2. Washington, DC: Government Printing Office. http://ws680.nist.gov/publication/get_pdf.cfm?pub_id=908938.

Gustin, Bill. 2010. "The Hazards of Grow Houses." *Fire Engineering* 163 (6). http://www.fireengineering.com/articles/print/volume-163/issue-6/Features/the-hazards-of-grow-houses.html.

Havel, Gregory. 2008a. "Construction Concerns: Breaching Walls." FireEngineering.com (blog), 12 August. http://www.fireengineering.com/articles/2008/08/construction-concerns-breaching-walls.html.

———. 2008b. "Construction Concerns: Knob and Tube Wiring." FireEngineering.com (blog), 13 November. http://www.fireengineering.com/articles/2008/11/construction-concerns-knob-and-tube-wiring.html.

Haynes, Hylton J. G. 2016. *Fire Loss in the United States during 2015.* Quincy, MA: National Fire Protection Association.

Hurley, Morgan, ed. 2016. *SFPE Handbook of Fire Protection Engineering.* 5th ed. New York: Springer.

Izydorek, Mark, Patrick Zeeveld, Matthew Samuels, and James Smyser. 2008. *Report on Structural Stability of Engineered Lumber in Fire Conditions.* Project No. 07CA42520. Northbrook, IL: Underwriters Laboratories. http://www.ul.com/global/documents/offerings/industries/buildingmaterials/fireservice/NC9140-20090512-Report-Independent.pdf.

Kennedy, Tom. 1991. "Energy Efficient Windows in Multiple Dwellings." In "Training Notebook," special issue, *Fire Engineering*, January.

Kerber, Steve. 2010. *Impact of Ventilation on Fire Behavior in Legacy and Contemporary Residential Construction.* Northbrook, IL: Underwriters Laboratories. http://www.ul.com/global/documents/offerings/industries/buildingmaterials/fireservice/ventilation/DHS%202008%20Grant%20Report%20Final.pdf.

———. 2013. "What Research Tells Us about the Modern Fireground." *Fire Rescue* 8 (7). http://www.firerescuemagazine.com/articles/print/volume-8/issue-7/strategy-and-tactics/what-research-tells-us-about-the-modern-fireground.html.

———. 2014. "Top 20 Tactical Considerations from Firefighter Research." Powerpoint presentation. Underwriters Laboratories Firefighter Safety Research Institute. https://www.dropbox.com/s/g1fcomhxdkccbdz/FDIC%202014%20Classroom%20-%20Top%2020.pptx.

Kerber, Stephen, Daniel Madrzykowski, James Dalton, and Bob Backstrom. 2012. *Improving Fire Safety by Understanding the Fire Performance of Engineered Floor Systems and Providing the Fire Service with Information for Tactical Decision Making*. Northbrook, IL: Underwriters Laboratories.

Kerber, Stephen, and Robin Zevotek. 2014. *Study of Residential Attic Fire Mitigation Tactics and Exterior Fire Spread Hazards on Fire Fighter Safety*. Northbrook, IL: Underwriters Laboratories. https://firenotes.ca/files/UL_FSRI_Attic_Final_Report.pdf.

Knapp, Jerry. 2013. "Rapid Fire Spread at Private Dwelling Fires." *Fire Engineering* 166 (10). http://www.fireengineering.com/articles/print/volume-166/issue-10/features/rapid-fire-spread-at-private-dwelling-fires.html.

———. 2015. "Modern House Fires Warrant Tactical Agility." *Fire Engineering* 168 (10). http://www.fireengineering.com/articles/print/volume-168/issue-10/features/modern-house-fires-warrant-tactical-agility.html.

Knapp, Jerry, and Christian Delisio. 1997. "Energy Efficient Windows: Firefighter's Friend or Foe?" Firehouse.com (blog), 1 July. http://www.firehouse.com/news/10544872/energy-efficient-windows-firefighters-friend-or-foe.

Knapp, Jerry, Tim Pillsworth, and Christopher Flatley. 2003. "Nozzle Tests Prove Fireground Realities, Part 2." *Fire Engineering* 156 (9): 71–76. https://nozzleforward dotcom1.files.wordpress.com/2013/03/nozzle-tests-prove-fireground-realities-part-2-2003.pdf.

Knapp, Jerry, and George Zayas. 2011. "The Dangers of Illegally Converted Private Dwellings." *Fire Engineering* 164 (6). http://www.fireengineering.com/articles/print/volume-164/issue-6/features/the-dangers-of-illegally-converted-private-dwellings.html.

Kutz, James R. 2012. "Firefighter Response to Natural Gas Leaks and Emergencies." FireEngineering.com (blog), May 15. http://www.fireengineering.com/articles/2012/05/firefighter-response-to-natural-gas-leaks-and-emergencies.html.

Layman, Lloyd. 1955a. *Attacking and Extinguishing Interior Fires*. 2nd ed. Boston: National Fire Protection Association.

———. 1955b. *Fire Fighting Tactics*. Reprint. Boston: National Fire Protection Association.

Madrzykowski, Daniel, Gary Roadarmel, and Laurean DeLauter. 1997. "Durable Agents for Exposure Protection in Wildland/Urban Interface Conflagrations." In *Thirteenth Meeting of the UJNR Panel on Fire Research and Safety, March 13–20, 1996*, edited by Kellie Ann Beall, 345–50. Gaithersburg, MD: Building and Fire Research Laboratory. http://fire.nist.gov/bfrlpubs/fire97/PDF/f97119.pdf.

Madrzykowski, Daniel, and R. Vettori. 2000. *Simulation of the Dynamics of the Fire at 3146 Cherry Road NE, Washington D.C., May 30, 1999*. NISTIR 6510. Gaithersburg, MD: National Institute of Standards and Technology.

Marsar, Stephen. 2010. "Survivability Profiling: How Long Can Victims Survive in a Fire?" FireEngineering.com (blog), 1 July. http://www.fireengineering.com/articles/2010/07/survivability-profiling-how-long-can-victims-survive-in-a-fire.html.

McDowell, Anthony. 2009. *The Wall of Fire: Training Firefighters to Survive Fires in Vinyl-Clad Houses*. Emmitsburg, MD: National Fire Academy.

Mittendorf, John. 1988. *Ventilation Methods and Techniques*. El Toro, CA: Fire Technology Services.

———. 2011. *Truck Company Operations*. 2nd ed. Tulsa: PennWell Corporation.

Moran, Lee. 2015. "Ark. Deputies Find Meth while Trying to Save Cats after Fire." *New York Daily News*, May 6. http://www.nydailynews.com/news/crime/ark-cops-find-meth-rushing-house-fire-save-cats-article-1.2212025.

Mowrer, Frederick. 1998. *Window Breakage Induced by Exterior Fires*. NIST-GCR-98-751. College Park, MD: Department of Fire Protection Engineering.

National Fire Data Center. 1995. *Three Firefighters Die in Pittsburgh House Fire—Pittsburgh, Pennsylvania*. Emmitsburg, MD: US Fire Administration.

———. 1999. *Establishing a Relationship between Alcohol and Casualties of Fire*. Arlington: TriData Corporation for US Fire Administration.

———. 2009. "Multiple-Fatality Fires in Residential Buildings." *Topical Fire Report* 9 (3). https://www.hsdl.org/?view&did=26259.

———. 2012. "Residential Building Fires (2008–2010)." *Topical Fire Report* 13 (2). https://nfa.usfa.fema.gov/downloads/pdf/statistics/v13i2.pdf.

———. 2015. "One- and Two-Family Residential Building Basement Fires (2010–2012)." *Topical Fire Report* 15 (10). https://www.usfa.fema.gov/downloads/pdf/statistics/v15i10.pdf.

———. 2017. "Civilian Fire Fatalities in Residential Buildings (2013–2015)." *Topical Fire Report* 18 (4). https://www.usfa.fema.gov/downloads/pdf/statistics/v18i4.pdf.

National Fire Data Center and National Fallen Firefighters Foundation. 2008. *Firefighter Fatalities in the United States in 2008*. Emmitsburg, MD: US Fire Administration.

———. 2011. *Firefighter Fatalities in the United States in 2011*. Emmitsburg, MD: US Fire Administration.

National Fire Protection Association. 2005. *NFPA 1410: Standard on Training for Initial Emergency Scene Operations*. Quincy, MA: National Fire Protection Association.

———. 2016. *NFPA 1710: Standard for the Organization and Deployment of Fire Suppression Operations, Emergency Medical Operations, and Special Operations to the Public by Career Fire Departments*. Quincy, MA: National Fire Protection Association.

National Institute for Occupational Safety and Health. 2003. *Volunteer Lieutenant Dies Following Structure Collapse at Residential House Fire—Pennsylvania*. NIOSH Report F2002-49. Atlanta: Department of Health and Human Services. https://www.cdc.gov/niosh/fire/reports/face200249.html.

———. 2005. "Preventing Injuries and Deaths of Fire Fighters due to Truss System Failures." NIOSH Alert.

———. 2007a. *Career Engineer Dies and Fire Fighter Injured after Falling through Floor while Conducting Primary Search at a Residential Structure Fire—Wisconsin*. NIOSH Report No. F2006-26. https://www.cdc.gov/niosh/fire/reports/face200626.html.

———. 2007b. *Volunteer Fire Fighter Dies after Falling through Floor Supported by Engineered Wooden-I Beams at Residential Structure Fire—Tennessee*. NIOSH Report F2007-07. Atlanta: Department of Health and Human Services. https://www.cdc.gov/niosh/fire/reports/face200707.html.

———. 2008. *Career Fire Fighter Dies in Wind Driven Residential Structure Fire—Virginia*. NIOSH Report F2007-12. Atlanta: Department of Health and Human Services.

———. 2009. *Volunteer Fire Lieutenant Killed While Fighting a Basement Fire—Pennsylvania*. Fatality Assessment and Control Evaluation Investigation Report No. F2008-08. Atlanta: Department of Health and Human Services.

———. 2012. *Career Fire Fighter Dies and Another Is Injured Following Structure Collapse at a Triple Decker Residential Fire—Massachusetts*. Fatality Assessment and Control Evaluation Investigation Report No. 2011-30. https://www.cdc.gov/niosh/fire/pdfs/face201130.pdf.

National Lightning Safety Institute. 2014. "Lightning Costs and Losses from Attributed Sources." http://lightningsafety.com/nlsi_lls/ListofLosses14.pdf.

Naum, Christopher. 2011. "Prince William County (VA) Fire Rescue Kyle Wilson LODD 2007; Is This on Your Radar Screen?" Firehouse.com (blog), 17 April. http://www.firehouse.com/blog/10459541/prince-william-county-va-fire-rescue-kyle-wilson-lodd-2007-is-this-on-your-radar-screen.

New York State Office of Fire Prevention and Control. 2012. *Recognizing Clandestine Drug Lab Operations: Student Manual*. Albany: New York State Office of Fire Prevention and Control.

Norman, John. 2012. *Fire Officer's Handbook of Tactics*. 4th ed. Tulsa: Fire Engineering/PennWell Corporation.

O'Donnell, Andrew. 1995. "Choosing a Nozzle." *Fire Engineering* 148 (9). http://www.fireengineering.com/articles/print/volume-148/issue-9/features/choosing-a-nozzle.html.

Occupational Safety and Health Administration. N.d. 1910: Occupational Safety and Health Standards, subpart I: Personal Protective Equipment, standard no. 1910.134: Respiratory Protection. https://www.osha.gov/pls/oshaweb/owadisp.show_document?p_table=STANDARDS&p_id=12716#1910.134(g)(4).

Office of Policy Development and Research. 2016. *2016 Characteristics of New Housing*. Department of Housing and Urban Development. https://www.census.gov/construction/chars/pdf/c25ann2016.pdf.

Peluso, Paul. 2010. "Backdraft Throws Arizona Firefighters." Firehouse.com (blog), 28 July. http://www.firehouse.com/news/10465231/backdraft-throws-arizona-firefighters.

Pillsworth, Tim, Jerry Knapp, Christopher Flatley, and Doug Leihbacher. 2007. "How Kinks Affect Your Fire Attack System." *Fire Engineering* 160 (10). http://www.fireengineering.com/articles/print/volume-160/issue-10/features/how-kinks-affect-your-fire-attack-system.html.

Port Gibson Fire Department. 2014. "The Two In, Two Out Rule." Last modified March 23. http://www.portgibsonfd.com/uploads/1/9/9/7/19973347/two_in_two_out.pdf.

Reeder, Forest. 2014. "Understanding the New SLICERS Acronym." *Fire Rescue Magazine* 9 (2). http://www.firerescuemagazine.com/articles/print/volume-9/issue-2/training-0/understanding-the-new-slicers-acronym.html.

Rice, Curtis, and Elvin Gonzalez. 2011. "Situational Awareness and 'Reading' a House." *Fire Engineering* 164 (2): 87.

Routley, J. Gordon. 1995. *Three Firefighters Die in Pittsburgh House Fire*. USFA-TR-078. Emmitsburg, MD: US Fire Administration.

Schnepp, Rob. 2009. "Where There's Fire—There's Smoke!" In "Cyanide and Carbon Monoxide: The Toxic Twins of Smoke Inhalation," supplement, *Smoke* 2:3–8.

Smith, James. 2002. *Strategic and Tactical Considerations on the Fireground*. Upper Saddle River, NJ: Prentice Hall.

Tenniswood, Bruce. 2009. "Bungalow Fires: Construction Dictates Tactics." *Fire Engineering* 162 (10). http://www.fireengineering.com/articles/2009/10/bungalow-fires-construction.html.

Underwriters Laboratories. 2013. "Innovating Fire Attack Tactics." *New Science Fire Safety*, Summer. http://docplayer.net/6007857-New-science-fire-safety-article-innovating-fire-attack-tactics-summer-2013-ul-com-newscience.html.

———. 2014a. "New Dynamics of Basement Fires." *New Science Fire Safety* 2:18–23. http://newscience.ul.com/wp-content/uploads/2014/04/NS_FS_Journal_Issue_21.pdf.

———. 2014b. "Study of the Impact of Fire Attack Utilizing Interior and Exterior Streams on Firefighter Safety and Occupant Survival." https://ulfirefightersafety.org/research-projects/impact-of-fire-attack-on-firefighter-safety-and-occupant-survival.html#more.

US Bureau of the Census. 2011. "Geography and Environment." In *Statistical Abstract of the United States: 2011*, 219–42. Washington, DC: US Bureau of the Census.

US Fire Administration. 2008a. *Fire-Related Firefighter Injuries in 2004*. N.p.: Federal Emergency Management Agency.

———. 2008b. *Residential Structure and Building Fires*. N.p.: Federal Emergency Management Agency.

US Fire Administration and National Fire Data Center. 2010. *A Profile of Fire in the United States 2003–2007*. 15th ed. Emmitsburg, MD: US Fire Administration.

Walsh, Donald. 2007. "Hydrogen Cyanide in Fire Smoke: An Unrecognized Threat to the American Firefighter." In "Perceptions, Myths, and Misunderstandings," supplement, *Smoke*, pp. 4–8.

Index

Symbols

2 in, 2 out 106
360-degree size-ups 8–9, 82–83, 97–98, 104

A

accountability on the fireground 105–6
  Electronic Firefighter Accountability System (EFAS) 106
  Personnel Accountability Report (PAR) 106
  search and rescue (SAR) and 136, 141
aggressive interior fire attack. *See* offensive fire attack
air flow and nozzle operation 162–63
alcohol
  burn injuries and 5
  house fires and 4–5
arson 3, 66–67, 195–96
attack strategies, fire. *See* fire attack strategies
attack teams 156–59
  training 160, 166
attic fires 165, 231–43. *See also* knee walls
  in balloon frames 29, 243
  case histories 231–32
  causes 231–32
  ceiling types and 62, 234
  egress and 232
  exterior firefighting at 241
  fire attack strategies 237–43
  flashover and 235
  fog streams and 240
  hazards 232–35
  hoseline placement 239–40
  ladders and 232–33, 236
  occupied attics and 242–43
  pulling ceilings at 232, 234–36, 239
  salvage and 240–41
  scenarios and discussion 235–37, 239–42
  smoke and 217
  statistics 231
  structural collapse and 236
  vertical ventilation and 181, 236–38, 239
automatic nozzles 155–56
Avillo, Anthony 33–34, 108–9, 136

B

backdraft 26, 54–55
  vs. flashover 54
backup firefighter (attack team) duties 158–59, 160
bad information 87–88, 139–40
bailout 173, 180
balloon frames 27–31
  attic fires and 29, 243
  basement fires and 27, 28, 73–74, 200
  case histories 29–32, 214–16
    hidden fires in 215–16
    identifying 27
    routes of fire spread 28–30, 89, 215–16
    ventilation of 215–16
    windows 28
basement fires 188–00
  balloon frames and 27, 28, 73–74, 200
  confinement of fire at 191–92
  construction methods and 200–201
  in duplexes 201
  exposures at 191
  extinguishment of 193–94
  hazards 197
  hoseline positioning at 190–91, 196, 199, 204–7
  offensive fire attack at 204
  platform construction and 200
  rescue at 190–91
  safety considerations 192
  scenario and discussion 204–6
  size-ups at 199
  statistics 188
  ventilation at 193, 199, 205
  visibility and 189
  windows, applying water through 196–97, 205
bathrooms
  location 8
  search and rescue (SAR) and 145
bedrooms
  location 8
  searching 128, 130, 213
biomimetic carbon monoxide (CO) detectors 246
boiling liquid expanding vapor explosion (BLEVE)
    hazards 63, 73
Borkowski, Barry 52–53
braced framing. *See* post-and-beam frames
Brannigan, Frank 18–19, 33
Brunacini, Alan 153
Bukowski, Richard 51
bungalows
  floor plans 15–16
  routes of fire spread 16
burn injuries 160
  alcohol and 5
  flashover and 48
businesses operating from homes 66, 201

C

cancer 178
candles 3
Cape Cod–style houses 11–12
carbon monoxide (CO) 245–48
  exposure symptoms 246, 248
  history of detectors 245–46

carbon monoxide (*continued*)
  investigating alarms 246–48
  pets and 246
  statistics 245–46
carcinogens 178
case histories
  attic fires 231–32
  balloon frames 29–32, 214–16
  cyanide poisoning 127
  exterior fire spread 56–61
  first-floor fires 214–16
  flashover and house fires 51–52
  legacy vs. lightweight construction 23–24
  lightning strikes 259
  lightweight construction and structural collapse 24–25
  natural gas emergencies 255–58
  outside search and rescue (SAR) team 133–35
  post-and-beam frames 33–34
  search rope 137
cause of fires 3–5, 55
  attic fires 231–32
  determining 90, 110–11, 195
ceilings and floors. *See also* collapse, structural; routes of
      fire spread
  adhesive failure and fire spread 38, 60
  attic floors, pulling 232, 234–36, 239
  ceiling temperatures 46, 124, 160
  collapse time on fire 25
  drop ceilings 188
  flashover and ceilings 17, 50–51, 62
  floor temperatures and structural collapse 25
  hidden fire and heat and 12, 30, 169, 218
  voids between and fire spread 38–39, 60–61
  weakened floors 26, 79, 188, 199–00
chauffeur firefighters 131
checklists for size-ups 84–85
chemical hazards 67, 69–70, 207–8
cigarettes
  fire fatalities and 5
  house fires and 4–5
clandestine drug labs 68–69, 209
Clark, William 50
CO. *See* carbon monoxide (CO)
COAL WAS WEALTH checklist 84
codes
  Code of Federal Regulations and natural gas 249
  electrical 258
collapse, structural 20, 33–35
  attic fires and 236
  case histories 24–25
  fire attack and 111–13
  firefighter fatalities and 24
  of floors 199
  floor temperatures and 25
  of garage doors 202–3
  joints and 22

joists and 198
  of lightweight/engineered construction 22–24, 125
  safety considerations 31
  thermal imaging cameras (TICs) and 24, 25, 199
  time to collapse of structural elements , 26
colonial revival houses 17
combination nozzles 159–60, 160–61, 179
combustion products 126–27, 178
command
  and control 101–7
  incident command system (ICS) 103–4
  posts 104
  transferring 105
complacency, danger of 6, 252, 257
conditions, actions, and needs (CAN) reports 104–5, 179
confinement (RECEO VS) 89, 110, 136, 196–97
  at basement fires 191–92
  at modern house fires 219–21
conserving property. *See* property conservation
construction methods of houses 19–42
  balloon frames 27–31
  basement fires and 200–201
  garages 203
  legacy 23
  lightweight/engineered construction 20–27
  modular 36–37
  platform 27, 125
  post-and-beam frames 32–36
  size-ups and 81–82
  stick-built 23
  terms, common 19
  World War II–era 15, 27–28
contemporary houses 12–14
converted private dwellings (CPDs) 92–95
  hazards 93–95, 144–45
  identifying 95
  search and rescue (SAR) at 144–46
cooking and house fires 3, 89, 134
CRAVE attack strategy 108–9
crime scenes 66–67
  evidence, preserving 72, 194–95
  gas explosions and 253
cyanide compounds 127

D
dangers of. *See also* hazards
  complacency 6, 252, 257
  fog streams 124–25, 163–64, 179–80
deaths and injuries statistics
  alcohol and 4–5
  from attic fires 231
  from basement fires 188
  burn injuries 5, 160
  from carbon monoxide poisoning 245
  from house fires 2–4, 125–26, 129
  from lightning strikes 258

from residential building collapses 24
smoke inhalation and 125
structural firefighting 5
defensive fire attack 112
defensive-offensive fire attack 114–15, 116
detached garages 203
exposures and 203
scenario and discussion 206–7
development of fire 49–50. *See also* flashover; life cycle of fire
ventilation and 151–52, 175–76, 184
documentation of house fires 107
door firefighter (attack team) duties 158–59
doors
backdraft and 54–55
egress and front doors 212
fire-rated 203
flashover and 172
front vs. rear for fire attack 212
garage 201–2
hoselines and 108, 116
rear 130–31
search and rescue (SAR) and 129–31, 135
size-ups and 9
windows as 180
dormers 11–12
double-wide houses 35–36
drop ceilings 188
drug labs 68–69, 209
Dunn, Vincent 36, 50, 140–41
duplex houses 15
basement fires in 201

E

EFAS (Electronic Firefighter Accountability System) 106
ego 44
egress
in attics 232
in basements 189
front doors and 212
protecting 206–7, 205–6, 173
searching areas of 128, 174
windows and 134, 141, 180
elderly people 126
electrical
codes 258
hazards 40, 63–64, 69, 233–34
malfunctions 3, 231
systems in houses 258–59
Electronic Firefighter Accountability System (EFAS) 106
energy efficient houses 65, 98
energy efficient windows 59–60, 65, 174–77
false vents and 65, 176
fire resistance of 175–77
engine companies. *See also* attack teams
front doors and access to buildings 130

search procedures of 136
engineered construction materials. *See* lightweight/engineered construction
English basements 190
ethyl mercaptan 250
evidence of crime scenes, preserving 72, 194–95
explosions and crime scenes 253
exposures (RECEO VS) 88–89, 110
at basement fires 191
definition 88
at detached garage fires 203
flow paths and 196
gas leaks and 255
water flow rate and 153
exterior 360s. *See* 360-degree size-ups
exterior firefighting 185, 196–97
at attic fires 241
at basement fires 205
at exterior fires 62, 237, 243
at garage fires 203
exterior fires 55–62, 243–44
becoming interior fires 237
case histories 56–61
sheathing and insulation and 58–59, 238
strategies 61–62
vinyl siding 56–59
extinguishment (RECEO VS) 89, 110
at basement fires 193

F

false alarms 87
false vents 65, 176
fans
attic 232
for positive pressure ventilation (PPV) 91, 182–83
FASTs (firefighter assist and search teams) 14, 106
FDNY (Fire Department of the City of New York) 127–28, 167, 245–46
Federal National Manufactured Housing Construction and Safety Standards Act of 1974 36
finger joints 22–23
firearms 68
fire attack strategies 107–14. *See also* RECEO VS; SLICERS
for attic fires 237–43
defensive 112
defensive-offensive 114–15, 116
front vs. rear door 212
knee walls and 238
in modern houses 219–20
offensive. *See* offensive fire attack
offensive-defensive 112–13
scenarios and discussion 101–2, 114–17
summary 114
Fire Department of the City of New York (FDNY) 127–28, 167, 245–46

fire development. *See* development of fire; flashover; life cycle of fire

firefighter assist and search teams (FASTs) 14, 106

firefighters
  backup firefighter (attack team) duties 158–59, 160
  chauffeur 131
  door firefighter (attack team) duties 158–59
  fatalities and structural collapse 24
  life safety of 124, 150. *See also* safety considerations
  nozzle operator duties 157–58, 220, 236
  as occupants 124, 174
  priorities of. *See* priorities of firefighting

Firefighting Conscience concept 80

Firefighting Principles and Practices 50–51

Fire Officer's Handbook of Tactics 128–29

fire-rated doors 203

fire reports as public records 107

first aid for carbon monoxide (CO) exposure 247

first-floor fires 212–16
  case history 214–16
  horizontal ventilation and 214
  scenario and discussion 212–14

flashover 47–55, 146
  at attic fires 235
  vs. backdraft 54
  burn injuries and 48
  case history 51–52
  ceilings and 17, 50–51, 62
  definitions 49–50
  front doors and 172
  house fires and 51–52, 121, 170
  insulation and 65
  in modern house fires 169–70
  pyrolysis and 50–51
  rollover and 50
  safety considerations 48, 54, 141
  search and rescue (SAR) and 121, 173
  smoke and 50, 51
  training 54
  ventilation and 50–51, 151, 174

Flatley, Christopher 152–53

floor plans, houses 8–18
  bungalows 15–16
  Cape Cod 11–12
  colonial revival 17
  contemporary 12–14
  four square 15
  McMansions 17–18
  ranch 9–11
  saltbox 14–15
  size-ups and 81–82
  two-and-a-half-story wood frame 11
  Victorian 16
  windows and 8–9

flow paths. *See* routes of fire spread

fog streams

  for attic fires 240
  dangers of using 124–25, 163–64, 179–80
  for ventilation 164
  history of use and misinterpretation 164
  vs. straight streams 179–80, 196

FOIA (Freedom of Information Act) 107

forcible exit 70–71

four-square houses 15

fraternity houses 4

Fredericks, Andy 156, 198

Freedom of Information Act (FOIA) 107

friction loss and water flow 154

furniture
  moving during a search 139–40
  protecting from damage 111

G

Gallagher, Kevin 36

garages 201–4
  construction 203
  detached 203, 206–7
  door safety 201–2
  exterior firefighting at garage fires 203
  hazards 201
  home businesses and 201
  hoseline placement at garage fires 206–8
  location in the home 202–3
  scenarios and discussion 206–9
  ventilation of garage fires 206–7

gas. *See* natural gas

Governors Island Experiments (GIE) 192

grade/slope hazards 64, 134

grow houses 69

gusset plates 22

Gustin, Bill 61, 104, 155, 179

H

Havel, Gregory 21, 233–34

hazards 65–71
  of attic fires 232–35
  of basement fires 197
  chemical 67, 69–70, 207–8
  of converted private dwellings (CPDs) 93–95, 144–45
  defensive fire attack and 112
  defensive-offensive fire attack and 114–15
  electrical 40, 63–64, 69, 233–34
  of garage fires 201
  grade/slope of site 64, 134
  of McMansions 17
  of modified housing 9, 30, 81
  natural gas 62, 73. *See also* natural gas
  propane 63, 73
  scenarios and discussion 45, 72–74
  small rooms 144

Healy, Thomas 178

heating ventilation and air conditioning (HVAC)
    systems 232, 259
hidden fires
  in balloon frames 215–16
  in ceilings 30
  in knee walls 243
  lightning strikes and 260
HIPAA laws 107
history of
  carbon monoxide (CO) detectors 245–46
  fog streams 164
hoarders 67–68
home businesses 66, 201
horizontal ventilation 90, 178–81
  advantages 180
  of first-floor fires 214
  hoselines and 179
  methods 180–81
  search and rescue (SAR) and 129, 180
hoselines. *See also* nozzles; water flow
  additional 108
  first, importance of 107–8, 116, 151
  friction loss in 154
  horizontal ventilation and 179
  kinks in 155–56, 159
  placement/positioning. *See* positioning hoselines
  pressure of 160
  protecting stairs with 108, 130, 212
  search and rescue (SAR) teams and 129, 207, 172
house fires
  causes 3–5, 55, 89, 134
  documentation 107
  fire attack strategies 107–14
  flashover and 51–52, 121, 170
  life cycle of 45–48, 170
  property loss and 1–2, 102, 188
  statistics 1–5, 126
  time of day and 3
  traditional tactics 220–21
  vs. nonresidential fires 2
  water amount needed to fight 152–53
houses
  construction methods. *See* construction methods of
      houses
  definition 3
  electrical systems 258–59
  energy efficient 65, 98
  floor plans 8–18
  fraternity 4
  modern. *See* modern house fires
  modifications, hazards of 9, 30, 81. *See also* converted
      private dwellings (CPDs)
  overcrowding 4, 82, 93, 145
  room groupings, typical 8–9, 82
HVAC (heating ventilation and air conditioning)
    systems 232, 259

I
ICs. *See* incident commanders (ICs)
ICS. *See* incident command system (ICS)
identifying
  balloon frames 27
  converted private dwellings (CPDs) 95
  routes of fire spread 196
IDR (instant detection and response) sensors 246
I-joists, 188, 26. *See also* lightweight/engineered construction
immediately dangerous to life or health (IDLH)
    atmospheres 106–7
Impact of Ventilation on Fire Behavior in Legacy and
    Contemporary Residential Construction *168–72*
improvised explosive buildings (IEBs) 254–55
incident commanders (ICs) 104–7
  cause of fire, determining 90, 195
incident command system (ICS) 103–7
  documentation 107
  transferring command 105
information, reliability of 87–88, 139–40
injuries. *See* deaths and injuries statistics
inside search and rescue (SAR) team 129–31, 197
  factors impacting success 130
  front vs. rear door for entry 130–31
  objectives 129
  tool selection 130
instant detection and response (IDR) sensors 246
insulation 33
  fire spread and 58–59
  flashover and 65
intelligence, preincident. *See* preincident intelligence
interior size-ups 10, 84
investigating
  carbon monoxide (CO) alarms 246–48
  cause of fires 195–96

J
joints in
  modular homes 40
  parallel cord trusses 22–23
  post-and-beam frames 33
joists 20
  balloon frames and 27
  collapse potential of 198
  I-joists , 188, 26
  platform construction and 27

K
Kerber, Steve 124, 168
kill boxes 251
kinks in hoselines 155–56, 159
kit homes 36. *See also* modular houses
knee walls 14–16, 242–43
  fire attack strategies 238
  fire spread and 12, 216, 238–39
  hidden fires in 243

knob-and-tube wiring 233–34
Kutz, James R. 255–58

L

LACoFD (Los Angeles County Fire Department) 239
ladders 134
   attics and 232–33, 236
   placement of 90, 180, 183
   roof operations and 181
   ventilation and 180
laminated veneer lumber (LVL) 20–21
law enforcement. *See* police officers
lawsuits 107
Layman, Lloyd 164
leadership 260–61
legacy construction 23
   collapse time on fire 26
   vs. lightweight construction 23–24
   windows 175–77
life cycle of fire 45–48, 170–71
   real vs. training situations 45–46
   temperature graphs 47, 49, 171
life safety 86, 103. *See also* safety considerations
   fire attack and 115
   firefighter 124, 150
lightning strikes 258–60
   case history 259
   hidden fires and 260
   statistics 258
lightweight/engineered construction 20–27. *See also* modern house fires
   basement fires and 200
   collapse time on fire 26
   contemporary houses and 13
   definition 20
   vs. legacy construction 23–24
   parallel cord trusses. *See* parallel cord trusses
   risk to firefighters, minimizing 26, 125
   structural collapse 22–24, 125
   ventilating 26
locations of
   bathrooms 8
   bedrooms 8
   fire victims 128, 131, 180
Los Angeles County Fire Department (LACoFD) 239
LVL (laminated veneer lumber) 20–21

M

manufactured houses 35–36. *See also* modular houses
marijuana 69
marriage walls 36, 39
matches 3
Maydays 106, 141
McGarry, Mac 155
McMansions 17–18, 169
   basements 189–90

floor plans 17–18
   hazards of 17
methane 250. *See also* natural gas
mobile homes 35
modern house fires
   confinement of 219–21
   fire attack strategies 219–20
   flashover risk 169–70
   life cycle 170–71
   scenario and discussion 216–20
   synthetic materials and 169, 221
   ventilation of 168–76
modular houses 35–41
   advantages 37
   construction methods 36–37
   inspections 41
   marriage walls 36, 39
   safety considerations 41
   void space and fire spread 38–40, 60
   wiring 40
moving
   fire victims 121, 143–44
   furniture during a search 139–40
   nozzle while operating 185, 220
multifamily houses 2, 82. *See also* converted private dwellings (CPDs)
mutual aid 184

N

National Firefighter Near Miss Reporting system 24
National Fire Incident Reporting System (NFIRS) 107, 195
National Fire Protection Association (NFPA) 1, 106
National Institute of Occupational Safety and Health (NIOSH) 5, 24, 26
National Institute of Standards and Technology (NIST) 192
natural disasters 2
natural gas 248–58
   case history 255–58
   Code of Federal Regulations 249
   example response 257–58
   explosions 249–50, 253–54
   exposures and gas leaks 255
   hazards 62, 73
   kill boxes 251
   meters 251–52
   procedures, developing 248–49, 254
   scenarios and discussion 252–55
NFIRS (National Fire Incident Reporting System) 107, 195
NFPA (National Fire Protection Association) 1, 106
NIOSH (National Institute of Occupational Safety and Health) 5, 24, 26
NIST (National Institute of Standards and Technology) 192
nongovernmental organizations (NGOs) 104
Norman, John 128–29
nozzles
   air flow and 162–63

automatic 155–56
characteristics of 160
combination 155–56, 160–61, 179
fog streams. *See* fog streams
moving vs. not moving 185, 220
operator duties 157–58, 220, 236
pressure and flow rates 155
reactions of 160–61
smoothbore 155–56, 162
varieties, comparison of 155, 161–62

O

objectives of
  search and rescue (SAR) 121
  size-ups 78–79
  ventilation 178
occupants
  in attics 242–43
  elderly 126
  firefighters as 124, 174
  survival profiles 124, 146, 190
Occupational Safety and Health Administration (OSHA) 106–7
odors
  natural gas 248, 250
  smoke 89
offensive-defensive fire attack 112–13
offensive fire attack 111–12, 149–65, 185
  at basement fires 204
  target flow 151–56
  team members and duties 156–59
on-scene reports 91–92, 97–98, 146, 195
oriented searches 122
oriented strand boards (OSBs) 58
outside search and rescue (SAR) team 131–36
  case history 133–35
  members 131
overcooling 183, 194
overcrowding in houses 4, 82, 93, 145
overhaul (RECEO VS) 90, 110–11
  ventilation during 178

P

panelized houses. *See* modular houses
parallel cord trusses 20–21, 188, 220
  joints 22–23
Personnel Accountability Report (PAR) 106
pets 68
  carbon monoxide (CO) and 246
Pillsworth, Tim 259
PIOs (public information officers) 104
platform construction 27, 125
  basement fires and 200
police officers 205
  command posts and 104
  information from 88, 133–34
porch roofs 131–33

positioning hoselines
  at attic fires 239–40
  at basement fires 190–91, 196, 198, 199
  from outside 185, 196–97
  at garage fires 206–8
  to protect egress 206–7, 235
positive pressure ventilation (PPV) 91, 182–83
  overcooling and 183
post-and-beam frames 32–36
  case history 33–34
  joints 33
  routes of fire spread 34–35
  tactical challenges 33–35
PPV. *See* positive pressure ventilation (PPV)
prefabricated houses. *See* modular houses
preincident intelligence 7–9, 18
  safety and 7, 78–79, 125
primary searches 122–23
priorities of firefighting 103, 119–20
  for engine companies 136
  evidence preservation 72, 194–95
products of combustion 126–27, 178
professionalism 194–95
propane hazards 63, 73
property conservation 102, 185, 194
  at attic fires 240
property loss 102, 188
  statistics about 1–2
public information officers (PIOs) 104
public records and fire reports 107
pumps
  friction loss and 154
  operators 107
  pressure and flow rates 155
pushing fire 219–20
  with fog streams 163–64
pyrolysis
  definition 47
  flashover and 50–51
  of vertical surfaces 61

R

radiant heat 47, 50
  defensive fire attack and 112
  gas leaks and 255
ranch houses 91–92, 146
  floor plans 9–11
rapid intervention crews/teams (RICs/RITs) 14, 106
RECEO VS 85–91, 96–97, 109–10. *See also* rescue; exposures; confinement; extinguishment; overhaul; ventilation; salvage
reports. *See also* studies
  conditions, actions, and needs (CAN) 104–5, 179
  fire 107
  on-scene 91–92, 97–98, 146, 195
  Personnel Accountability Report (PAR) 106

rescue (RECEO VS) 86, 109–10
  at basement fires 190–91
  RECEO VS vs. SLICERS 198
  reliable information and 88
  size-ups and 86
residential
  fires. *See* house fires
  property 3
RICs/RITs (rapid intervention crews/teams) 14, 106
risk-taking 102, 120–21
  ventilation and 179
Rockland County (NY) Fire Training Center 45
  flashover simulator 50–51
rollover 50
roofs
  operations on 181–82
  porch roofs 131–33
  size-ups of 181
  ventilation of 181–82, 215
room groupings in houses, typical 8–9, 82
routes of fire spread 115
  in balloon frames 28–30, 89, 215–16
  in bungalows 16
  ceiling adhesive failure 38–39, 60
  exposures and 196
  exterior to interior 237
  identifying and controlling 196
  insulation 58
  knee walls 12, 216, 238–39
  in modular homes 38–40, 60
  in post-and-beam frames 34–35
  vinyl siding 56–59
  void spaces 33–35, 38–40, 60–61
  in wood frames 140

S
safety considerations 102
  backdraft 55–56
  breaching walls 70–71
  electrical hazards 63–64
  flashover 48, 54, 141
  garage doors 202–3
  grow houses 69
  modular homes 41
  natural gas 252–55
  overhaul 110
  preincident intelligence and 7, 78–79
  roof operations 181–82
  search and rescue (SAR) 140–41, 145–46
  structural collapse 31
  vent, enter, isolate, and search (VEIS) 174
Safety Tactics Operations Plans (STOP) 19
saltbox houses 14–15
salvage 90, 111, 194–96
  at attic fires 240–41
SAR. *See* search and rescue (SAR)

scenarios and discussion
  attic fires 235–37, 239–42
  basement fires 204–6
  command, control, and fire attack strategies 101–2, 114–17
  first-floor fires 212–14
  garage fires 206–9
  hazards of house fires 45, 72–74
  modern house fires 216–20
  natural gas leaks 252–55
  search and rescue (SAR) 119–20, 146–48
  size-ups 77–78, 95–99
search and rescue (SAR) 107–10. *See also* vent, enter, isolate, and search (VEIS)
  accountability and 136, 141
  considerations for employing 113, 115, 122–25, 207
  at converted private dwellings (CPDs) 144–46
  egress, areas of 128, 174
  engine companies, procedures for 136
  flashover and 121, 173
  front vs. rear door for inside team entry 130–31
  hoselines and 129, 207, 172
  inside team 129–31, 197
  moving furniture during 139–40
  moving victims 121
  objectives 121
  oriented searches 122
  outside team 131–36
  primary vs. secondary searches 122–23
  safety considerations 145–46
  scenario and discussion 119–20, 146–48
  supporting 124–25
  tool selection 130–31
  training 121, 125
  ventilation and 124, 129, 180
  victim profiles 123
search rope 137–39
  case history 137
secondary searches 122–23
security bars 70
sheathing and insulation and fire spread 58–59, 238
single-family dwellings (SFDs) 8. *See also* house fires; houses
single-room occupancies (SROs) 92–93, 144. *See also* converted private dwellings (CPDs)
single-wide homes 35–36
size-ups 8–9, 77–99
  arrival on scene 81–82
  at basement fires 199
  checklists 84–85
  common house characteristics and 81–82
  construction methods and 81–82
  contemporary houses and 12
  doors and 9
  exterior. *See* 360-degree size-ups
  exterior fires and 61

floor plans and 81–82
interior 10, 84
methods 79–84
objectives of 78–79
on-scene reports 91–92
questions to ask 81, 86–87, 89
reliability of sources 87–88
rescue and 86
of roofs 181
scenarios and discussion 77–78, 95–99
smoke and 83, 123, 172
SLICERS 85, 195–98. *See also* RECEO VS
ventilation and 198
smoke
attic fires and 217
backdraft and 55
components 126
fatalities caused by 125
flashover and 47, 51
odors 89
scenario and discussion 208
size-ups and 83, 123, 172
thermal imaging and 141
smoothbore nozzles 161, 162
social media 107
soffits
exterior firefighting and 239
fire spread and 57, 115, 217
solid streams. *See* straight streams
soot 126
sounding floors 199
Spaulding, Mike 52
spread of fire. *See* routes of fire spread
SROs (single-room occupancies) 92–93, 144. *See also* converted private dwellings (CPDs)
stairs
basement 189, 199
protecting with hoselines 108, 130, 212
for search and rescue (SAR) 135
statistics about
attic fires 231
basement fires 188
carbon monoxide (CO) 245–46
fire victims 1, 3–4, 126, 129
house fires 1–5, 126
house sizes 169
lightning strikes 258
property loss 1
structural firefighting deaths 5
vinyl siding 56
steam 124–25, 163–64
stick-built houses 23, 125
inspections 41
vs. modular homes 36
STOP (Safety Tactics Operations Plans) 19

straight streams 183
vs. fog streams 179–80, 196
structural collapse. *See* collapse, structural
studies
Governors Island Experiments (GIE) 192
Impact of Ventilation on Fire Behavior in Legacy and Contemporary Residential Construction *168–72*
Study of Residential Attic Fire Mitigation Tactics and Exterior Fire Spread Hazards on Fire Fighter Safety *237–38*
suicides 66–67, 195
Sullivan, James 240
surprises
construction methods and 33
size-ups and 87
survivability profiling 124, 146, 190
symptoms of carbon monoxide (CO) exposure 246, 248
synthetic building materials 169

T
target water flow 151–56, 165
temperature graphs of fire life cycle 47, 49, 171
tempered glass windows 175
terminology
construction methods 19
occupancy 80
terrorism 71–72, 209
thermal imaging cameras (TICs)
floor collapse and 25, 26, 199
grow houses and 69
overview 141–42
smoke and 141
third-party damage 255
time of day
civilian fatalities and 129
house fires and 3
tools for search and rescue (SAR) teams 130–31
training
attack teams 160, 166
fires vs. real fires 46–47
flashover 54
for the future 261
moving fire victims 143–44
NIOSH reports and 24
search and rescue (SAR) 121, 125
transferring command 105
trusses
collapse of 188
in modular houses 40
parallel cord. *See* parallel cord trusses
in stick-built houses 23
triangular 20
turnout gear 221
two-and-a-half-story wood frame houses 11
types of nozzles 155, 161–62

U

Underwriters Laboratories (UL) 25–26, 124, 168
US Fire Administration (USFA) 2
utilities (gas, water, etc.) hazards 62–63, 188–89, 197. *See also* natural gas
utility companies and crews 104
  gas leaks and 253–55
  procedures, sharing 254

V

varieties of nozzles 155
varities of nozzles 161–62
vent, enter, isolate, and search (VEIS) 98–99, 102–3, 198, 213–14
  at attic fires 235
  for Cape Cod houses 11
  for contemporary houses 12–13
  safety advantages of 174
  search and rescue (SAR) and 134–36
  vs. vent, enter, and search (VES) 103
ventilation 85, 168–84. *See also* horizontal ventilation; vertical ventilation
  of attic fires 236, 238, 239
  backdraft and 26, 55, 181
  of balloon frames 215–16
  of basement fires 193, 199, 205
  cancer and 178
  fire development and 151–52, 175–76, 184
  flashover and 50–51, 151, 174
  fog streams and 164
  of garage fires 206–7
  ladders and 180
  of lightweight construction 26
  methods 180–81, 182
  of modern home fires 168–76
  objectives of 178
  overhaul and 178
  positive pressure 91, 182–83
  risks of 179
  search and rescue (SAR) and 124, 180
  timing of 177–79, 185, 199, 238
  windows and 12, 180, 193
ventilation-limited fires 49, 135
  attic fires as 238
  basement fires as 193, 205
Ventilation Methods and Techniques 90
vertical ventilation 181–82
  of attic fires 181, 236–38, 239
  backdraft and 181
  considerations for employing 90–91
  methods 182
  safety of roof operations 181–82
  timing of 236, 238
victims of fire
  elderly 126
  locations, probable 128, 131, 180

  moving 121, 143–44
  profiles 128–29, 146, 190
  statistics about 1, 3–4, 126, 129
Victorian houses 16
  vs. two-and-a-half-story wood frame houses 11
vinyl siding and fire spread 56–59
visibility at basement fires 189
void spaces and fire spread 33–35
  attics and knee walls and 242
  basement fires and 200
  between floors 38–39, 60–61
  in modular houses 38–40, 60

W

walk-out basements 190
walls
  breaching 70–71
  fires in 34–36, 39, 215–16
  marriage 36, 39
war relics 69
water damage 194. *See also* property conservation
water flow
  exposures and 153–54
  friction loss and 154
  inadequate 152–55
  targets 151–56, 165
  testing 153–54
water-reactive substances 69
water supply
  establishing 81
  running out, misplaced fear of 61
webbing for victim removal 143–44
Wetherhold, Dennis 255–58
windows
  applying water through 196–97, 205, 219
  in balloon frames 28
  basement 192–93
  as doors 180
  dormers 11–12
  egress and 134, 141, 180
  energy efficient. *See* energy efficient windows
  fire resistance, legacy vs. modern 175–77
  as fire stops 29
  floor plans and 8–9
  in saltbox houses 14–15
  tempered glass 175
  ventilation and 12, 90, 180, 193
wiring
  knob-and-tube 233–34
  lightning strikes and 259
  in modular homes 40
World War II housing 15, 27–28

Z

Zevotek, Robin 237–38